Clinical Trial Modernization

As the pharmaceutical industry navigates this new era of technological innovation, the integration of AI, big data, and advanced analytics into clinical trials holds immense potential to transform drug development. *Clinical Trial Modernization: Technological, Operational, and Regulatory Advances* provides a comprehensive overview of the current trends, challenges, and opportunities in modernizing clinical trials, offering a roadmap for stakeholders in this evolving field.

This book serves as a valuable resource for professionals, researchers, and regulators, providing actionable insights into the future of clinical trials and their critical role in bringing new therapies to market faster and more effectively.

Harry Yang, Ph.D., is Vice President of Biometrics at Recursion Pharmaceuticals. With over 25 years of experience in small molecule, biologics, and cellular immunotherapy development, his expertise spans the therapeutics areas of transplantation, vaccines, autoimmune and inflammatory disease, rare disease, and oncology. Dr. Yang is well-versed in innovative clinical trial design, regulatory submissions, and integration of real-world data, AI, and machine learning in drug R&D. He is a prolific author, having published 9 books and over 130 articles and book chapters covering critical statistical, scientific, and regulatory topics in drug R&D.

Liang Zhao, Ph.D., is Professor of Bioengineering and Therapeutic Sciences at University of California, San Francisco, following his role as the Director of Division of Quantitative Methods and Modeling (DQMM), Office of Research and Standards, Office of Generic Drugs, CDER/FDA. Dr. Zhao has introduced a broad array of innovative tools in drug deliveries, bioequivalence assessment, and pharmacometrics. He has published 100+ articles and book chapters and is the recipient of the 2023 Gary Neil Prive for Innovation in Drug Development from ASCPT.

Chapman & Hall/CRC Biostatistics Series

Series Editors: Mark Chang, *Boston University, USA*

Case Studies in Innovative Clinical Trials
Edited by Binbing Yu and Kristine Broglio

Value of Information for Healthcare Decision Making
Edited by Anna Heath, Natalia Kunst, and Christopher Jackson

Probability Modeling and Statistical Inference in Cancer Screening
Dongfeng Wu

Development of Gene Therapies
Strategic, Scientific, Regulatory, and Access Considerations
Edited by Avery McIntosh and Oleksandr Sverdlov

Bayesian Precision Medicine
Peter F. Thall

Statistical Methods for Dynamic Disease Screening and Spatio-Temporal Disease Surveillance
Peihua Qiu

Causal Inference in Pharmaceutical Statistics
Yixin Fang

Applied Microbiome Statistics
Correlation, Association, Interaction and Composition
Yinglin Xia and Jun Sun

Association Models in Epidemiology
Study Design, Modeling Strategies, and Analytic Methods
Hongjie Liu

Likelihood Methods in Survival Analysis with R Examples
Jun Ma, Malcolm Hudson, and Annabel Webb

Biostatistics for Bioassay
Ann Yellowlees and Matthew Stephenson

Power and Sample Size in R
Catherine M. Crespi

For more information about this series, please visit: www.routledge.com/Chapman--Hall-CRC-Biostatistics-Series/book-series/CHBIOSTATIS

Clinical Trial Modernization

Technological, Operational, and Regulatory Advances

Harry Yang and Liang Zhao

Designed cover image: Shutterstock_2166598585

First edition published 2025
by CRC Press
2385 NW Executive Center Drive, Suite 320, Boca Raton FL 33431

and by CRC Press
4 Park Square, Milton Park, Abingdon, Oxon, OX14 4RN

CRC Press is an imprint of Taylor & Francis Group, LLC

ISBN: 9781032123608 (hbk)
ISBN: 9781032127538 (pbk)
ISBN: 9781003226086 (ebk)

DOI: 10.1201/9781003226086

Typeset in Palatino
by Newgen Publishing UK

Contents

Foreword

A comprehensive treatise on the exciting advances in clinical trial science is timely and will serve to both educate and stimulate further innovation. Thus, it is with great enthusiasm that I introduce *Clinical Trial Modernization*, a remarkable contribution to the field authored by Drs. Harry Yang and Liang Zhao. This book comes at a pivotal time, as the landscape of clinical trials is undergoing unprecedented transformation, driven by innovation, necessity, and a collective drive to better serve patients and society. As someone who has spent his entire career in clinical pharmacology and regulatory science, I can comfortably say that this work addresses critical knowledge gaps and offers a roadmap for the future of clinical research.

It has been my great privilege to witness the evolution of clinical pharmacology and regulatory science over the last 50 years. Yet, I believe the changes in these domains that are now underway are among the most consequential. *Clinical Trial Modernization* not only captures the uniqueness of this moment but also provides a practical guide for navigating it.

For decades, clinical trials have been the cornerstone of drug development and regulatory decision-making, yet the processes that underpin them have often been criticized as slow, expensive, and, at times, misaligned with the realities of modern healthcare needs. What Drs. Yang and Zhao have achieved in this book is to capture not only the challenges inherent in modernizing clinical trials but also the opportunities. They present practical and visionary approaches to rethinking trial design, novel implementation, and regulatory alignment.

One of the most striking aspects of this book is its focus on patient-centricity. For far too long, the perspectives and experiences of patients have been underrepresented in the design and execution of clinical trials. By emphasizing patient engagement, real-world evidence, decentralized and inclusive trial designs, the authors highlight a much-needed shift toward trials that better reflect the diversity and complexity of real-world healthcare. This shift is not merely aspirational—it is essential if the system aims to serve all patients equitably.

Importantly, Drs. Yang and Zhao also address the increasing roles of technology and advanced data analytics in reshaping clinical trials. From wearable devices and remote monitoring to artificial intelligence, Bayesian inference, and pharmacometrics, the technological advances detailed in this book are nothing short of transformative. These innovations hold the potential to increase trial efficiency, reduce costs, and expand access to previously underserved populations. The authors' discussion of decentralized trials, in particular, is timely, pro-patient and thought-provoking, offering insights into how technology can bring trials closer to patients' lives.

Of equal importance is the thoughtful exploration of regulatory innovation presented in this work. By encouraging adaptive trial designs, novel endpoints, model-based procedures, and employment of real-world data, *Clinical Trial Modernization* reshapes the traditional regulatory frameworks that have guided drug development for decades. Drs. Yang and Zhao articulate a compelling vision for how regulatory science is evolving to meet the demands of these new paradigms without compromising rigor, transparency, or patient safety. Their expertise shines through in their nuanced understanding of the challenges and opportunities facing regulatory agencies, sponsors, and researchers alike.

Beyond technology and regulatory frameworks, *Clinical Trial Modernization* also tackles the operational complexities of modern trials. Issues such as data integration, interoperability, and workforce training are presented with clarity and depth, underscoring the multifaceted nature of the challenges we face. The authors provide actionable procedures and insights, drawing on their own extensive experience to guide stakeholders in navigating these complexities.

Ethical considerations, a cornerstone of clinical research, are not overlooked in this work. The authors remind us that modernization must never come at the expense of ethical principles. Equity, transparency, and the safety of participants must remain non-negotiable pillars as we move forward. This balance between innovation and ethical rigor is one of the book's greatest strengths.

As I reflect on the breadth of topics covered in this book, I am struck by the authors' ability to weave together technical expertise, practical solutions, and a vision for the future. Drs. Yang and Zhao's deep understanding of the challenges and opportunities in clinical trial modernization makes this book an invaluable resource for a wide audience, from researchers and clinicians to policymakers and regulators.

I am reminded of the words of Louis Pasteur: "Fortune favors the prepared mind." This book prepares us for the future, preparing the minds of readers with the knowledge and tools to embrace change and improve the way we conduct clinical trials. I am thrilled to contribute this foreword and excited for the conversations and advancements that this important work will inspire.

Dr. Carl Peck
San Luis Obispo, CA

Preface

High attrition rates in drug development are well-documented. Many drugs that show promise in early studies ultimately fail in late-stage development, with approximately 60–70% of Phase II trials and 30–34% of Phase III trials proving unsuccessful. Overall, about 90% of new drug candidates do not make it past clinical development. Major causes for failure include lack of efficacy, safety concerns, and the challenge of identifying optimal dosing where the benefits outweigh the risks. Recruitment issues also significantly impact the productivity of clinical trials. There is a growing consensus that the current clinical trial process is broken and in need of reform.

Reducing costs and improving efficiency in drug development require identifying unpromising candidates early and directing resources to those with the highest likelihood of success. However, high attrition is challenging to mitigate due to clinical trial decisions often being made with incomplete information and significant uncertainties, compounded by the need to act quickly amid competitive pressures.

Recent advancements in innovative trial designs offer new opportunities to optimize clinical trials and support timely decision-making. For example, adaptive designs allow for timely adjustments of design features based on early trial results, and enrichment strategies enhance trial effectiveness by focusing on populations more likely to benefit. Additionally, the integration of digital technologies and AI is transforming clinical trials, utilizing data from sources such as electronic health records (EHRs), medical claims, disease registries, and wearable devices. When effectively harnessed, these data provide valuable insights for optimizing trial design, improving clinical operations, and guiding clinical decisions.

Despite these advancements, there is a notable lack of systematic discussion and illustrative examples on integrating statistical design strategies with digital innovations. This book aims to address this gap by providing a comprehensive reference for practitioners interested in applying these integrative approaches to advance clinical development. It explores how modern innovations are reshaping clinical trials, offering solutions to longstanding challenges, and enhancing the efficiency of drug development.

This book comprises eight chapters that explore a wide range of topics related to clinical trials, including statistical and technological innovations, as well as the associated challenges and opportunities.

Chapter 1 examines the role of statistical innovations, big data, AI, and predictive analytics in reshaping clinical trial design and execution. While these advances promise to optimize study design, streamline operations, reduce costs, and increase success rates, they also present new challenges that must be carefully navigated.

Chapter 2 delves into the complexities of managing digital data flows in clinical trials. Despite the widespread adoption of platforms like Clinical Trial Management Systems (CTMS) and Electronic Data Capture (EDC), the disconnected nature of these systems creates inefficiencies. The chapter highlights how unified data solutions can address these issues and improve trial precision.

In Chapter 3, the focus shifts to clinical operations, a critical component in advancing new therapies. AI and other advanced technologies are increasingly being used to optimize processes, from trial feasibility assessments to patient recruitment and retention. Real-world examples demonstrate how these innovations are expediting clinical trials.

Chapter 4 expounds the concept of Quality by Design (QbD) in clinical trials, advocating for a proactive, risk-based approach to ensure data integrity and trial reliability. The chapter presents case studies on how AI and machine learning can enhance oversight and improve participant protection.

Chapter 5 tackles the optimization of clinical trials, focusing on the strategic, scientific, and operational elements that influence trial success. It discusses how to leverage AI analytics and novel sources of data to design more precise and efficient clinical trials, leading to better outcomes.

Chapter 6 explores model-informed drug development, a quantitative approach that uses data from both external sources and in-house clinical studies to inform decision-making, with a particular focus on various statistical frameworks for facilitating go and no-go decisions after Phase II proof of concept trials.

Data analysis takes center stage in Chapter 7, where AI and machine learning are revolutionizing the way safety and efficacy data are processed. The chapter details how these technologies automate tasks, handle complex datasets, and generate meaningful insights that guide decision-making.

Finally, Chapter 8 looks at the regulatory landscape, examining how agencies like the FDA are adopting AI and other digital technologies to modernize clinical trials. The chapter highlights recent regulatory initiatives that aim to streamline drug development, accelerate approvals, and improve overall efficiency.

As the pharmaceutical industry navigates this new era of technological innovation, the integration of AI, big data, and advanced analytics into clinical trials holds immense potential to transform drug development. This book provides a comprehensive overview of the current trends, challenges, and opportunities in modernizing clinical trials, offering a roadmap for stakeholders in this evolving field.

We hope this book will serve as a valuable resource for professionals, researchers, and regulators, providing actionable insights into the future of clinical trials and their critical role in bringing new therapies to market faster and more effectively.

We thank David Grubbs, Senior Editor at Chapman & Hall/CRC, for the opportunity to write this book. We are also grateful to the four anonymous reviewers for their valuable feedback and comments on the book proposal. Lastly, the views expressed in this book are solely those of the authors and do not necessarily reflect those of their respective companies or institutions.

Harry Yang
Gaithersburg, Maryland
Liang Zhao
San Francisco, California

1

Transforming Clinical Trials through Harnessing Statistical and Technological Breakthroughs

1.1 Introduction

In drug research and development, clinical trials (CTs) stand as indispensable pillars, providing essential evidence of drug safety and efficacy pivotal for regulatory approval. CTs are time-consuming, costly, and labor-intensive. In recent years, this arduous process has been further compounded by the frequent setbacks encountered, often emerging late in the drug development cycle. Literature reveals that merely about 10% of drugs progressing to the CT stage ultimately obtain approval from the U.S. Food and Drug Administration (FDA). The failures of the remaining 90% are frequently attributed to challenges such as lack of efficacy, safety issues, suboptimal patient cohort selection, ineffective recruiting strategies, and inadequate infrastructure to support complex trials. The pharmaceutical industry is on the brink of a transformative phase in CTs, propelled by advancements in innovative trial design and conduct, powered by big data and digital technologies including artificial intelligence (AI), predictive analytics, cloud computing, and sensor technology. These innovations are opening doors to novel approaches in trial design and execution. Collectively, they empower enhanced insights and flexibility in trial design, cohort selection and trial administration, and evidence synthesis. Moreover, they enable a deeper understanding of the fundamental mechanisms influencing efficacy, side effects, and interindividual variations. The integration of AI presents a promising avenue to streamline various facets of CTs. From optimizing patient recruitment to enhancing data analysis, pattern recognition, and early identification of potential adverse events, AI holds the potential to expedite drug development processes, potentially curbing costs and enhancing patient outcomes. However, like any nascent technology, the implementation of AI in CTs is not without its share

DOI: 10.1201/9781003226086-1

of hurdles and limitations. In this chapter, we discuss the opportunities, and potential implications of big data and AI integration in CTs.

1.2 Clinical Development

1.2.1 Evidence Generation Process

Clinical development refers to the process of testing a new medical intervention, such as a drug, in humans to determine its safety and efficacy. Paracelsus, a Swiss physician, in the 15th century once said, "Every substance is a poison; it is only a matter of the dose." This encapsulates the inherent risk of clinical development. Since there are no known theoretical models that can reliably predict the effects of drugs, clinical development is largely an empirical effort, relying on trial and error through experimentation.

Clinical development is a complex, lengthy, and resource-intensive process. The process has been discussed by many authors, including Lipsky and Sharp (2001). Figure 1.1 presents a diagram of clinical development.

Upon completion of preclinical development, the drug is ready to be studied in human subjects through CTs. The sponsor must submit an Investigational New Drug Application (IND) with the FDA or Investigational Medicinal Product Dossier (IMPD) with European Medicines Agency (EMA) if a CT is to be conducted in one or more European Union Member States. IND and IMPD are requests for FDA and EMA authorization to administer the investigational drug to humans. Both filings contain data from nonclinical studies, quality, manufacture, and control of the investigational drug as well as comparator(s) if applicable. In addition to the regulatory approval, the intended clinical study must also be endorsed by the Institutional Review Board (IRB) at the sites where the trial is conducted.

In clinical development, the usual practice isn't about trying to "boil the ocean." Instead, it's about taking methodical steps, addressing one set of inquiries in phase at a time, and determining the subsequent course of

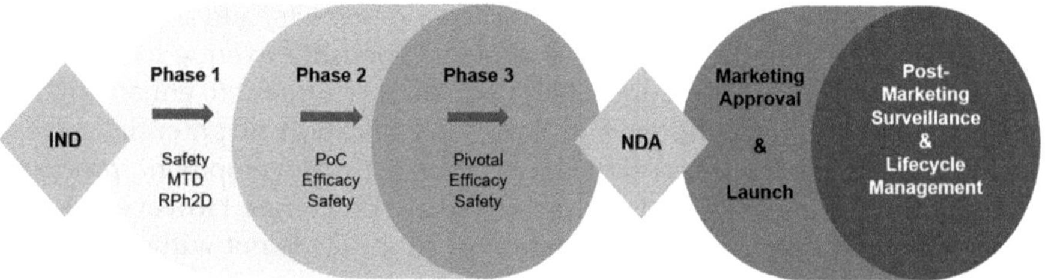

FIGURE 1.1
Clinical development process.

action. As such, clinical development comprises three stages, each with distinct objectives, commonly referred to as Phase I, II, or III CTs with a focus on clinical pharmacology and early safety, efficacy evaluation in targeted patient populations, and confirmation of drug's safety and efficacy, respectively. CTs are often carried out utilizing a mechanism in which patients in treatment group(s) or a control group are randomly assigned. These trials are usually referred to as RCTs. Randomization is used to rule out the effect of potential confounding factors and ensure comparable patients across the group (Barton 2000). Patients in the studies are closely followed to ensure treatment adherence. Additionally, individuals, including the medical monitors from sponsors, patients, and investigators, are blinded about the patient treatment, and the way data are analyzed is pre-specified. Taken together, these measures ensure internal validity of the study design and conclusions.

The use of RCTs to generate evidence of drug safety and efficacy is widely accepted and is likely to remain the gold standard for developing safe and effective drugs. This is despite the long-standing acknowledgment of the significant investment and high risks involved for pharmaceutical companies (Askin et al. 2023; Harrer et al. 2019). Evidence generated from RCTs is integrated and submitted to regulatory agencies such as the U.S. FDA for marketing approval. Potentially, Phase IV studies, often termed post-approval trials, may be required after marketing approval. It is aimed at gaining better understanding of either potential long-term adverse effects or rare adverse events associated with the product.

1.2.2 Clinical Trials

In the following, a brief description of various phases of CTs and other key elements of clinical development is provided.

1.2.2.1 Phase I Clinical Trial

The primary objective of a Phase 1 study is to determine the safety profile and to study the pharmacokinetic property of the drug. For drugs with moderate toxicity, the trial is normally carried out using healthy male volunteers. However, for cytotoxic agents, Phase 1 trials are usually conducted in the target patient populations to minimize unnecessary exposures of the drugs in healthy volunteers. Dose-escalation designs may be used. They typically begin with a low dose of the drug predicted from the animal studies and progressively escalate to higher doses if the drug is well tolerated. A range of doses is explored, and the maximum tolerated dose (MTD) is determined. The study also collects pharmacodynamic and pharmacokinetic data to address important questions such as side effects, therapeutic effects, and ADME (absorption, distribution, metabolism, and excretion). The knowledge garnered from this stage of development, including the safe dosing range, is used to guide the next phase of clinical development.

1.2.2.2 Phase II Clinical Trial

After a MTD is identified from Phase I trials, Phase II studies are carried out with the primary focus on demonstrating the efficacy of the drug and finding an optimum dosing regimen. The studies are usually conducted in patients who have an illness or condition that the drug is intended to treat. These are relatively larger trials with several hundred patients. Phase II trials can be further divided into Phase IIa and Phase IIb studies. The former are typically single-arm trials to screen out inefficacious drugs. The latter usually contain treatment arms of the drug at different dose levels and dosing schedules and a control arm. They are often randomized trials to allow for accurate assessment of treatment effects. At the conclusion of the Phase II trials, researchers expect either to have identified the effective dose, route of administration, and dosing range for Phase III trials or decide to terminate the clinical development of the drug.

1.2.2.3 Phase III Clinical Trial

Phase III trials, also known as pivotal or confirmatory trials, are conducted in a much larger number of patients to substantiate safety and efficacy findings from the previous studies in support of market approval of the drug. These studies are often lengthy, multi-center, possibly global, and are consequently very costly to complete. Most Phase III trials are randomized, double-blind, and multi-armed with a comparator. To gain marketing approval by regulatory approval, typically two Phase III trials are required.

1.2.2.4 Phase IV Clinical Trial

For a new drug, Phase IV trials are conducted for various purposes. They may be used to assess long term effect of the drug. They may be also carried out to determine risk and benefit in specific subgroup of patients. Further, a Phase IV trial may be conducted to support market authorizations in different regions or countries and expand product label.

1.2.2.5 Translational Research

In recent years, a great deal of emphasis has been placed on translational research to expedite and shorten the time of "bench-to-bedside" of a new medicine, by harnessing knowledge of both basic science and clinical research. Translational research leverages data generated from advanced technologies, such as next generation gene sequencing, proteomics, bioinformatics, and robotics, to identify drug candidates that have greater potential to be developed into novel drugs. In addition, it also promotes incorporation of learnings from the clinical setting in drug discovery. For instance, the analysis of patient samples might help identify a subgroup that responds better to the treatment. Thus, a companion diagnostic tool based on genomic or proteomic markers may ensure more targeted therapy development and enhance the probability of success. It is for these reasons that translational research techniques have been widely utilized to advance new drug development.

1.2.3 Target Product Profile and Clinical Development Plan

For a clear line of sight in clinical development, two pivotal documents – the Target Product Profile (TPP) and the Clinical Development Plan (CDP) – are essential and often developed before getting the first study off the ground. The TPP outlines the envisioned attributes of a potential new drug, while the CDP, including many CT strategies for a pharmaceutical or biotech product, from its initial stages to potential regulatory approval and beyond, and serves as a blueprint for clinical programs. A detailed discussion of TPP and CDP is provided in Chapter 6.

1.2.4 Regulatory Requirements

1.2.4.1 Key Regulatory Milestones in the United States

Due to its inherent risk, CTs are strictly regulated. As a result, bringing safe and effective drugs to the market involves conducting CTs and demonstrating drug safety and efficacy in accordance with regulatory requirements. The regulation of drug development has evolved significantly over the past century, driven by scientific advancements, public health concerns, and the need to ensure the safety and efficacy of pharmaceutical products. This evolution has been marked by several key regulatory milestones presented in Figure 1.2. In the following, we discuss key regulatory milestones in drug development in the United States.

In the early 20th century, there were minimal regulations governing drug development and marketing. The tragic consequences of unregulated products, such as the widespread use of toxic substances in medications, led to growing public outcry and demands for government intervention.

One of the earliest regulatory milestones in the United States was the passage of the Pure Food and Drug Act in 1906. This legislation aimed to prevent the adulteration and misbranding of food and drugs, requiring accurate labeling and banning the sale of misbranded or adulterated products. While it represented a significant step forward in consumer protection, it lacked the teeth to regulate drug efficacy.

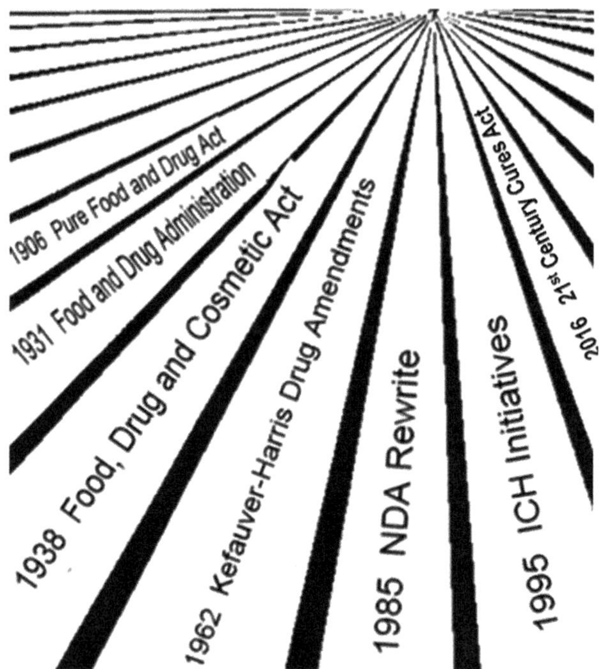

FIGURE 1.2
Significant regulatory milestones in the United States.

One of the most significant events shaping drug regulation was the thalidomide tragedy in the late 1950s and early 1960s. Thalidomide, a drug prescribed for morning sickness, caused severe birth defects in thousands of infants whose mothers had taken the medication during pregnancy. This crisis prompted the U.S. Congress to pass the Kefauver-Harris Amendments in 1962, requiring drug manufacturers to prove the effectiveness of their products before marketing them and mandating stricter oversight by the FDA.

The FDA was officially established in 1938, consolidating various regulatory functions related to food, drugs, and cosmetics. Over the years, the FDA's role in regulating drug development expanded significantly, with a focus on ensuring the safety, efficacy, and quality of pharmaceutical products. In the United States, the FDA requires that an NDA/BLA includes integrated summaries of preclinical and clinical study results. In addition, the application should also contain information on chemistry, manufacturing, and control (CMC), and proposed labeling (FDA 1999). The NDA/BLA is reviewed by the FDA. After comprehensive review, the FDA may approve the drug product or request additional data or studies.

The FDA has played a crucial role in fostering innovation in CT design. The inception of the Critical Path Initiative in 2004 marked a pivotal moment, demonstrating the agency's commitment to advancing medical research. Subsequently, the passage of the 21st Century Cures Act underscored the FDA's dedication to facilitating streamlined pathways for treatment breakthroughs.

In recent times, the FDA has been proactive in addressing challenges within contemporary oncology drug development by issuing a series of guidelines tailored to industry needs. Notably, one such guideline focuses on master protocols, while another centers on expansion cohorts in first-in-human CTs. Although these guidelines are currently in draft form, they signify the FDA's concerted effort to provide comprehensive and timely guidance to stakeholders. This commitment is aimed at expediting the development of impactful new therapies while aiding sponsors in navigating the intricacies of the approval process.

As drug development became increasingly globalized, efforts to harmonize regulatory standards and streamline the approval process gained traction. Initiatives such as the International Conference on Harmonization (ICH) brought together regulatory authorities from around the world to develop common guidelines for the conduct of CTs, drug safety evaluation, and marketing authorization.

1.2.4.2 Regulatory Advances in Europe and Other Regions

Although the regulatory requirements for clinical development and premarketing approval of a new drug and review vary, there are common elements in new drug applications (NDAs) and biologics license applications (BLAs). For marketing authorization applications in Europe and other regions, similar requirements are mandated by regulatory authorities. While the regulatory agencies have the ultimate authority to approve or disapprove the applications, they may solicit the opinion of an independent advisory committee, consisting of experts in the field.

1.2.4.3 Evolving Regulatory Landscape

The evolution of drug regulation is likely to continue as new technologies, such as gene editing and personalized medicine, reshape the landscape of drug development. Regulatory agencies will need to adapt to these changes by incorporating innovative approaches to assessment and approval while maintaining rigorous standards for safety and efficacy. Collaboration between industry, academia, regulatory authorities, and patient advocacy groups will be essential in overcoming the challenges and hurdles that lie ahead in the quest to develop safe and effective treatments for patients worldwide.

1.3 Challenges of Clinical Trials

1.3.1 High Attrition

In recent years, the pharmaceutical industry has grappled with an unprecedented productivity challenge. Despite numerous innovations in biomedical research offering abundant opportunities for the detection, treatment, and prevention of serious illnesses, a significant proportion of candidate drugs that initially showed promise in early research failed in the later stages of clinical development. Approximately, 60–70% of Phase II trial failed to meet their primary endpoints, while about 30–40% Phase III turned out to be negative. Overall, about 11–19% of experimental drugs were able to transition from Phase I to marketing approval for a lead indication (Feijoo et al. 2020; Kola and Landis 2004; DiMasi et al. 2010; Arrowsmith 2012; Hay et al. 2014).

While overall research and development (R&D) spending surged to unsustainable levels, the number of new drug approvals plummeted significantly – with as many as 30% of drugs failing to obtain regulatory approval (Arrowsmith 2012). A survey conducted across nine major pharmaceutical companies revealed that, on average, fewer than one new drug per company received approval from the U.S. FDA annually during the period from 2005 to 2010. In stark contrast, annual R&D spending had soared to a staggering $60 billion by the close of 2010 (Bunnage 2011). Furthermore, the development costs for innovative drugs have escalated to as much as $1.8 billion, and the median development cycle has extended to approximately 13 years compared to the previous 9 years. Concurrently, many companies are grappling with challenges such as numerous blockbuster drugs reaching patent expiration and a sharp decline in revenue growth. Reports indicate that between 2002 and 2018, revenue losses due to generic drug competition are estimated to exceed $290 billion (Winquist 2012).

Extensive research has been conducted to comprehend the underlying causes of high attrition rates (Kola and Landis 2004; Paul et al. 2010; Bunnage 2011; Winquist 2012). Kola and Landis (2004) analyzed the underlying attrition issues using data from ten major pharmaceutical companies. It was found that while attrition stemming from adverse pharmacokinetic and bioavailability outcomes has notably decreased, issues related to efficacy and safety remain largely unchanged, accounting for over 80% of the attrition. Other published findings further suggest that attrition rates due to lack of efficacy vary across therapeutic areas and are notably higher in fields where animal models inadequately predict clinical responses. Two notable examples are CNS (Central Nervous System) and oncology, both exhibiting high attrition rates in late-phase trials (Kola and Landis 2004). Other factors potentially contributing to the elevated attrition rates include the absence of predictive biomarkers to identify patients more likely to respond to innovative drugs, unacceptable safety profiles, and inefficient trial designs.

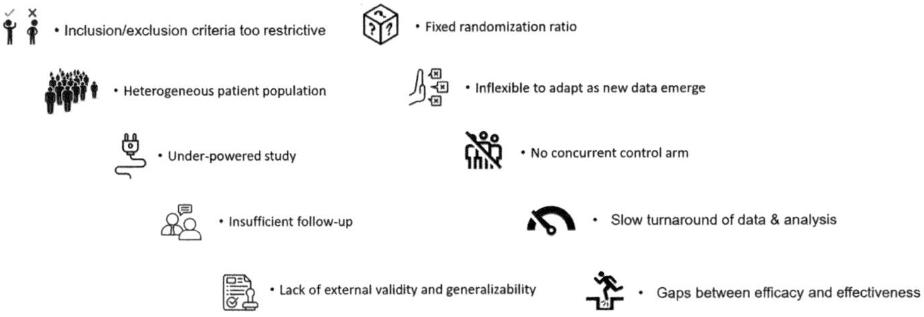

FIGURE 1.3
Limitations of traditional clinical trial designs.

1.3.2 Suboptimal Study Design

Traditional CTs were primarily tailored for mass-market drugs such as COVID-19 vaccines for broad patient populations. When dealing with intricate therapies targeting smaller and more diverse patient populations, these designs fall short in terms of analytical power to demonstrate efficacy due to the heterogeneity of both diseases and patients. They also lack adaptability, because of the inability of adjusting ongoing trials based on emerging data, discrete and fixed points at which transitions from one phase study to another are made. The selection of study populations and endpoints and associated effect sizes tend to be heavily dependent on key opinion leaders' views as opposed to being driven by patient data. This often leads to too restrictive eligibility criteria, inaccurate estimate of effect sizes, leading to under-powered studies. Figure 1.3 depicts several other limitations of the traditional trial designs.

1.3.3 Inefficient Trial Conduct

1.3.3.1 Manual and Labor-Intensive Process

The traditional CT is a manual and labor-intensive process, encountering numerous bottlenecks (Taylor et al. 2020). For example, while the development of many CT systems requires key information from the study protocol, the current practice involves manually compiling and curating information from the protocol to disseminate it among various stakeholders (DiCicco et al. 2023). This laborious and repetitive manual process not only results in much rework but also is error-prone.

Among the various manual processes, setting up a database and electronic data collection tools stands out as a prime example of the limitations inherent in manual curation, particularly during study start-up. Despite efforts to streamline processes, data collection tools are often configured as digital replicas of paper forms, perpetuating inefficiencies and causing delays in initiating CTs. Because of its manual nature, constructing a database takes 12–14 weeks. Frequent protocol amendments exacerbate the situation, resulting in further delay of key timelines.

Addressing these challenges necessitates a paradigm shift toward embracing digital solutions that facilitate seamless integration, automation, and analytics-driven optimization of study protocols and data collection processes. By embracing digital-first approaches and leveraging advanced technologies like machine learning (ML) and NLP, the clinical research community can overcome longstanding barriers and usher in a new era of efficiency and innovation in CT management.

1.3.3.2 Ineffective Patient Selection, Recruitment, and Retention

Ineffective patient selection, recruitment, and retention, together with lack of a technical infrastructure to cope with the complexity of CTs, are contributing to high trial failure rates and raising the costs of research and development (Harrer et al. 2019). As CTs become focused on testing innovative drugs in smaller patient populations, the inclusion and exclusion criteria becoming increasingly intricate. Selecting participants who precisely meet these criteria presents a formidable challenge. Selection bias may lead to underrepresentation of certain groups of patients, leading to study findings that are not generalizable (Mayo et al. 2017; Sen et al. 2017). It also has the potential to result in under- or over-enrollment, thus increasing the overall cost of CTs.

Recruiting participants for CTs stands as a perennial challenge, significantly impacting the setup and execution of these studies. The repercussions of inadequate recruitment reverberate throughout the trial process, inflicting substantial delays and escalating financial burdens. A myriad of factors contributes to the intricacies of recruitment, including the inherent complexity of trial protocols, limited awareness of the trial among potential participants, apprehensions stemming from emotional concerns about participation, and the pervasive issue of competing studies vying for the same participant pool. In essence, the recruitment phase of CTs encapsulates a multifaceted interplay of logistical, psychological, and procedural challenges.

Patient retention in CTs poses a significant challenge due to the extensive demands placed on participants, including frequent laboratory work, numerous on-site visits, and prolonged follow-up periods. These requirements can lead to participant fatigue and inconvenience, particularly for those who need to manage these commitments alongside their daily responsibilities and personal lives. The physical and emotional strain of repeated medical procedures, the time-consuming nature of travel to trial sites, and the long duration of follow-up can discourage continued participation. Additionally, unforeseen personal or health issues may arise, further complicating retention.

1.3.4 Lack of Generalizability of Study Findings

Use of RCTs for demonstrating drug safety and efficacy has been the gold standard for assessing a drug's safety and efficacy. The evidence generation process of RCT is consistent with regulatory expectations. However, there are several drawbacks concerning the RCT methodology. First of all, the outcomes from RCTs may lack external validity as RCTs are often conducted under strictly controlled experimental conditions which are different from routine clinical practice. Consequently, RCTs often provide an estimate of the efficacy of the drug rather than the true measure of effectiveness in the real world (Black et al. 1996), resulting in a gap between the efficacy demonstrated in the CTs and effectiveness of the drug in real-world use. Factors that contribute to the gap include patient adherence, age, comorbidities, concomitant medications, etc. Because of these variations, findings from RCTs may not always translate into the performance of the product in the practical setting. Secondly, due to rising medical costs, healthcare decision-making by payers has increasingly relied on the balance of cost and benefit of a new treatment. However, despite the demand for demonstration of the value of medical products to justify payment, traditional RCTs offer little information because of the gaps in efficacy-effectiveness. The situation worsens for drug products which are approved based on single-arm studies, surrogate endpoints, or short-term outcomes. Lastly, since not every patient responds to the same treatment in the same way owing to heterogeneity of the patient population, it is of interest to both the patient and prescribing physician to understand potential effects of the treatment on an individual patient. In the traditional CTs, the efficacy and safety of a treatment is demonstrated by comparing the average outcomes of the treatment and control groups. Therefore, it does not provide patient-specific assessments of efficacy and safety. Evidence generated from traditional clinical research fails to guide patients, physicians, and health systems for real-world decisions, which has been noted by many researchers.

1.4 Paradigm Shift in Clinical Development

1.4.1 Patient Centricity

To effectively address the challenges in clinical development, there has been a notable shift toward a patient-centric approach (Yang 2023a). This evolution acknowledges that not all patients respond uniformly to the same drug, thus emphasizing the need to tailor treatments to individual patient profiles. By incorporating deep insights about patients into CTs, the precision and effectiveness of these studies are significantly improved, leading to better patient outcomes and more efficient use of resources (Yang 2023b).

A key element of this patient-centric approach is the identification of biomarkers that can predict individual responses to specific treatments. Predictive models based on genomic, proteomic, and other molecular data can be used to identify patients that are more likely to benefit from certain therapies. This approach not only enhances therapeutic efficacy but also minimizes the exposure of patients to potentially ineffective or harmful treatments (Collins and Varmus 2015). Moreover, such an approach facilitates adaptive trial designs, including platform trials, which allow for the evaluation of multiple interventions within a single study framework, thereby increasing the speed and efficiency of drug development (Saville and Berry 2016).

Drug developers are also focusing on detecting ineffective or unsafe drugs at an earlier stage. This proactive strategy aims to prevent patients from being subjected to drugs that offer no benefit or pose significant risks. Early detection is achieved through the integration of advanced analytics, real-time monitoring, and interim data analysis. These tools enable the continuous assessment of treatment efficacy and safety, allowing for timely modifications to the trial design or termination of underperforming drug arms (Thall et al. 2007).

The viability of this patient-centric strategy relies heavily on optimizing CT designs and automating clinical operation processes. Automation technologies, such as AI and ML, are being employed to streamline data collection, analysis, and reporting. These technologies reduce human error, enhance data accuracy, and accelerate decision-making processes (Topol 2019a, 2019b). Additionally, decentralized CTs, which utilize telemedicine and remote monitoring, are becoming increasingly popular. These trials enhance patient engagement and retention by reducing the burden of frequent site visits, thus ensuring a more diverse and representative patient population (Dorsey and Topol 2020).

The shift toward a patient-centric approach in clinical development represents a significant advancement in the field. By tailoring treatments to individual patient profiles, utilizing innovative trial designs, and leveraging automation technologies, there is great potential to improve the precision and efficiency of CTs. This evolution not only improves patient outcomes but also accelerates the development of new therapies, ultimately transforming the landscape of clinical research.

1.4.2 Innovative Clinical Trial Designs

In recent years, novel CT designs such as enrichment design, adaptive design, and master protocols have emerged, significantly enhancing the effectiveness and efficiency of CTs and ultimately accelerating clinical development. These designs are tailored to improve the success rate of trials by adapting treatments to patient profiles and dynamically responding to emerging data from ongoing trials or external sources. Regulatory agencies have increasingly supported these innovative designs, as reflected in their guidelines and reflection papers (FDA 2019a, 2019b, 2019g, 2023a; EMA 2007; PMDA 2007). Enrichment designs focus on selecting patient subpopulations most likely to benefit from a treatment, thereby increasing the trial's sensitivity and reducing required sample sizes. Adaptive designs allow for modifications to the trial procedures based on interim results, enhancing agility and resource utilization. Master protocols, including platform trials, facilitate the simultaneous examination of multiple therapies or diseases within a single study framework, offering unparalleled efficiency by allowing for the addition of new treatments at different times. These innovative approaches provide high-quality data to support regulatory and reimbursement decision-making, ultimately benefiting patients by expediting the development of effective therapies. In the following sections, we briefly describe these novel designs and their applications throughout the lifecycle of CTs. Further details of these design and associated regulatory advances are provided in Chapter 8.

1.4.2.1 Enrichment Design

Many drugs have failed CTs, not because they are inherently ineffective, but rather due to our inability to identify the right patient population. Enrichment in CTs entails the proactive utilization of patient characteristics to refine the study population, enhancing the likelihood of detecting a drug effect, if present, compared to an unselected cohort. This approach has garnered increasing interest, particularly in targeted therapy development. There are several types of enrichment. Prognostic and predictive enrichment strategies identify patients with higher disease progression risks or greater response likelihood to treatments. These strategies enhance trial efficiency and permit the use of smaller patient populations, resulting in substantially better benefit-risk for patients (FDA 2019a, 2019f).

Enrichment strategies have been effectively employed in achieving drug approval. Notable examples include Trastuzumab for patients with overexpressed HER-2-neu receptors, Imatinib for c-kit positive GIST tumors, and Vemurafenib for melanoma with the BRAF mutation. These cases highlight the practical application of biomarker-based enrichment in the successful development and approval of targeted therapies. These methods hinge on the identification of single biomarkers indicative of either the patient's prognosis or response to therapeutic intervention. However, developing markers that account for inter-dependence of patient and disease characteristics calls for more innovative approaches (Yang 2024).

1.4.2.2 Adaptive Design

Adaptive design stands at the forefront of trial optimization, offering a dynamic framework that allows for modifications to trial parameters based on accumulating data (Thorlund et al. 2018). By incorporating interim analyses and adaptive decision-making algorithms, adaptive designs enable researchers to respond dynamically to emerging insights and make modifications to trial designs, such as the termination of ineffective treatment arms or the enrichment of a study population. This flexibility enhances trial efficiency and increases the likelihood of successful outcomes by maximizing the use of available data (Mahlich et al. 2021). The method is also consistent with the patient-centric principles. Broadly implementing adaptive trials could mitigate many unforeseen risks that hinder the progress of effective drugs and unnecessarily lengthen development timelines. For instance, adaptive methods can consolidate trial dosing information within a single two-year combined Phase II/III study, a process that might otherwise span several years and require three or more consecutive conventional trials. This approach not only accelerates development but also reduces the total sample size by utilizing the same patients across multiple stages. Additional benefits include flexible allocation rates, early termination, and the reassessment of sample sizes, patient subgroups, or specific treatments, all of which enhance trial efficiency.

1.4.2.3 Bayesian Method

Bayesian methods offer a robust framework for integrating prior knowledge with observed data, enabling more informed decision-making in both clinical development and regulatory review and approval. In CTs, the Bayesian approach is increasingly recognized for its versatility and effectiveness across a wide range of applications (Ionan et al. 2023; Travis et al. 2023). One of the key strengths of Bayesian methods is their ability to facilitate adaptive decision-making by continuously updating probability distributions based on accumulating evidence (Yang and Novick 2019). This real-time adaptability allows for more flexible and responsive trial designs, which can adjust to new information as it becomes available.

By incorporating diverse sources of information, such as historical data, expert opinions, and current trial outcomes, Bayesian designs optimize trial efficiency. This integration of prior knowledge helps in refining the estimates of treatment effects and improves the precision of statistical inferences (Berry et al. 2015). For instance, if early results in a trial indicate that a particular treatment is highly effective, Bayesian methods can increase the probability of assigning future patients to this treatment arm, thus enhancing the ethical aspects of the trial by potentially providing more patients with beneficial therapies (Thall et al. 2007).

Moreover, the Bayesian approach enhances decision-making accuracy in clinical research. By providing a probabilistic framework for hypothesis testing and parameter estimation, it allows researchers to quantify uncertainty more effectively and make more nuanced decisions regarding trial continuation, modification, or termination (Saville and Berry 2016). This leads to more reliable and robust conclusions, ultimately improving the overall quality of clinical research.

1.4.2.4 Dose Optimization

Dose optimization strategies are essential in CT optimization, as they aim to identify the optimal dosage regimen that maximizes therapeutic efficacy while minimizing adverse effects. Traditional fixed-dose approaches often fail to account for interindividual variability in drug response, resulting in suboptimal dosing recommendations that may not be effective or safe for all patients. By contrast, modern dose optimization techniques, such as adaptive dose-finding algorithms and model-based dose escalation, provide a more sophisticated and personalized approach.

Adaptive dose-finding algorithms are particularly valuable as they allow for real-time adjustments based on accumulating trial data. These algorithms can dynamically modify dosing levels according to individual patient responses and emerging safety profiles (Cheung and Chappell 2000). This method enhances the precision of dose selection and ensures that patients receive a dosage that is tailored to their specific needs, ultimately improving therapeutic outcomes.

Model-based dose escalation, another advanced technique, leverages mathematical and statistical models to predict the optimal dose-response relationship. This approach uses preclinical data, historical trial data, and ongoing patient data to inform dose adjustments (Piantadosi 1997). By systematically exploring various dose levels, model-based strategies can identify the most effective and safest dose more efficiently than traditional methods.

When drug concentration-response relationships are the primary focus for dose optimization, randomized concentration-controlled clinical trials (RCCTs), introduced by Peck in 1989, offer a powerful alternative to randomized dose-controlled trials (RDCTs) (Peck 1990; Sanathanan, Peck et al. 1991; Sanathanan and Peck 1991; Sanathanan and Peck 1993). RCCTs are particularly advantageous for drugs such as immunosuppressants, antiepileptics, and oncology agents, where precise dosing is crucial for balancing efficacy and safety. By closely controlling drug concentrations, RCCTs significantly reduce the risk of toxicities associated with pharmacokinetic variability (Grahnén and Karlsson 2001; Kraiczi et al. 2003). Although their implementation requires additional resources and expertise, RCCTs provide a valuable framework for enhancing the precision and reliability of clinical trial outcomes, ultimately supporting optimal dose selection. Incorporating these advanced dose optimization strategies into CTs not only improves patient outcomes but also enhances trial efficiency. Tailoring treatment regimens to individual patient characteristics reduces the risk of adverse effects and increases the likelihood of achieving therapeutic goals. Furthermore, these strategies can streamline the drug development process by identifying optimal doses more quickly, reducing the duration and cost of CTs.

Overall, dose optimization techniques represent a significant advancement in CT methodology. By addressing interindividual variability and utilizing adaptive and model-based approaches, researchers can develop more effective and safer dosing regimens. This not only enhances the quality of clinical research but also accelerates the development of new therapies, ultimately benefiting patients and healthcare systems alike.

1.4.2.5 Master Protocol

Master protocols represent a comprehensive framework to trial optimization, facilitating the evaluation of multiple interventions or diseases within a single overarching framework. These protocols provide a flexible infrastructure for assessing diverse treatments and patient populations simultaneously, streamlining the drug development process and reducing redundancy. By accommodating adaptive design features, Bayesian methods, and dose optimization strategies, master protocols optimize trial efficiency and accelerate the generation of actionable data, thereby advancing the field of clinical research. These master protocols are particularly beneficial in oncology, where using biomarkers to predict patient response to therapy has become standard practice (Hirakawa et al. 2018). However, master protocols also have potential in other therapeutic areas with multiple treatment options or diseases that can be categorized into several subtypes (EFPIA 2020). Recent examples in the non-oncology areas include CTs for Alzheimer's Disease (Bateman et al. 2017; Schneider et al. 2003) and infectious diseases (Dodd et al. 2016; Scott et al. 2015). Despite their benefits, these complex trials pose significant challenges, underscoring the need for guidance and best practices among industry and other stakeholders.

1.4.2.6 Platform Trial

Platform trials, a special kind of master protocol, represent a scalable and flexible approach to optimizing CTs, allowing for the assessment of multiple interventions within a common infrastructure. By seamlessly integrating new treatments and modifications during active enrollment, platform trials maximize efficiency and statistical power while minimizing resource consumption and timeline constraints (Saville and Berry 2016). This adaptive approach accelerates the identification of promising therapies and enhances patient access to cutting-edge treatments, revolutionizing the landscape of clinical research.

An integral facet of platform trial design is adaptive randomization. This dynamic feature empowers the adjustment of randomization ratios among treatment arms based on interim results, optimizing patient assignment to the most promising interventions. Consequently, patient outcomes often improve as more participants receive potentially effective treatments (Berry et al. 2015). Beyond individual benefits, adaptive randomization bolsters trial efficiency by streamlining patient recruitment and achieving desired outcomes with a reduced participant cohort (Thall et al. 2007).

1.5 Powering Clinical Trial Innovation through Big Data and AI

The pharmaceutical industry has undergone two pivotal transformations in recent years (Yang and Yu 2021). The first is the proliferation and application of real-world data (RWD) from diverse sources, such as electronic health records (EHRs) and disease registries, powered by advancements in cloud storage and GPU computing. The second transformation is the integration of information technologies, including AI and ML, across the entire product development lifecycle. This convergence has fundamentally altered how new medicines are discovered, developed, approved, and delivered to patients, presenting numerous opportunities for CT innovation.

The digital transformation has enabled data to be increasingly connected and interrogated for insights. Data are no longer viewed as a one-off solution to specific inquiries. Instead, they are now perceived as assets that can address numerous questions and serve as a continuous aid in decision-making and problem-solving throughout clinical development (Topol 2019). For example, EHR data can provide the insight of nature history of a disease in an individual patient over time. By using ML and data mining techniques, phenotyping algorithms can identify and characterize diseases, their subtypes, and clusters of comorbidities using EHRs, thereby enabling personalized healthcare (Kirby et al. 2016). In addition, predictive models built upon EHR data further enable patient-centric healthcare. Of equal importance is that EHR data are poised to play a critical role in evidence-based medicine, including regulatory approval and label expansion, clinical practice guidelines, and reimbursement policies. This shift in perspective has been crucial in harnessing the full potential of big data and AI.

Figure 1.4 illustrates data from diverse sources that can be used to optimize the design and conduct of a CT.

One of the primary benefits of integrating big data and AI in CTs is the ability to enhance patient recruitment and retention. By analyzing vast amounts of RWD, AI algorithms can identify and predict eligible patient populations more accurately, ensuring a more targeted and efficient recruitment process

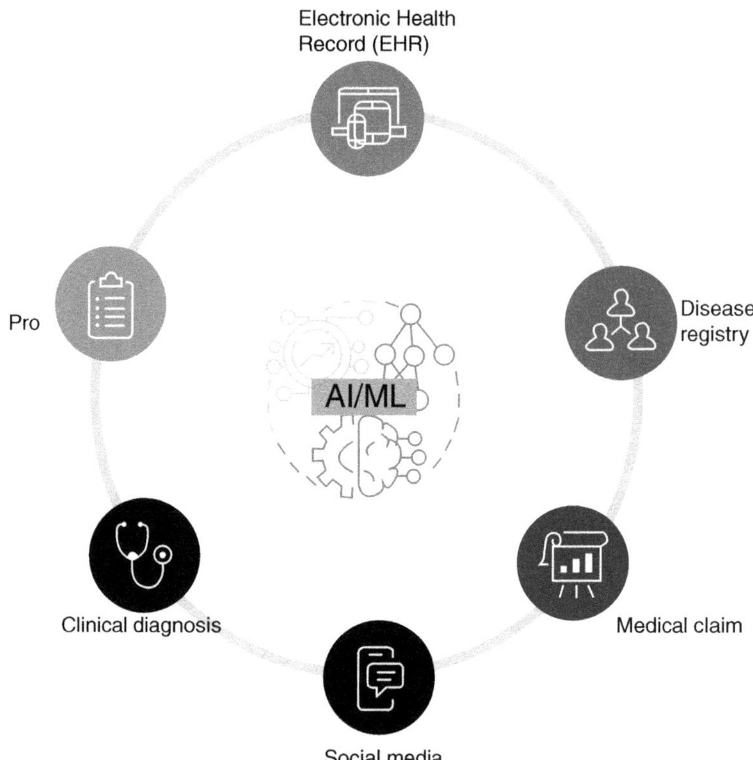

FIGURE 1.4
Integration of data from diverse sources and AI in advancing clinical trial design and conduct.

(Sperrin et al. 2020). Additionally, AI-driven insights can personalize patient monitoring and engagement strategies, thereby improving retention rates and overall trial adherence.

Furthermore, AI and ML technologies enable more sophisticated trial designs and adaptive methodologies. For instance, AI can be used to optimize dosing regimens, predict patient responses, and identify potential adverse events before they occur, leading to safer and more effective CTs. These technologies also facilitate real-time data analysis, allowing for quicker adjustments and more agile trial management.

The continuous analysis of connected data also improves the accuracy and reliability of trial outcomes. By integrating EHRs, genomic data, and other RWD, researchers can achieve a more comprehensive understanding of treatment effects across diverse patient populations. This holistic view aids in the identification of biomarkers, the validation of therapeutic targets, and the personalization of treatments, ultimately enhancing the efficacy and safety of new medicines (Collins and Varmus 2015).

1.5.1 Real-World Data

RWD are collected from diverse sources outside the confines of conventional randomized controlled trials (RCTs). These data often consist of observational outcomes in heterogeneous patient populations. Because RWD are not gathered in well-controlled experimental settings and come from diverse origins, they tend to be unstructured, heterogeneous, complex, and inherently variable. Key sources of RWD include EHRs, claims and billing activities, product and disease registries, patient-related activities in outpatient or in-home settings, and health-monitoring systems. Moreover, advances in digital technologies have enabled the capture of RWD through social media and wearable devices, offering opportunities to provide a holistic picture of an individual's health status (Sherman et al. 2016).

The integration of digital technologies into healthcare facilitates better adherence and disease management by allowing patients to engage with their own data. Wearable devices, for instance, can continuously monitor vital signs, activity levels, and other health metrics, feeding this information back to healthcare providers for real-time insights into a patient's health (Wang et al. 2019). Social media platforms also offer a novel source of health data, capturing patient experiences, treatment outcomes, and adverse events in a timely manner (Weinstein et al. 2017). This vast and diverse data reservoir supports a shift in healthcare from a "drugs for disease" model to "human engineering," where care is personalized based on patient and disease characteristics.

However, the unstructured, noisy nature of RWD presents challenges. Data often come in free text, making it essential to use advanced analytics and natural language processing (NLP) to extract meaningful information (Murphy et al. 2007). The key to realizing the potential of RWD lies in using high-quality data, valid statistical methods, and advanced analytics to synthesize information from large and diverse datasets.

1.5.2 Real-World Evidence

Real-world evidence (RWE) is clinical evidence regarding the usage and potential benefits or risks of a medical product, derived from the analysis of RWD. RWE complements and enhances the evidence obtained from RCTs, providing crucial information for decision-making throughout the product lifecycle (FDA 2017a). The validity of RWE depends not only on the quality of the RWD but also on the robustness of the study design and the appropriateness of the statistical analysis (Makady, Ham et al. 2017; Makady, van Veelen et al. 2017).

At the outset of drug discovery, RWE serves as a valuable tool for researchers to comprehend patient demographics, disease prevalence, and the effectiveness of existing standards of care. When coupled with genomic data, RWE holds the potential to unveil biomarkers that inform targeted drug development strategies for CTs (Kim and Scialli 2011). Moreover, RWD and RWE play pivotal roles in enhancing trial design and operations, facilitating aspects such as study feasibility assessments, site selection, and patient recruitment. Innovative trial designs, including single-arm trials bolstered by synthetic control arms derived from RWE or pragmatic studies aimed at generating evidence of comparative effectiveness, are enabled by these data sources.

Beyond the confines of traditional RCTs, RWE plays a critical role in regulatory evaluations, informing decisions on label expansions, payor coverage, drug pricing strategies, and the management of drug supply chains and inventories (Corrigan-Curay et al. 2018). While RWE has undeniably revolutionized the landscape of new medicine development, realizing its full potential hinges on the availability of high-quality RWD and the capacity to extract meaningful insights from complex and heterogeneous datasets. Additionally, clarity in regulatory policies regarding the utilization of RWE for registrational purposes is paramount.

Despite the extensive literature on diverse applications and use cases of RWE, there remains a notable absence of a comprehensive resource systematically covering the latest developments in the field. Addressing this gap is imperative to foster deeper understanding and broader adoption of RWE-driven approaches within the healthcare and pharmaceutical industries.

1.5.3 Impact of RWD on Clinical Trial Design

The synergy between RWD and digital technologies such as AI and ML has catalyzed the shift from the traditional drug-centric model toward a more patient-focused approach and holds immense potential to streamline processes and improve precision. This amalgamation facilitates a more patient-centric approach in clinical development. By leveraging RWD and advanced analytics, researchers can identify patient subpopulations, predict treatment responses, and detect adverse events earlier in the drug development process. This not only enhances the efficiency of CTs but also minimizes risks for patients by ensuring that they receive treatments tailored to their individual characteristics and needs.

Moreover, the integration of RWD into clinical development facilitates comparative effectiveness research, enabling healthcare providers and payers to make more informed decisions about treatment options. This comparative evidence is particularly valuable in disease areas targeted by high-priced specialty drugs, where stakeholders seek assurance of both clinical efficacy and economic value. By incorporating RWE into decision-making processes, healthcare stakeholders can optimize treatment selection and resource allocation, ultimately leading to better patient outcomes and more efficient healthcare delivery.

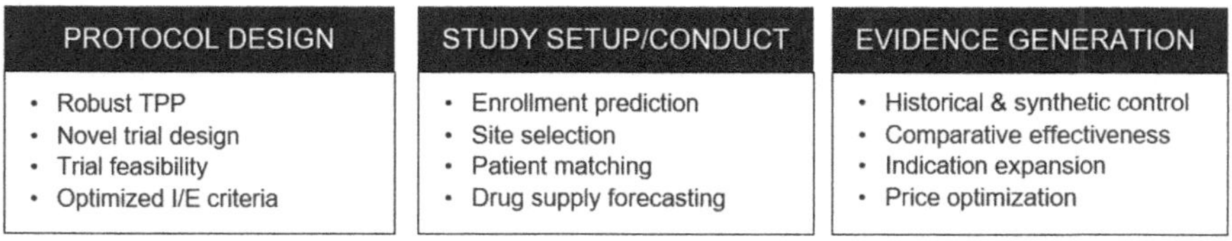

PROTOCOL DESIGN	STUDY SETUP/CONDUCT	EVIDENCE GENERATION
• Robust TPP • Novel trial design • Trial feasibility • Optimized I/E criteria	• Enrollment prediction • Site selection • Patient matching • Drug supply forecasting	• Historical & synthetic control • Comparative effectiveness • Indication expansion • Price optimization

FIGURE 1.5
Applications of RWD in clinical development enabled by AI and machine learning.

A broad range of AI and RWD-enabled solutions are highlighted in Figure 1.5.

1.5.3.1 Optimizing Trial Design

RWD is utilized for actionable insights throughout a drug product development lifecycle. Applications of RWD can bring about fresh insights about disease prevalence, treatment pathways, resulting robust TPPs, enabling novel trial designs that personalize patient treatment using information from wearable devices, identification of high-performing sites, and right patients to match a drug for improved trial efficiency. It also allows for generation of evidence of comparative effectiveness of a product to support regulatory approval and product launch, and market access.

RWE has traditionally served as a cornerstone for regulatory agencies, providing insights into the long-term safety profiles and rare adverse events associated with marketed drug products. However, it is only in recent years that stakeholders within the healthcare ecosystem have begun to recognize the transformative potential of RWE, integrating it as a pivotal driver of outcomes- and value-based strategies throughout the drug development lifecycle. This shift has been catalyzed by advancements in digital technologies, notably cloud data storage, and sophisticated analytics such as AI and ML, which have enabled the seamless integration and interrogation of RWD sourced from diverse repositories.

The evolving landscape of laws, regulations, and mounting pressure on pharmaceutical companies to substantiate the value propositions of their products to healthcare stakeholders have compelled drug developers to prioritize the establishment of comprehensive RWE strategies and capabilities. Furthermore, the recent trend toward personalized healthcare and the adoption of value-based coverage and payment policies in drug development have further propelled the utilization of RWE as a guiding force.

RWE encompasses data derived from settings beyond the confines of traditional CTs, drawing from sources including EHRs, administrative and claims databases, disease and product registries, as well as patient-generated data collected through in-home settings or wearable devices. Demonstrated by numerous successful applications, RWE has emerged as a transformative force across all facets of drug research, development, and commercialization.

1.5.3.1.1 Eligibility Criteria

RWD hold immense potential in optimizing inclusion and exclusion criteria for CTs by providing insights into patient demographics, disease prevalence, and treatment response patterns across diverse populations. By leveraging EHRs, disease registries, and other sources, researchers can identify specific patient subgroups that are most likely to benefit from investigational therapies while minimizing risks associated with adverse events or non-response. This approach allows for a more nuanced selection of trial participants based on real-world characteristics and biomarkers, thereby enhancing the relevance and efficiency of CT designs (Sherman et al. 2016). Moreover, integrating genomic and proteomic data from RWD further refines patient stratification strategies, enabling personalized medicine approaches that align treatment interventions with individual patient profiles and disease phenotypes (Collins and Varmus 2015; Topol 2019). By optimizing inclusion/exclusion criteria through RWD, researchers can increase the likelihood of trial success and accelerate the development of novel therapies tailored to meet the needs of specific patient populations.

1.5.3.1.2 External Control

RWD play a crucial role in creating external controls for CTs by providing a comparator group derived from real-world patient data outside the trial setting. This approach, often termed synthetic control arms, leverages data from EHRs, disease registries, and other sources to construct a historical cohort that closely matches the characteristics of the trial's intervention group. By using RWD to define baseline demographics, disease progression metrics, and treatment outcomes, researchers can simulate the expected outcomes of a trial in the absence of active intervention. This method not only facilitates more efficient and ethical trial designs by reducing the need for placebo or untreated control groups but also enhances statistical power and generalizability of trial results (Sherman et al. 2016). Moreover, advancements in ML and statistical methodologies allow for the adjustment of synthetic control arms over time to reflect changes in clinical practice and patient management, further improving the accuracy and reliability of comparative effectiveness studies (Rosenbaum 2020). By integrating RWD to create external controls, clinical researchers can achieve more robust evaluations of new treatments while minimizing ethical concerns and accelerating the translation of research findings into clinical practice.

1.5.3.1.3 In Silico Trial, Digital Twin

In silico trials utilize advanced simulations to create virtual populations that closely mimic real patient populations. Through these virtual populations, researchers can simulate drug effects, predict efficacy, understand potential side effects, and optimize dosing strategies. This approach significantly enhances the efficiency and effectiveness of the drug development process (Viceconti et al. 2021).

The integration of *in silico* trials into drug development offers numerous benefits, including increased efficiency, cost-effectiveness, and predictive accuracy. By conducting virtual experiments in silico, researchers can expedite the screening of potential drug candidates, prioritize promising leads for further preclinical and clinical evaluation, and reduce the time and resources required for traditional trial enrollment and follow-up. Additionally, in silico trials facilitate iterative refinement of therapeutic interventions based on real-time feedback, accelerating the pace of therapeutic innovation and improving the likelihood of successful outcomes.

In silico trials leverage sophisticated computational techniques to create virtual populations that closely resemble real patient cohorts. These virtual populations are characterized by various demographic, physiological, and clinical parameters, allowing researchers to simulate the complexities of human biology and disease progression with remarkable fidelity. By integrating data from diverse sources, such as EHRs, genomic databases, and CT repositories, in silico trials enable researchers to construct comprehensive models that capture the intricacies of disease pathophysiology and treatment response.

One primary application of in silico trials is simulating drug effects within virtual patient populations. By modeling the pharmacokinetic and pharmacodynamic properties of a drug, researchers can simulate its interactions with biological targets, predict efficacy across diverse patient subgroups, and anticipate potential therapeutic outcomes. These simulations provide invaluable insights into the mechanisms of action underlying drug response, enabling researchers to optimize treatment regimens and identify promising candidates for further clinical evaluation.

Beyond predicting efficacy, in silico trials also facilitate the assessment of potential side effects and safety profiles associated with drug interventions. By incorporating data on drug metabolism, toxicity pathways, and adverse event profiles, researchers can simulate the likelihood and severity of adverse reactions within virtual patient populations. These simulations enable early identification of safety concerns, guiding decisions regarding dose selection, patient monitoring, and risk mitigation strategies.

Another key advantage of in silico trials is their ability to optimize dosing strategies for investigational drugs. By simulating the pharmacokinetics of different dosing regimens across diverse patient populations, researchers can identify optimal dosage levels, frequency, and treatment duration to maximize therapeutic efficacy while minimizing the risk of adverse events (Rostami-Hodjegan 2012). These simulations enable precision dosing tailored to individual patient characteristics, enhancing therapeutic outcomes and patient safety.

Despite the tremendous promise of in silico trials, several challenges remain. There is a need for robust validation of computational models, standardization of simulation methodologies, and integration of diverse data sources (Viceconti and Hunter 2016). Regulatory acceptance and adoption of in silico trial results also pose hurdles that must be overcome to realize their full potential in drug development. Nonetheless, ongoing advances in computational biology, ML, and AI hold promise for addressing these challenges and ushering in a new era of precision medicine fueled by *in silico* experimentation.

1.5.3.1.4 Precision Medicine

Many drugs have failed CTs not because they are inherently ineffective, but due to the challenge of identifying the right patient population. AI-powered approaches are revolutionizing patient selection through CT enrichment strategies. Prognostic and predictive enrichment strategies leverage AI to identify patients with higher risks of disease progression or a greater likelihood of responding to treatments. These strategies enhance trial efficiency and permit the use of smaller patient populations, resulting in substantial benefits (Collins and Varmus 2015).

The application of AI and ML to identify markers for trial enrichment represents a burgeoning area of research. When employed judiciously, enrichment designs have the potential to significantly bolster the efficiency of drug development and advance the principles of precision medicine. This involves tailoring treatments to individuals based on a comprehensive array of clinical, laboratory, genomic, and proteomic factors (Topol 2019). By doing so, AI-driven enrichment strategies can more accurately predict which patients will benefit from specific interventions, thus optimizing clinical outcomes.

The widespread adoption of EHR systems has facilitated access to previously inaccessible patient data, enabling electronic phenotyping to identify individuals with specific conditions or outcomes. Electronic phenotyping reduces population heterogeneity, although its application is not inherently geared toward achieving prognostic or predictive enrichment. By integrating AI and ML, researchers can mine EHR data to identify subtle patterns and correlations that are indicative of potential treatment responses.

RWD have emerged as valuable resources for determining prognostic indicators or baseline characteristics for prognostic enrichment. Predictive markers derived from real-world sources are increasingly informing enrichment trial design, particularly in fields like oncology, where personalized approaches are paramount (Sherman et al. 2016). Integration of genomic data with real-world clinical outcomes holds promise for uncovering biomarkers predictive of therapeutic response and disease resistance, thereby refining drug development strategies (Kim and Scialli 2011).

However, predictive enrichment necessitates the use of more complex models to effectively characterize and assess disease progression. AI and ML models, capable of handling vast and diverse datasets, are instrumental in developing these complex predictive models. By continuously learning from new data, these models can improve their predictive accuracy over time, further enhancing trial outcomes.

Enrichment design lies at the nexus of translational research and precision medicine, representing a pivotal component of modern CT paradigms. This approach not only accelerates the drug development process but also ensures that new therapies are tailored to the needs of specific patient subgroups, thereby improving overall treatment efficacy. For a comprehensive exploration of this topic, refer to Huang and Chiu (2023).

1.5.3.1.5 Individualization of Dosing

Real-time collection of patient data from diverse sources, including electrocardiograms (ECG) and wearable devices, when coupled with advanced analytics, presents a valuable opportunity to gain timely insights into patient health status. These insights pave the way for the development of personalized medicine, as highlighted by Zheng et al. (2013). By harnessing such data, clinicians can optimize treatment strategies tailored to the individual patient's needs.

The application of reinforcement learning holds significant promise in shaping individualized treatment regimens, as demonstrated by research conducted by Zhao et al. (2009) and Zhang et al. (2019). This approach also facilitates the selection of personalized optimal combination therapies, as illustrated by the work of Liang et al. (2018). Particularly in scenarios where multiple treatment options are available, reinforcement learning methods prove to be highly effective, as noted by Godfried (2018).

In recent years, the applicability of reinforcement learning has been extensively explored across various medical settings, including intensive care units. These studies underscore the potential of reinforcement learning algorithms to revolutionize treatment paradigms, offering tailored solutions that optimize patient outcomes based on real-time data insights.

1.5.3.1.6 Pragmatic Studies

Pragmatic studies, also known as pragmatic clinical trials (PCTs), represent a critical avenue for leveraging RWE to address concerns regarding the quality of RWD and the validity of findings. Unlike traditional RCTs, PCTs employ less constrained study designs and encompass broader patient populations, making them an effective bridge between RWD sources and RCTs. By adhering to fundamental principles of patient randomization within controlled trial settings and pre-specified follow-up protocols, PCTs offer a more inclusive and representative approach compared to typical RCTs. Importantly, PCTs do not demand strict adherence to study protocols, mirroring the variability of real-world clinical practice.

When appropriately designed, PCTs have the capacity to generate robust evidence that informs regulatory and payer decision-making processes. Notable examples, such as the Salford Lung Studies (SLS), have demonstrated the utility of PCTs in evaluating treatment effectiveness and safety within real-world settings (Vestbo et al. 2016; Woodcock et al. 2017). For example, the SLS assessed the efficacy and safety of fluticasone furoate in COPD patients by collecting EHR data from various healthcare settings over a 12-month period. The study, which randomized 2,799 patients to different treatment arms, garnered acceptance from regulatory agencies such as the EMA and fulfilled post-approval commitments. Furthermore, Webster and Smith (2019) advocated for the integration of RWE in disease management, citing its significance in chronic myeloid leukemia (CML) treatment. Their research emphasizes how RWE provides insights into early treatment milestones and patient perspectives, thereby enhancing treatment optimization efforts.

Pragmatic studies offer a promising avenue for leveraging RWE to address concerns regarding RWD quality and the validity of findings. As demonstrated by the SLS and other notable examples, these studies play a vital role in informing clinical decisions and generating insights that benefit various stakeholders across the healthcare ecosystem.

1.5.3.1.7 Drug Repurposing and Label Expansion

In recent years, drug repurposing and label expansion have emerged as promising strategies for accelerating therapeutic innovation and expanding the clinical utility of existing drugs. By leveraging CT data, RWD, and RWE, researchers can identify new indications, therapeutic combinations, and patient populations for approved drugs, thereby maximizing their clinical impact and enhancing patient care.

Drug repurposing offers significant advantages, including reduced development costs and shorter timeframes to clinical application compared to traditional drug discovery. This is because many safety and pharmacokinetic profiles of repurposed drugs are already well-documented, enabling faster transitions through the CT phases. For example, repurposed drugs might skip straight to Phase II and III trials if earlier phases have already established their safety (Austin et al. 2021; Jonker et al. 2023).

Furthermore, drug repurposing can address unmet medical needs, especially in fields like oncology and infectious diseases, by finding new therapeutic uses for drugs originally developed for other indications. For example, the antidiabetic drug metformin has been repurposed to exhibit anticancer effects (Kulkarni et al. 2023). Additionally, the repurposing strategy has been pivotal in identifying treatments for diseases like COVID-19, where existing drugs were evaluated for efficacy against the virus (Sperry and Ingber 2024).

Overall, the integration of comprehensive data sources and innovative research methods in drug repurposing and label expansion significantly contributes to more efficient drug development processes and better healthcare outcomes.

CT data serve as a cornerstone for drug repurposing and label expansion efforts, providing valuable insights into the safety, efficacy, and pharmacological profile of approved drugs across different patient populations and disease contexts. By reanalyzing existing trial data and conducting secondary analyses, researchers can identify unexpected therapeutic effects, off-target interactions, and novel indications for approved drugs. For example, the antihypertensive drug minoxidil was originally developed for the treatment of hypertension but was later repurposed for the management of male pattern baldness based on observed hair regrowth in CT participants.

RWD derived from routine clinical practice, EHRs, and patient registries offer complementary insights into drug effectiveness, safety, and tolerability in real-world settings. By analyzing RWD in conjunction with CT data, researchers can generate RWE to support label expansion and regulatory decision-making. For example, the antidiabetic drug metformin was initially approved for the treatment of type 2 diabetes mellitus based on CT data. Subsequent real-world studies utilizing EHRs demonstrated its potential benefits in reducing cardiovascular risk and improving overall survival in diabetic patients, leading to label expansions for these indications.

Aspirin: Originally approved as an analgesic and anti-inflammatory agent, aspirin has undergone extensive label expansion over the years. CT data supported its use for secondary prevention of cardiovascular events, including myocardial infarction and stroke. Furthermore, RWE confirmed its benefits in reducing the risk of colorectal cancer and mortality in certain patient populations, leading to additional label indications.

Thalidomide: Despite its infamous history as a teratogenic agent, thalidomide has found new therapeutic applications in the treatment of multiple myeloma and leprosy. CTs demonstrated its immunomodulatory and antiangiogenic properties, leading to its approval for these indications. RWE further validated its efficacy and safety in diverse patient populations, supporting its continued use in clinical practice.

Gabapentin: Originally approved for the management of epilepsy and neuropathic pain, gabapentin has been repurposed for various off-label indications, including fibromyalgia and restless legs syndrome. CT data provided initial evidence of its efficacy in these conditions, while real-world studies corroborated its therapeutic benefits and safety profile in real-world settings.

Amantadine: Initially developed as an antiviral agent for the treatment of influenza, amantadine has been repurposed for the management of Parkinson's disease and multiple sclerosis. CTs demonstrated its efficacy in alleviating motor symptoms and reducing disease progression in these neurological disorders. RWE further supported its long-term benefits and tolerability in patients, leading to expanded label indications.

Drug repurposing and label expansion represent valuable strategies for maximizing the clinical utility of approved drugs and addressing unmet medical needs. By leveraging CT data, RWE, and RWD, researchers can identify new indications, therapeutic combinations, and patient populations for existing drugs, thereby enhancing patient care and driving therapeutic innovation. Ongoing collaboration between academia, industry, and regulatory agencies is essential to support these efforts and realize the full potential of drug repurposing and label expansion in improving patient outcomes.

1.5.3.2 Smart Clinical Operations

Despite being fundamental to drug development, CTs are predominantly driven by manual processes, from protocol development and database construction to study setup, drug supply management, data cleaning, analysis, reporting, and regulatory submission. In recent years, AI and ML, particularly deep learning, have significantly improved CTs by automating labor-intensive processes. Across various therapeutic domains, AI tools have revolutionized the integration of diverse datasets encompassing demographic details, laboratory findings, imaging results, and other-omics data. By amalgamating these disparate sources, AI facilitates precise patient matching against complex inclusion criteria, thereby ensuring appropriate recruitment into CTs. This impact is notably pronounced in fields such as metabolic disorders, where AI-driven analyses of extensive datasets have enhanced equitable trial access. In addition to improving patient matching, advanced AI applications play a pivotal role in boosting awareness about CTs and augmenting recruitment efforts.

A critical component of AI's success in recruitment lies in the adoption of standardized eligibility criteria language, which ensures seamless interoperability across different systems. Effective AI tools must accurately interpret and process inputs to function optimally. To refine eligibility screening, a hybrid approach integrating structured data with insights from NLP of patient reports holds immense promise. Recent advancements in summarizing eligibility criteria from extensive CT databases underscore the feasibility and utility of these methodologies.

Furthermore, AI offers a compelling solution to alleviate the operational burden of pivotal CTs. AI-driven algorithms can expedite internal investigator selection, streamline site initiation processes, and accelerate recruitment efforts. This enhanced efficiency promises to elevate the overall execution of CTs by expediting participant enrollment. Moreover, the potential for AI to automate trial recommendations could further improve patient selection by disseminating trial information to a broader audience through public CT platforms.

1.6 Clinical Trials of the Future

The landscape of CTs, powered by big data and digital technologies, is rapidly advancing, with numerous exciting trends and innovations set to revolutionize CT methodologies. These developments demonstrate a strong dedication to leveraging state-of-the-art technologies and novel CT designs to enhance the precision, efficiency, and patient-centered nature of clinical research (Padmaja et al. 2024). As the field progresses, the integration of advanced data analytics, AI, and digital health tools will become increasingly prominent, enabling more tailored and adaptive trial designs. These tools not only streamline processes but also ensure that patient experiences and outcomes are prioritized. Furthermore, innovations such as decentralized trials, real-time data monitoring, and enhanced patient engagement platforms are transforming the way trials are conducted, making them more accessible and flexible. Ultimately, these trends signify a transformative shift toward more sophisticated, agile, and patient-focused clinical research practices, promising significant improvements in the quality and impact of future medical discoveries.

1.6.1 Empowering Participants through Digital Engagement

One of the key tenets of the future of CTs is a renewed focus on patient-centricity. In this framework, the role of patients has changed from passive participants to active partners. As such, patient perspectives become an integral part of CT design and conduct. Digital tools and platforms empower trial participants by providing them with greater control over their healthcare journey. From matching patients to trial, remote participation, data collection, and monitoring, and virtual consultations to patient-reported outcomes, electronic informed consent, analysis of their own data, digital engagement initiatives enhance patient convenience, autonomy, and adherence, ultimately improving trial outcomes and retention rates. In addition, such decentralized CT approach also improves patient diversity and inclusivity, particularly for underserved populations and those with limited mobility or geographical constraints.

1.6.2 Seamless and Real-Time Data Capture and Integration

The future of CTs involves seamless and timely integration of data from diverse sources including EHR, EDC, data from wearable devices and sensors. Real-time data capture enables continuous monitoring of patient health metrics and trial endpoints, providing researchers with timely insights into treatment efficacy, safety, and tolerability. By leveraging wearable devices, biosensors, and mobile health apps, RWD streams can be seamlessly integrated into CT workflows, enabling adaptive trial designs, dynamic patient stratification, and personalized treatment algorithms. The seamless integration of EHR and EDC holds the promise of generating evidence of drug safety and efficacy using programmatic studies where data are collected in less controlled setting and broader population (Vestbo et al. 2016; Woodcock et al. 2017).

1.6.3 Advanced Predictive Analytics

The future of CTs is poised for greater utilization of predictive analytics, driven by the evolving power of ML algorithms. These sophisticated algorithms will enhance the accuracy of predictions regarding patient responses to treatments, enabling researchers to discern intricate patterns within complex datasets. As a result, the design of CT protocols will become not only more personalized but also more predictive. This evolution will empower researchers to tailor treatments more precisely to individual patient needs, potentially improving outcomes and making trials more efficient. The integration of these advanced predictive, from site selection and patient recruitment to patient engagement and compliance.

1.6.4 Automation

The continued adoption of AI and ML enables automation of CT processes. The future of CTs will witness end-to-end automation. Starting from study protocol design, CT system development, to data collection, cleaning, review, analysis and reporting, and clinical study report writing, and regulatory submissions, the entire process is facilitated by intelligent tools with little manual support. By leveraging automation based on AI and ML, researchers can streamline trial operations, improve data quality, and accelerate the pace of therapeutic innovation.

1.6.5 Regulatory Innovation

Regulatory agencies play a crucial role in shaping the future of digital trials by providing guidance, frameworks, and incentives to foster innovation while ensuring patient safety and data integrity. Initiatives such as the establishment of FDA's Center for Clinical Trial Innovation (C3TI) (FDA 2024a) and Digital Health Center of Excellence (FDA 2020a), and the subsequent publication of Digital Health Innovation Action Plan (FDA 2020b), and the European Medicines Agency's Regulatory Science Strategy to 2025 (2020) aim to support digital adoption, streamline regulatory processes, and enhance collaboration between stakeholders. The future of CTs witnesses the continued adoption regulatory frameworks to the rapid evolution of CTs.

1.6.6 Collaborative Ecosystems

The future of CTs relies on collaborative ecosystems that bring together stakeholders from across the healthcare ecosystem, including pharmaceutical companies, technology providers, regulatory agencies, healthcare providers, and patient advocacy groups. By fostering interdisciplinary collaboration and knowledge sharing, these ecosystems accelerate the development and adoption of digital solutions, driving transformative changes in clinical research.

1.7 Concluding Remarks

While RCTs are anticipated to remain as the gold standard for clinical development in the immediate future, in the foreseeable future, the landscape of CTs will undergo a profound transformation to a patient-centric paradigm propelled by advances in innovative CT design and convergence of big data and digital technologies. CTs will incorporate flexible scheduling, decentralized trial methods, and supportive measures to maintain participant engagement and compliance throughout the study period. Patient retention will be increased through a decentralized process where trial data can be collected with wearable wireless devices and sensors in patients' home comfort. CTs will be seamlessly integrated into healthcare systems, leveraging routine care delivery procedures to generate a majority of the requisite data, thus minimizing reliance on study-specific case report forms. Additionally, safety monitoring will undergo automation, with real-time data monitoring supplanting traditional adverse event reporting by sponsors. The utilization of programmatic studies and real-world data/evidence (RWD/RWE) will increasingly serve as accelerators for evidence generation in support of marketing approvals, aligning with the pharmaceutical industry's shift toward a patient-centric paradigm. This transition will manifest in the evolution of TPPs to encompass patient care needs, with digital tools becoming integral components thereof. Furthermore, ongoing trial data will be readily shared with patients, physicians, and payers, empowering informed treatment decisions.

Moreover, the integration of AI-enabled automation will permeate every facet of clinical operations, spanning from patient eligibility assessment and consent to treatment, to data collection, cleaning, and reporting of trial outcomes. This comprehensive automation will extend to remote monitoring and the advent of siteless trials, reshaping the landscape of clinical development. Crucially, all stakeholders will be actively engaged throughout the entire clinical development process, ensuring collective participation in realizing the full potential of these transformative technologies.

These advancements will revolutionize design, execution, and subsequent analysis of trial outcomes by seamlessly integrating data from diverse sources. A pivotal objective is the personalization of medicine, achieved not only through the deployment of AI-driven diagnostic technologies for disease detection and treatment but also through the application of ML methodologies, such as reinforcement learning, to optimize treatment protocols tailored to individual patient needs over time.

However, the transition to a patient-centric approach poses several challenges that must be addressed. Historically, patient experience data were often overlooked in CTs and rarely included in product labels. To fully embrace a patient-centric model, stakeholders must redefine study designs, endpoints, and regulatory frameworks to incorporate patient-reported outcomes and RWE. This shift necessitates the adoption of novel methodologies and technologies, such as wearable devices and remote monitoring tools, to collect and analyze patient data effectively.

Furthermore, the patient-centric approach may require investment in smaller patient populations, which may not yield immediate returns. This financial burden could strain healthcare providers and insurers, particularly in the absence of reimbursement mechanisms for personalized care services and technologies. To overcome these challenges, stakeholders must collaborate to develop sustainable funding models and reimbursement policies that incentivize the adoption of patient-centric practices while ensuring equitable access to innovative therapies for all patients. Lastly, to fully realize the potential of these dynamic approaches, regulatory policies must adapt accordingly.

2

Digital Data Flow for Clinical Trials

2.1 Introduction

Clinical trials involve complex systems, processes, data from diverse sources, and various stakeholders. Seamless data flow across different systems, studies, and programs is essential for quality data, timely decision-making, regulatory compliance, and ultimately, success of the trials. Despite the broad adoption of technological platforms such as Clinical Trial Management System (CTMS), Electronic Data Capture (EDC) system, and wearable devices, deriving insights from aggregated data has been a significant challenge. Today, many clinical trial systems and processes are disconnected, and data siloed. The data flow of clinical trials is fraught with manual effort, rework, and inefficiency, impeding the ability for timely data insight to inform decision-making. These challenges are intensifying with the advent of new technologies that enable the generation of much more data from disparate sources at a greater speed, and the growing emphasis on patient centricity in clinical trials. This chapter begins with a discussion of various CTMSs and the traditional data flow in the lifecycle of a clinical trial. Following that, it highlights specific challenges posed by the current manual-driven, document-centric, and disconnected data practices. Subsequently, it discusses how to unlock the value of data through unified data solutions and advanced analytics to enhance clinical trial precision and efficiency from study start-up to study close-out.

2.2 Clinical Trial Systems

A clinical trial is a multifaceted undertaking, involving development of various systems and processes for managing study conduct, data collection, curation, integration, and analysis, and collaboration among multiple stakeholders of

DOI: 10.1201/9781003226086-2

varying roles and responsibilities. Systems commonly used in clinical trials for managing data and information gathered at each stage include CTMS, Electronic Trial Master File (eTMF), EDC, Interactive Response Technology (IRT), Laboratory Information Management System (LIMS), Safety Database, Electronic Patient Reported Outcome (ePRO) systems, and Electronic Health Record (EHR) (Alexsoft 2020; Blumenthal and Tavernner 2010).

2.2.1 Clinical Trial Management System

CTMSs serve as comprehensive platforms for effective planning, executing, and monitoring clinical trials. These systems offer functionalities for protocol management, patient recruitment, site management, budgeting, and regulatory compliance (Zhang et al. 2022). CTMS software enables centralized management of trial-related documents, facilitates communication among stakeholders, and provides real-time visibility into trial progress. Integration with EDC and other data management systems enhances data flow and ensures data consistency across different trial components.

2.2.2 Electronic Trial Master File

eTMF systems streamline the management and organization of essential trial documents, including regulatory submissions, investigator agreements, and study protocols. These systems offer centralized repositories for storing, accessing, and tracking trial documentation throughout the trial lifecycle. eTMF platforms enhance document version control, facilitate document collaboration, and support regulatory inspections and audits. Harmonization with other clinical trial systems ensures data consistency and compliance with regulatory requirements.

2.2.3 Electronic Data Capture System

In contrast to CMTS that is focused on project management aspects of a clinical trial, EDC systems are aimed for patient data collection and management (Krishnankutty et al. 2012). EDC platforms facilitate the direct entry of clinical trial data into electronic databases, streamlining data collection, reducing transcription errors, and enabling real-time data monitoring. These systems typically feature user-friendly interfaces for data entry, validation checks to ensure data accuracy, and audit trails to track data modifications. Moreover, EDC systems support remote data entry by investigators and can integrate with other trial management systems for seamless data flow.

2.2.4 Interactive Response Technology

IRT, also known as Interactive Voice/Web Response Systems (IVRS/IWRS), facilitates randomization, drug assignment, and subject management in clinical trials (Alexsoft 2020). These systems enable automated patient screening, randomization, and drug dispensation based on predefined algorithms and trial protocols. IRT platforms enhance data accuracy by minimizing manual errors in subject assignment and drug allocation. Moreover, they provide real-time access to trial data, enabling investigators to make informed decisions and ensure protocol compliance.

2.2.5 Laboratory Information Management System

LIMS play a crucial role in managing and tracking laboratory data generated during clinical trials. These systems streamline sample tracking, test scheduling, result reporting, and data analysis in laboratory settings. LIMS software integrates with EDC and CTMS platforms to ensure seamless data exchange between clinical sites and central laboratories. By automating laboratory workflows and data management processes, LIMS enhances efficiency, data quality, and regulatory compliance in clinical trials.

2.2.6 Safety Database

The safety database plays a crucial role in clinical trials, serving as a centralized repository for all safety-related data collected throughout the study. Its primary purpose is to ensure the integrity, accuracy, and completeness of safety information, facilitating the monitoring and evaluation of adverse events (AEs) and other safety parameters. The development of a safety database involves several key steps, starting with the design phase, where the database structure is established to capture specific data points relevant to the trial. This includes defining data fields, coding AEs according to standardized medical dictionaries like MedDRA (n.d.), and implementing validation rules to ensure data quality. Integration with EDC systems allows real-time data entry and immediate access for safety monitoring teams. Additionally, robust security measures are incorporated to protect patient confidentiality and comply with regulatory requirements. The database supports ongoing safety assessments by enabling data analysis and reporting, thereby playing an integral role in safeguarding participant health and ensuring regulatory compliance throughout the clinical trial process.

2.2.7 System of Patient Reported Outcomes

The Patient-Reported Outcome (PRO) system is a pivotal component in clinical trials, designed to capture direct input from patients regarding their health status, treatment experiences, and quality of life. This system empowers patients by allowing them to report symptoms, side effects, and overall well-being, providing valuable insights that may not be fully captured through clinical assessments alone (Alexsoft 2020). The development of a PRO system begins with meticulous planning to identify the relevant domains and constructs to be measured, ensuring alignment with the trial's objectives and regulatory guidelines. The system is then built using user-friendly electronic platforms such as web-based interfaces or mobile applications, which facilitate easy and accurate data entry by patients. Key features include intuitive design, multilingual support, and reminders to enhance compliance and data completeness. Data collected through the PRO system are integrated into the broader clinical trial database, enabling comprehensive analysis alongside clinical and laboratory data. This integration supports a holistic evaluation of the treatment's impact, providing a richer, patient-centered perspective on efficacy and safety outcomes. Ensuring robust security and privacy measures, the PRO system also adheres to regulatory standards, protecting patient information while delivering critical data to inform clinical decision-making and improve patient care.

2.2.8 Electronic Health Record

EHR systems capture, store, and manage patient health information, including medical history, medications, laboratory results, and diagnostic images (Blumenthal and Tavenner 2010; Jha et al. 2009). In clinical trials, EHR integration facilitates access to real-time patient data, enhances patient safety, and supports clinical decision-making. By linking EHR data with trial databases, researchers can identify eligible patients, monitor AEs, and assess trial outcomes more efficiently. EHR interoperability with other clinical trial systems promotes data standardization and facilitates regulatory compliance.

2.3 Data Flow

In the conventional clinical trial framework, the flow of data among different clinical systems and across the trial lifecycle is illustrated in Figure 2.1 (Khin et al. 2020). The solid arrows in the figure represent the most common data flows between systems, while the dashed arrows indicate the various potential pathways that developers can choose from when integrating data systems. The data flow encompasses crucial stages such as data collection, curation, integration, analysis, reporting, and dissemination. The central focus of this process is the systematic collection of data of patients enrolled in the clinical trial, followed by rigorous cleaning, analysis, and reporting of findings to support regulatory submissions. These data are often referred to clinical data.

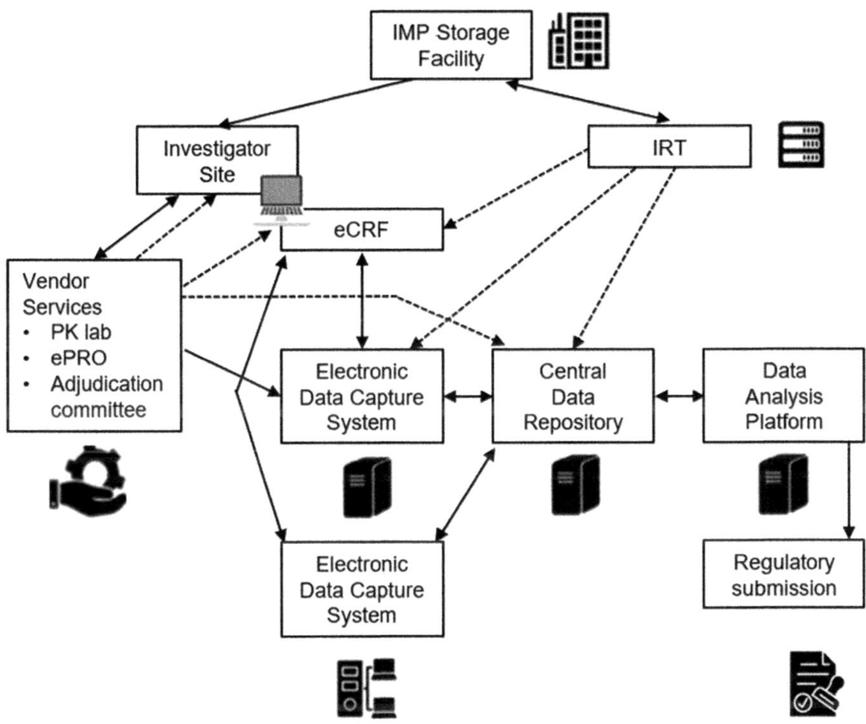

FIGURE 2.1
General data flow in clinical trials. (Adapted from Khin 2020.)

2.3.1 Clinical Data

2.3.1.1 Data Collection

Initially, data collection begins at clinical sites where investigators or study coordinators gather and record patient information, treatment details, and observations either in source documents or on paper Case Report Forms (CRFs). These documents captured various data points such as patient demographics, baseline characteristics, AEs, and clinical outcomes. Once completed, the data in these documents are manually transcribed into electronic databases. This transcription process is labor-intensive and error-prone, often leading to data discrepancies that needed to be resolved through additional review. Following data entry, a data cleaning phase ensues, where data managers review the records for inconsistencies, missing values, and outliers. Queries are then generated and sent back to clinical sites for resolution, creating a loop of feedback and corrections. Periodic monitoring visits are conducted by the trial sponsor to verify source data against CRFs.

2.3.1.2 Data Integration

Raw data from clinical sites, local, central, or specialty laboratories are extracted and loaded to a central repository, where they are curated and integrated for analysis. This process involves several critical steps, including mapping raw data into the Study Data Tabulation Model (SDTM) and Analysis Data Model (ADaM) datasets (Zaidi et al. 2020).

2.3.1.2.1 Mapping Raw Data into SDTM

The SDTM provides a standardized structure for submitting clinical trial data to regulatory authorities. This model ensures consistency and clarity, enabling regulatory reviewers to assess the data efficiently. The mapping process involves several stages:

1. **Data Transformation**: Raw clinical data are transformed into standardized domains such as Demographics (DM), AEs, and Laboratory Test Results (LB). Each domain captures specific aspects of the trial data, facilitating systematic organization.
2. **Validation**: The transformed data are validated to ensure compliance with SDTM standards. This step involves checking for format errors, consistency issues, and adherence to predefined controlled terminology (CT).
3. **Documentation**: Detailed documentation is created to describe the mapping process, transformation rules, and any assumptions made. This documentation is essential for regulatory submission and for future reference during data review.

2.3.1.2.2 Mapping Raw Data into ADaM

The ADaM is designed to support efficient generation of statistical analyses. It provides a framework for creating analysis datasets that are structured to facilitate reproducible and transparent statistical analysis. The mapping process to ADaM involves:

1. **Creation of Analysis Datasets**: Based on the analysis plan, specific datasets are created to address key research questions. These datasets include variables derived from raw data, such as derived efficacy and safety endpoints.
2. **Analysis Metadata**: Metadata is created to document the derivation methods and analytical processes applied. This metadata ensures that the analyses are transparent and reproducible.
3. **Consistency Checks**: The analysis datasets undergo rigorous consistency checks to ensure they align with the corresponding SDTM datasets. This step verifies that the transformations and derivations were performed correctly.

2.3.1.3 Analysis and Reporting

Following data integration, trial data undergo rigorous analysis to derive meaningful insights and evaluate trial outcomes. Statistical analysis software such as SAS, R, and other tools are utilized to analyze clinical data, identify trends, and assess treatment effects.

Data analysis encompasses descriptive statistics, inferential analyses, and subgroup analyses aimed at supporting efficacy and safety assessments. The results of these analyses are compiled into clinical study reports (CSRs) (ICH 1995) and regulatory submissions for dissemination to regulatory authorities, investigators, and study sponsors. CSRs provide a comprehensive summary of the study design, methodology, statistical analyses, and findings. These reports are essential for regulatory review and approval processes.

2.3.1.4 Data Dissemination and Sharing

The final stage of data flow involves disseminating trial data to relevant stakeholders and sharing findings with the scientific community. Regulatory submissions, including New Drug Applications (NDAs) and Marketing Authorization Applications (MAAs), provide comprehensive summaries of trial data and analyses to regulatory authorities for product approval. These submissions include detailed information on the safety, efficacy, and quality of the investigational product (FDA 2021e).

Additionally, publication in peer-reviewed journals, conference presentations, and public databases enable broader dissemination of trial results and facilitate scientific collaboration and knowledge sharing. Sharing data with the scientific community not only advances medical research but also enhances transparency and trust in clinical trial outcomes.

2.4 Challenges of Conventional Clinical Trial Framework

In the conventional clinical trial paradigm, operational data, clinical data, and data from various sources are collected separately and entered into different systems using diverse formats and nomenclatures. This fragmented approach necessitates manual curation, transformation, and visualization of data using disparate standalone solutions. The lack of a cohesive data framework impedes seamless data sharing and integration, leading to inefficiencies in data management and analysis throughout the trial lifecycle. For instance, serious AE data are typically entered into the safety database for timely reporting and concurrently transcribed into EDC systems for clinical data management. However, without a common data fabric, safety reviewers and clinical trial staff often review and clean data in separate systems, periodically reconciling discrepancies. Integration of data from diverse sources, such as laboratory results and imaging data, for safety assessment proves cumbersome, requiring manual reconciliation due to varied formats. Standardizing these data sources becomes imperative to ensure accuracy and facilitate comprehensive analysis.

2.4.1 Disconnected Systems

The development and management of various systems for clinical trials require specialized technical expertise that often goes beyond the capabilities of a single organization. Therefore, it is a common practice for clinical trial sponsors to outsource these activities to a variety of vendors or contract research organizations (CROs). As a result, these systems are often built by multiple teams across different vendors, using information manually extracted from the study protocol. Such a fragmented and synchronization approach can lead to a disjointed infrastructure where different systems, even those intended for the same drug and the same indication at various phases, may operate on entirely separate platforms. The disparities in the systems used can create challenges in data integration and consistency. For example, data collection processes for a drug's Phase I trial may be managed by one vendor using a specific system, while data for its Phase II trial might be collected by another vendor using a different system. These variations can complicate the process of aggregating and analyzing data across the study lifecycle. There is a constant need and effort to harmonize data across these disconnected systems.

2.4.2 Document-Centricity

The conventional clinical trial process is often impeded by its document-centric nature, particularly in how study documents such as the clinical trial protocol are utilized. These documents, while crucial for outlining study design, methodology, objectives, and operational procedures, exist as static electronic documents (DiDicco and Evans 2023). This static nature complicates their use, as it necessitates manual extraction, transformation, and loading (ETL) processes to make the data useful for various purposes throughout the trial lifecycle. For example, key information needed for system development and implementation must be manually extracted from the protocol, a process that is labor-intensive and time-consuming. Different teams, including those involved in data management, clinical operations, and regulatory compliance, often need to independently extract the same information, leading to significant redundancy and potential discrepancies across various systems and processes (Figure 2.2). Moreover, as clinical trial protocols frequently undergo amendments, all teams must revisit and re-extract necessary information. This repetitive cycle further delays progress and increases the potential for errors, impacting the trial's overall efficiency and data integrity.

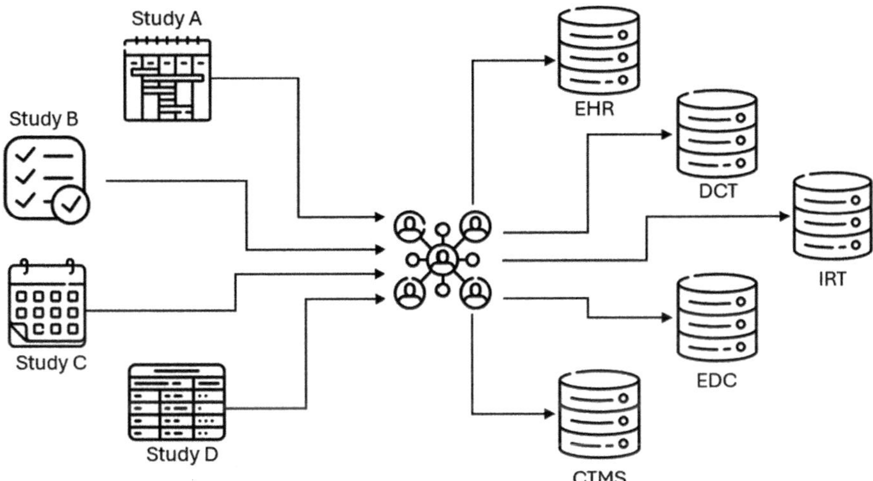

FIGURE 2.2
Manual process of extracting key information for building various clinical trial systems. (Adapted from TransCelerate n.d.)

To address these challenges, there is a growing need for a more dynamic and integrated approach to protocol management. Leveraging technologies such as structured, machine-readable protocols can significantly enhance efficiency. These technologies enable automated data extraction, reducing the need for manual transcription and minimizing errors. By creating a centralized repository of protocol data that can be accessed and utilized by all teams, redundancy of efforts can be eliminated, and consistency across systems can be maintained.

Moreover, implementing a more integrated protocol management system facilitates real-time updates and synchronization across all trial components whenever the protocol is amended. This ensures that all teams are working with the most current information, enhancing coordination and reducing the risk of discrepancies.

Transitioning from a static, document-centric approach to a dynamic, integrated protocol management system is essential to overcoming current inefficiencies. This shift not only streamlines the clinical trial process but also enhances data accuracy, reduces redundancy, and ultimately improves the quality and speed of clinical research.

2.4.3 Inefficient Data Flow

The disconnection in the development of clinical systems results in the use of disparate data standards, formats, and templates. This fragmentation creates significant barriers for data integration and analysis, limiting the ability to fully realize the potential of the data collected. The diversity in data standards and formats necessitates the laborious process of mapping data to a unified set of standards. Currently, this process is largely manual and time-consuming, often requiring extensive effort to ensure data from different sources can be harmonized and accurately integrated. Additionally, any changes in the underlying source data can further complicate this effort, introducing new inconsistencies and requiring ongoing adjustments.

Additionally, similar to study documents, most of the data gathered during a clinical trial are captured in static formats, which are not conducive to machine reading and automation. For instance, patient eligibility is assessed based on data from multiple sources, some of which are embedded in unstructured text and reside in different systems. Validating such data across various formats also requires manual checks or the use of software programs followed by manual evaluation of the outputs. This manual process creates significant bottlenecks, hindering the seamless flow of data through the trial process. The challenges posed by static data are further highlighted by the time it takes to initiate a trial. It has been reported that from the finalization of the study protocol, it typically takes a median of 150 days to start a trial (Karia and Taylor 2020). The reliance on static data formats necessitates extensive manual effort for extraction, validation, and analysis, leading to operational inefficiencies and potentially causing quality and compliance issues. The static nature of the data makes it difficult to analyze and extract insights efficiently. Without dynamic, machine-readable data formats, leveraging advanced analytics and automation tools is nearly impossible. This limitation hampers the ability to quickly derive actionable insights from the data, delaying critical decision-making processes and impacting the overall effectiveness of the clinical trial. To enhance efficiency and data quality, it is crucial to transition toward more dynamic, structured data formats that facilitate automated data processing and real-time analysis.

Moreover, the inability to create and reuse libraries of CRFs, data checks, and analytics across studies and programs results in redundant efforts. Different analytics tools, built separately by different teams for various purposes, lead to inefficiencies and inconsistencies. Without a unified approach to data management, implementing predictive analytics and artificial intelligence/machine learning (AI/ML) becomes nearly impossible. These advanced analytical methods require a robust data foundation, which can only be achieved by first connecting siloed data.

A set of principles known as FAIR – Findability, Accessibility, Interoperability, and Reusability – offers a framework for addressing these challenges (Wilkinson et al. 2016). Creating FAIR-compliant data should be a priority for any data science organization. By ensuring that data are findable, accessible, interoperable, and reusable, organizations can lay the groundwork for more efficient data integration and utilization.

Findability ensures that data can be easily located by both humans and machines. This involves the use of standardized metadata and persistent identifiers, which make data discoverable.

Accessibility guarantees that once data are located, they can be accessed under clearly defined conditions. This includes the use of standardized protocols for data retrieval and access.

Interoperability focuses on enabling data exchange and use between different systems. This requires the use of common data standards and vocabularies, facilitating seamless data integration and analysis.

Reusability emphasizes the need for data to be well-documented and richly described, allowing it to be reused in future research. This involves the creation of detailed metadata and adherence to community standards.

Adopting FAIR principles can transform how clinical trial data are managed, making them more conducive to advanced analytics and machine learning. By breaking down data silos and establishing a robust data foundation, organizations can enhance their ability to generate actionable insights from diverse data sources.

2.4.4 Rework

The fragmentation among various clinical trials systems across different studies and programs significantly hampers the ability to reuse essential data components. This includes standard CRFs, table shells, listing figures, and analytics for data monitoring, cleaning, and analysis. Each clinical trial often operates within its own isolated system, leading to inefficiencies and inconsistencies. These disparate systems prevent seamless integration and sharing of data, resulting in duplicated efforts and increased costs. Moreover, the lack of standardization complicates the process of data cleaning and analysis, as researchers must adapt to different formats and methodologies for each study. Streamlining these processes through unified systems or standardized protocols would enhance data interoperability, allowing for more efficient data reuse. This would not only save time and resources but also improve the accuracy and reliability of clinical trial outcomes by facilitating more consistent and comprehensive data analysis. Consequently, addressing this disconnect is crucial for advancing clinical research and optimizing the overall effectiveness of clinical trials.

2.4.5 Data Analysis Challenge

The current landscape of siloed data in clinical trials poses significant challenges for data integration and analysis. Overcoming these challenges requires a concerted effort to standardize data formats and adopt principles like FAIR. By doing so, the potential of clinical trial data can be fully realized, paving the way for more efficient and insightful research outcomes.

Integrating data from disparate sources is crucial to realizing the full value of big data in clinical trials. For instance, developing machine learning models to predict patient responders requires integrating data collected from laboratories and clinics across multiple studies. However, even seemingly homogeneous data present integration challenges. These challenges begin with cross-platform normalization and extend to meta-analysis methods, multiple testing issues, and new logical and statistical complexities, which only increase with greater data heterogeneity. Such complexities underscore the importance of establishing a standardized approach to data integration to enable advanced analytics and machine learning applications.

High-quality clinical trial data serve as the backbone for addressing study objectives, supporting marketing approval. Unfortunately, data collected at clinical sites and entered in various clinical trial systems including EDC are often teemed with errors and missingness. While an EDC system normally utilizes automated edit checks to prevent the entry of inaccurate information, the scope of those checks is limited, and cross eCRF checks infeasible. The current manual review and cleaning of clinical data are time-consuming and labor-intensive. It is unsustainable for large late-stage trials, which may enroll hundreds or thousands of patients. The lack of data standards makes it difficult to automate the data review process. The presence of siloed data further diminishes the opportunity to drive insights from diverse data sources, as the lack of integration hinders comprehensive data analysis and interpretation.

For the analysis of these data to draw meaningful conclusions, it should consider the volume, variety, velocity and veracity of the integrated data. The large volume and high-dimensional, longitudinal, structured and unstructured nature of these data make the classical statistical methods such as multivariate regression analysis inept at elucidate robust evidence. AI and machine learning have been shown to provide workable solutions that tap into the large volume of clinical data from a variety of sources, some of which gathered continuously at real-time speed, to provide accurate clinical insights. However, fragmented data scattering across disconnected systems in vastly different formats makes it challenging to train AI and machine learning models to support clinical activities.

2.4.6 Cost

Considerable costs are associated with building CTMSs. According to the report by Karia and Taylor (2020), on average six or more systems are built and used for a typical clinical trial. Clinical protocols often undergo several amendments due to emerging data or changing regulatory thinking. Each amendment may require updates of systems. An average clinical protocol amendment can cost $250–450K due to the same document-centric issue. Other cost-related concerns also exist. For example, the medical community is worried about costly dependence on technology companies that own the proprietary technology of clinical trial systems and the potential for them to monopolize the hardware and software required for interoperability. Open-source technology may have the potential to ease the high-cost concerns and inflexibility associated with proprietary technologies. However, the downside is that open-source technologies often lack evidence of validation.

2.5 Digital Transformation

2.5.1 Impact of Digital Technologies

This rise of digital technologies, coupled with advanced analytics powered by AI and machine learning, has created a wide range of new data opportunities for clinical trials. With the increasing adoption of these technologies, there is a greater chance for digitizing clinical protocol, streamlining the design and creation of clinical trial systems, connecting siloed data, and automating data monitoring and analysis for effective trial oversight and decision-making. Applications of digital technologies and machine learning provide broad and better data, in real-time, and even in real-life settings, thus furthering more patient-centric clinical trials.

2.5.2 Holistic Data Solution

To streamline the clinical trial process and maximize the potential of diverse data gathered throughout the study, a holistic approach to constructing interconnected clinical trial systems and ensuring seamless data flow is essential. This comprehensive method involves three critical components: data architecture including systems, tools, and processes for data collection, curation, and integration, advanced analytics, and digital data flow (DDF) (Figure 2.3).

Data architecture involves designing a robust framework to collect, store, and manage data effectively. This ensures data integrity, security, and accessibility, which are crucial for reliable analysis and decision-making. Advanced analytics leverages sophisticated statistical and machine learning techniques to interpret complex datasets. By applying these methods, researchers can identify patterns, predict outcomes, and generate actionable insights, significantly enhancing the efficiency and accuracy of the trial. DDF emphasizes the importance of integrating digital technologies to facilitate real-time data collection and transmission. This component ensures that data from various sources are seamlessly aggregated and processed, minimizing delays and

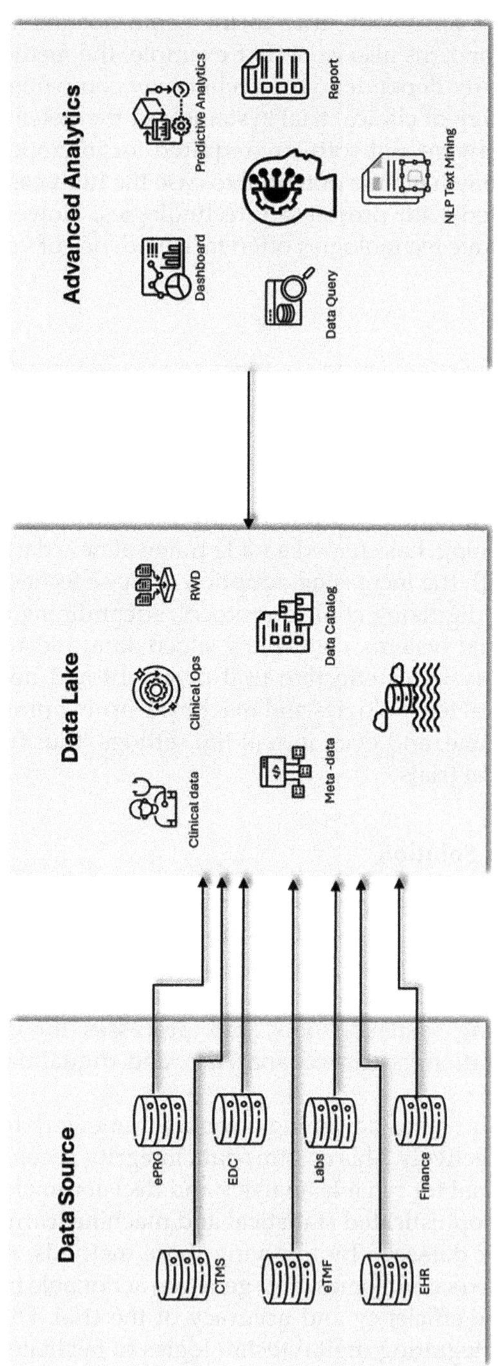

FIGURE 2.3
An end-to-end data solution to clinical trials.

errors associated with manual data handling. Together, these elements form a cohesive system that optimizes clinical trial processes, enabling more efficient, accurate, and insightful research outcomes.

2.5.2.1 Data Architecture

Data architecture is the cornerstone of efficient clinical trials. It encompasses various clinical trial systems and includes models, policies, and standards governing data collection, storage, organization, management, and security. Data extracted from these systems are integrated, mined, and analyzed in a central repository, enabling comprehensive and accurate insights. The digitization of these data, including documents into machine-readable formats, facilitates the development of AI-enabled analytics, thereby enhancing the ability to uncover patterns and predict outcomes.

Increasingly, cloud-based platforms are being utilized as repositories due to their numerous advantages. These platforms are cost-effective because they can be scaled up or down to fit a company's business needs, allowing for efficient resource management (Li and Tang 2021). Cloud platforms also enhance cross-functional collaboration and productivity by providing centralized access to data and tools. Furthermore, such platforms can significantly reduce maintenance efforts as system security, disaster recovery, software updates, and system monitoring are managed by the service providers.

Another important consideration in data architecture is its suitability for embedding and deploying analytic tools and AI applications across the enterprise. A well-designed data architecture ensures seamless integration and operational efficiency of these advanced technologies, which are pivotal for harnessing the full potential of big data and advanced analytics. Identifying and implementing the best-fitting data architecture are therefore crucial for delivering the promise of big data in clinical trials, enabling researchers to conduct more efficient, accurate, and insightful studies.

2.5.2.2 Advanced Analytics

Analytics has been transformative in revolutionizing how data are analyzed and critical knowledge for business decisions is extracted. AI-powered analytics, in particular, have demonstrated the ability to interrogate multiple data sources and extract insights from complex datasets more rapidly and, often, more accurately than traditional data specialists. These systems can process vast amounts of data in real time, identifying trends and patterns that might be missed by human analysts.

The development of user-friendly interfaces is essential for the widespread deployment and accessibility of these advanced analytics tools. Such interfaces ensure that employees across various locations and functions within a company can utilize the insights generated, fostering a data-driven decision-making culture. For instance, a well-designed dashboard can provide executives with real-time updates on key performance indicators, while operational staff can receive actionable insights pertinent to their daily tasks.

Developing effective analytics solutions requires a thorough understanding of the business's needs and the data sources available. It is crucial to align the analytics goals with the company's strategic objectives to ensure relevance and impact. Additionally, for addressing specific questions, companies must balance the use of external capabilities and internal tool development to remain cost-effective. Leveraging external expertise can provide access to cutting-edge technologies and methodologies, while internal development ensures customization and alignment with unique business processes.

Striking this balance is crucial for maximizing the return on investment in analytics. By carefully selecting which tools and capabilities to develop internally versus externally, companies can optimize their analytics infrastructure for both efficiency and effectiveness, ultimately driving better business outcomes.

2.5.2.3 Digital Data Flow

Key to delivering a holistic end-to-end solution for clinical trial design, analysis, start-up, conduct, and close-out is the digitization of data flow throughout the entire lifecycle of clinical trials. This digitization ensures that data are consistently captured and stored, in accordance with the FAIR principles. By implementing DDF, researchers can streamline processes, reduce errors associated with manual data entry, and facilitate seamless integration of various data sources.

Furthermore, digitization supports advanced analytics and AI applications, which can provide deeper insights and predictive capabilities that are not possible with traditional methods. This enables more informed decision-making, faster identification of trends, and more efficient allocation of resources. Cloud-based platforms play a crucial role in this process by offering scalable, secure, and cost-effective solutions for data management. These platforms allow for real-time data sharing and collaboration across different locations and teams, further enhancing the trial's efficiency and productivity.

Moreover, a digitized data flow ensures compliance with regulatory standards by maintaining accurate and up-to-date records that can be easily audited. This comprehensive approach not only accelerates the trial process but also improves the overall quality and reliability of the data, ultimately leading to better clinical outcomes.

2.6 Initiatives in Digital Data Flow

Various organizations have spearheaded initiatives aimed at laying the groundwork for digital transformation in clinical research.

2.6.1 ICH Guidance M11

In 2022, ICH working group published ICH Guidance M11 Clinical Electronic Structured Harmonized Protocol, also known as CeSharP (EMA 2022). The guideline provides a template and outlines technical specifications with a special emphasis on content definition for electronic exchange and design for content reuse. The publication of the guideline represents a pivotal step toward standardization and interoperability.

2.6.2 TransCelerate Common Protocol Template

Similarly, in 2016, TransCelerate published the TransCelerate Common Protocol Template (CPT). The CPT offers a harmonized and simplified format for clinical trial protocols, designed to enhance comprehension for study sites and global regulatory authorities. Its purpose is to facilitate the automation of numerous clinical processes while adhering to industry data standards. The CPT incorporates a standardized structure, suggested text, and regulatory approved endpoint definitions, providing flexibility for users to adopt with minimal adjustments. Originally introduced in 2015, the CPT has continuously evolved to meet evolving needs in the field. Fully aligned with CDISC's Clinical Data Standards, the guideline has facilitated industry-wide adoption, mitigating compliance challenges associated with evolving guidelines.

2.6.3 Unified Study Definition Model

Owing to their ongoing collaborations, in 2023 the Clinical Data Interchange Standard Consortium (CDISC) and TransCelerate have developed and released a framework, called Unified Study Definitions Model (USDM), based on the principle of data interoperability among various clinical systems. The USDM, accompanied by a Study Definitions Repository (SDR), forms the backbone for digital documentation of protocol-specified information, enabling seamless exchange and interoperability across diverse platforms (DiCicco and Evans 2023).

2.6.4 TransCelerate Digital Data Flow Initiative

In 2018, TransCelerate and other stakeholders launched a DDF initiative, aimed at catalyzing digital transformation in clinical trial and breaking the protocol documentation paradigm to enable seamless data flow (TransCelerate n.d.).

2.6.4.1 Shift from Document to Digital Data

The TransCelerate's DDF initiative is intended to transform the current document-based clinical trial process into a digital paradigm for clinical trials across the industry. It intends to modernize clinical trials by enabling a digital

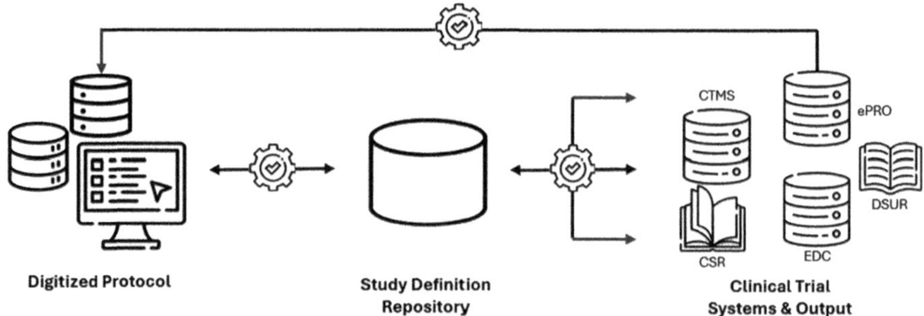

FIGURE 2.4
Future state of clinical data flow. Digitalized one-to-many process; Digital paradigm for protocol creation, with fully automated data flow and interoperability between systems. (Adapted from TransCelerate n.d.)

workflow to allow for automated creation of study assets including study protocol and configuration of study systems to support clinical trial conduct and execution. One of the key objectives of the DDF initiative is to replace the conventional clinical protocol with Study Design Metadata, spanning the entire study lifecycle. By standardizing data formats, DDF envisions automated and parallel data flow between systems, ensuring end-to-end traceability, consistency, efficiency, quality, and compliance (Figure 2.4).

The DDF intends to use vendor-agnostic, open-source solutions to enable rapid and seamless adoption by the pharmaceutical industry, ensuring scalability, flexibility, and adaptability to evolving requirements. Additionally, the DDF initiative is also intended to achieve several other objectives as outlined in Table 2.1 (TransCelerate n.d.).

2.6.4.2 Key Components of DDF

Key components of the DDF framework consist of both upstream and downstream systems interconnected through a Study Definition Repository (SDR), which leverages data and technology standards to ensure seamless integration. The vision of the DDF is to digitize clinical protocols, thereby improving data flow and facilitating efficient data exchange across various systems. To realize this vision, protocols are created using a specialized Study Builder tool. This tool enables the creation of a SDR, where elements of the study protocol are represented in machine-readable and executable formats. These elements are then made readily accessible across multiple systems through Application Programming Interfaces (APIs).

The successful implementation of DDF is pivotal for achieving automation, interoperability, and reuse throughout the study lifecycle. By standardizing the study protocols in a digital format, the DDF ensures that data can be consistently and accurately shared between different systems, reducing

TABLE 2.1

Objectives of DDF

Objective	Description
Digital Protocols	Moving from document-based protocols to digital formats simplifies the clinical trial process, enhancing efficiency, accuracy, and transparency.
Connectivity of Data and Processes	Establishing seamless connectivity between data and processes enables end-to-end lineage and automated information flows, fostering better decision-making and improved outcomes.
Advanced Analytics	By digitizing protocols and establishing lineage to downstream systems, DDF enables advanced analytics to optimize study design, patient engagement, operational efficiencies, and compliance, leading to better decision-making and outcomes.
Open and Flexible Solutions	Adopting a vendor-agnostic approach allows for the creation of customized solutions, scalability, and collaboration, facilitating easy adaptation to changes and fostering innovation in a plug-and-play mode.
Interoperability	DDF adheres to development principles aimed at promoting interoperability, efficiency, and agility in clinical trials:
Agile Development	Embracing an iterative, collaborative approach to software development facilitates flexibility and responsiveness to changing requirements.
Dynamic Alignment to Standards	DDF supports dynamic alignment to standards by providing a flexible framework for data exchange adaptable to different systems and technologies while adhering to CDISC standards.

redundancy and minimizing errors. This not only streamlines the process of data collection and analysis but also enhances the ability to monitor and manage clinical trials in real-time. Moreover, the DDF's standardized approach promotes greater collaboration and efficiency in clinical research, ultimately accelerating the development of new treatments and therapies. In essence, the DDF represents a significant advancement in the quest to modernize and optimize clinical trial processes (Figure 2.5).

The key components of the DDF framework include the Source/Upstream System, SDR, USDM, API, and CT. These components collectively facilitate streamlined data exchange and interoperability within clinical trials.

The Source/Upstream System serves as the initial point of data collection, ensuring that all relevant information is captured accurately. The SDR acts as a central hub where study protocols are stored in a machine-readable format, promoting consistency and ease of access. The USDM provides a standardized template for defining study elements, enabling uniformity across different trials and systems.

APIs play a crucial role by allowing various software applications to communicate and share data seamlessly, ensuring that information flows smoothly between the Source/Upstream System and the SDR. Finally, CT ensures that

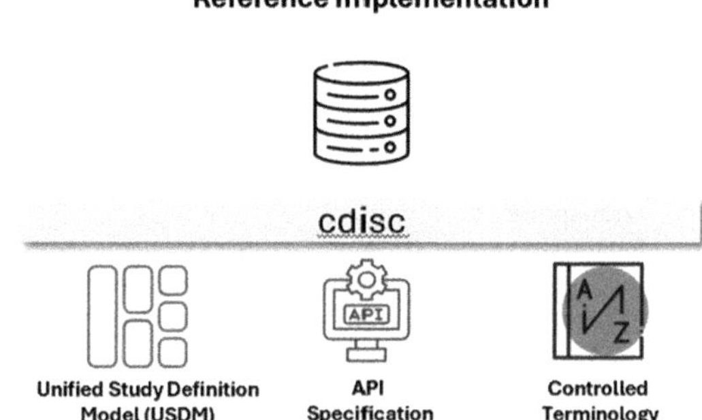

FIGURE 2.5
Study definition repository (SDR) reference implementation. (Adapted from TransCelerate n.d.)

TABLE 2.2
Key Components of DDF

Component	Description
Upstream System	The source system, comprising a protocol authoring tool, captures major protocol components using standards from the Study Definition Repository (SDR).
Downstream System	CTMS, eTMF, EDC, IRT, and other systems supporting clinical trial execution are all downstream system that consume information from the study protocol.
Study Definition Repository (SDR)	Serving as a middle layer, SDR efficiently manages information flow between study definitions/protocols and execution systems, supporting downstream applications via APIs.
Unified Study Definitions Model (USDM)	Developed by CDISC, USDM fosters data interoperability, enabling sharing and reuse of trial data across studies and organizations, and aligns with CDISC data standards.
Application Programming Interface (API)	APIs facilitate seamless integration between systems by transferring data based on predefined mappings, ensuring parallel execution of study activities and maintaining traceability.
Controlled Terminology (CT)	Standardized lists of values within CT streamline data selection within USDM, promoting consistency and interoperability across different systems.

all data elements are described using consistent and standardized language, reducing ambiguity and enhancing data clarity.

Together, these components of the DDF framework enable efficient data integration, improving the accuracy and reliability of clinical trial data. This streamlined approach not only reduces redundancies but also accelerates the research process, paving the way for faster development and approval of new therapies (Table 2.2).

2.7 Adoption of Digital Technologies in Clinical Trial

2.7.1 Technology Providers

Technology providers play a pivotal role in ensuring the maintenance of data standards through robust governance models enforced via underlying technology. Implementing vendor-agnostic, open-source solutions enable rapid and seamless adoption by the pharmaceutical industry, ensuring scalability, flexibility, and adaptability to evolving requirements. In the past few years, a significant number of pharmaceutical companies, academic institutions, government agencies are partnering up with technology start-up companies to the utility of integrating various digital platforms into clinical trials (Mitsi et al. 2022). Zhang et al. (2022) presented a case study where blockchain technology is utilized to construct a CTMS spanning all stages of clinical trials. The system incorporates several key features, including a sharable trial master file system, an expedited recruitment and simplified enrollment process, a secure and reliable EDC system, a reproducible data analytics platform, and a streamlined payment and reimbursement system with traceability. In a pilot study by Govil and Vundela (2019), a recommender system based on machine learning algorithms was discussed. The system creates a preliminary list of CRFs from a global library of standard CRFs, based on information extracted from the protocol.

There are many technology companies actively developing integrated data and analytics solutions to clinical trials. A select list of these start-ups is presented in Table 2.3.

Using AI to streamline DDF, Deloitte develops clinical management tools that create structured, standardized, and digital data elements from a range of input documents and sources (Karia and Taylor 2020). Once integrated with the existing core clinical trials systems, Deloitte's products promulgate the sharing of data and metadata common to all the systems. Various other startups are investigating and developing tools and platforms to enable the adoption of digital health technologies (DHTs). An AI-powered medical writing tool is developed by Feroze (2021). The tool autonomously extracts and consolidates information from source documents such as protocols, SAPs, and in-text data, organizing them into the relevant sections in accordance with the ICH E3 guidelines for CSRs. This automation significantly reduces the workload for medical writers. Additionally, the tool features workflow and dashboard modules that are customizable to meet the specific requirements of sponsors.

2.7.2 Adoption of Digital Technologies

In the past decade, there has been increasing utilization of DHTs in clinical trials and the trend is likely going up. DHTs have unlocked numerous new

TABLE 2.3

Select List of Technology Solution Providers

Company	Focus Area	Technology Platform	Reference
Deloitte	Clinical trial process automation	AI and ML-solutions: Study builder, CRF builder, automated SDTM conversion, ADaM conversion, SAP and TLF creation, and CSR writing	Deloitte
Taimei Technology	EDC	AI, NLP for EDC build, and creation of edit checks	Taimei
Indegene	EDC	Automation of EDC build and medical coding	Indegene
Deep 6 AI	Patient matching	AI, NLP for mining unstructured medical data to identify eligible patients	Deep 6 AI
Antidote Technologies	Patient recruitment	Smart search for studies via AI	Antidote Technologies
Paradigm	Cancer patient matching	AI analyzing clinical and pathology reports to find best trial matches for patients	Paradigm
Saama Technologies	Recruitment, and data curation and analysis	ML algorithms for analyzing clinical data and optimizing trial operations	Samma Technologies
Clindata Insight	Data analysis	AI tools for data cleaning, normalization, and analysis	Clindata Insight
ZYLIQ	Clinical study report (CSR)	Automation of CSR writing	Zyliq
Medidata Solutions	Platform data solution	AI tools for optimizing trial design, recruiting, and engaging patients, and managing clinical data.	MEDIDATA
ThoughtSphere	Data lake, data curation and integration, analytics	Integrated clinical data solutions, including platform streamlining data flow and clinical operations	ThoughtSphere
Yonalink	Platform data solution	AI-powered EDC, HER-to-EDC data streaming, ePROS, eCOA, eConsent, etc.	Yonalink
CDISC	Standards-setting	Developing standards for medical research data to enable information system interoperability to improve medical research and related areas of healthcare.	CDISC
TransCelerate BIOPHARMA	Clinical trial innovation and acceleration	Standards-setting, best practice, innovative solution in clinical trials	TransCelerate

data opportunities for clinical trials. These technologies encompass a wide range of tools, including software such as mobile health apps, hardware like wearable devices and sensors, and telemedicine platform solutions. In neurology trials, DHTs have demonstrated the ability to provide more accurate and comprehensive data from real-life settings, enhancing the relevance and applicability of trial outcomes.

The integration of DHTs into clinical trial design has facilitated more patient-centered research by enabling continuous monitoring and real-time data collection. This shift allows researchers to capture a holistic view of patient experiences and health outcomes, leading to more nuanced insights and personalized interventions. Additionally, the use of DHTs supports real-world data-driven decisions, which are crucial for developing effective and safe treatments. As advances in DHTs continue to evolve, their impact on clinical trials will likely expand, promoting more efficient, accurate, and patient-focused research.

However, despite these advances the intake of such technologies is uneven across various aspects of clinical trials. As presented in Figure 2.6 (TransCelerate 2023), the use of EDC systems for data management has become a common practice, there have been a significant adoption of technology systems to match patients with studies and connect site with various data sources. What is clearly lagging in technology intake are digitization of study protocol, broad utilization of wearable devices and sensors to collect key study endpoints and remote study visits. Other challenges also exist in the adoption of DHT in clinical trials (Mittermaier et al. 2023), including inaccessibility of DHTs due to limited access to the Internet or spare technology literacy (Reiner et al. 2019; Saeed and Masters 2021), lack of regulatory framework for validation and evaluation digital endpoints (Goldsack et al. 2020; Godfrey et al. 2020), and participant authentication and data reliability (Sonla 2007; Zola Matuvanga et al. 2021; Bottcher 2022). Although

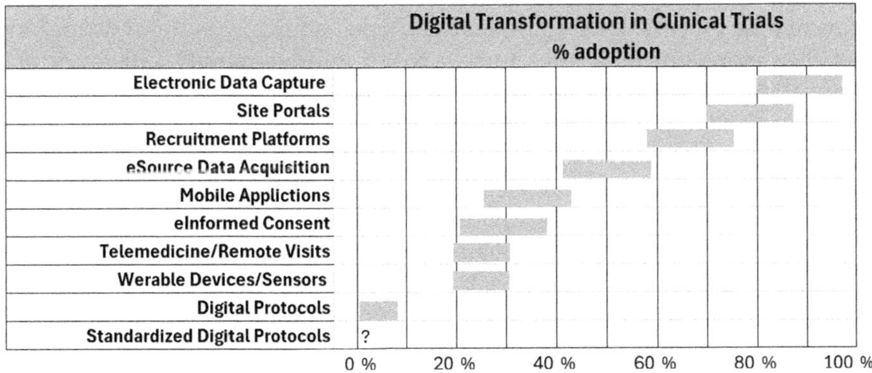

FIGURE 2.6
Adoption of DHTs in clinical trials. (Adapted from TransCelerate 2023.)

some of these issues, such as participant identity, can be addressed as new technologies emerge (Zhao et al. 2020; Zeng et al. 2017) broad applications of these methods require special considerations of participants' privacy (Baig and Eskeland 2021).

2.8 Concluding Remarks

Clinical trials involve intricate systems, multifaceted processes, data from diverse sources, and various stakeholders. Ensuring seamless data flow across different systems, studies, and programs is crucial for maintaining data quality, facilitating timely decision-making, achieving regulatory compliance, and ultimately ensuring the success of the trials. Despite the widespread adoption of technological platforms, deriving actionable insights from aggregated data to facilitate timely decision-making remains a significant challenge. The current disconnects among many clinical trial systems and processes result in siloed and under-utilized data. The data flow in clinical trials is often burdened with manual effort, rework, and inefficiency, which impedes timely insights and decision-making. These challenges are compounded by the increasing volume of data generated by DHTs, leading to additional complexities and new challenges in clinical trials.

DHTs play a pivotal role in enabling DDF and driving digital transformation in clinical trials. DHTs empower investigators to gather continuously diverse data in real-world scenarios, enabling the capture of data types previously unattainable. Moreover, they have demonstrated the ability to expedite patient recruitment and enhance both existing and novel endpoints through derived measures. By facilitating the creation of digital protocols, ensuring connectivity between systems, promoting advanced analytics, and providing open and flexible solutions, technology providers significantly optimize the clinical trial process. DDF's adherence to open-source, vendor-agnostic, and agile development principles further foster interoperability, efficiency, and innovation in clinical research.

The utilization of DHTs in clinical trials has seen a significant rise over the past decade, with ongoing growth and development (Baker et al. 2021). It is crucial to ensure the equitable deployment of DHTs in trials, aiming to bridge rather than widen the digital divide. This may necessitate substantial social investment from trial sponsors, supported by governmental guidance. The utilization of DHTs in clinical trials holds the potential to revolutionize the landscape, paving the way for a virtual era of distributed clinical trials.

3

Enhancing Clinical Operations Efficiency and Effectiveness

3.1 Introduction

In today's rapidly evolving healthcare landscape, the efficiency and effectiveness of clinical operations play a critical role in driving the development and delivery of new therapies to patients in need. From trial feasibility assessment to patient recruitment, adherence, and retention, every aspect of the clinical trial process presents unique challenges and opportunities for improvement. As the demand for innovative treatments continues to grow, stakeholders across the pharmaceutical and biotech industries are increasingly turning to artificial intelligence (AI) and advanced technologies to streamline operations, accelerate timelines, and optimize resource allocation. These technological advances enable a patient-centric approach to recruit and engage participants and broader collaboration among stakeholders to improve access to clinical trials, alleviate the patient burden, increase the diversity of participants, and accelerate approval of breakthrough therapies. This chapter begins with a brief discussion about the role of clinical operations in drug development, and evolution of the profession in that process. This is followed by the exploration of various strategies and technologies aimed at enhancing clinical operations efficiency, drawing on real-world examples and case studies to illustrate best practices and emerging trends. Embracing new technologies, methodologies, and best practices, Clinical Operations can unlock new opportunities in an increasingly dynamic and competitive landscape of clinical development.

3.2 Clinical Operations

3.2.1 Clinical Operations Process

Clinical trials are crucial for advancing medical treatments, but their success hinges on meticulous planning, execution, and oversight. Clinical Operations

DOI: 10.1201/9781003226086-3

is a multidisciplinary team comprising Project Managers, Study Managers, Clinical Research Associates (CRAs), Data Managers, Quality Associates, Drug Supply Personnel, and other supporting staff. These professionals are responsible for ensuring that trials are conducted ethically, scientifically, and within designated timelines. By adhering to best practices, Clinical Operations staff can navigate the complexities of clinical research and achieve meaningful advancements in healthcare (Spilker 1991).

The Clinical Operations process encompasses a wide array of activities spanning from the planning to the close-out of clinical trials. This process is visualized in Figure 3.1. By adhering to a holistic and structured approach, the Clinical Operations team ensures that clinical trials are conducted efficiently, ethically, and to the highest scientific standards, ultimately contributing to the advancement of medicine and patient care.

Below, we describe the Clinical Operations process in detail, highlighting the roles and responsibilities that drive the clinical program forward.

3.2.1.1 Study Planning

The process starts with study planning. During this phase, the Clinical Operations team plays a crucial role in laying the groundwork for successful downstream activities related to study start-up, conduct, and close-out. Closely collaborating with study team members, including clinical scientists, medical monitors, and biostatisticians, the Clinical Operations team provides input to the study protocol and informed consent form (ICF) from a trial execution standpoint (Friedman et al. 2010).

The team also develops assumptions regarding drug supply, dosage, and administration to provide initial guidance for planning purposes. This involves identifying potential vendors and contract research organizations (CROs) based on expertise, capabilities, and past performance, initiating the process of qualification (Spilker 1991). An initial budget is drafted, considering various study components such as site fees, vendor costs, and drug supply expenses.

Another key responsibility of this team is devising a strategy for selecting suitable clinical trial sites based on criteria such as patient population, infrastructure, and investigator experience. They establish estimates for patient enrollment rates and study timelines to guide planning efforts and resource allocation. Clinical trials can be affected by many unforeseen events, such as the COVID-19 pandemic. To manage these risks, the team proactively identifies potential risks to the study and develops mitigation strategies to ensure smooth operations and minimize disruptions (ICH 2023). It also evaluates resource requirements, including personnel, equipment, and facilities, to support study activities effectively.

The team is responsible for forming and briefing the study management team on their roles, responsibilities, and timelines for the upcoming study,

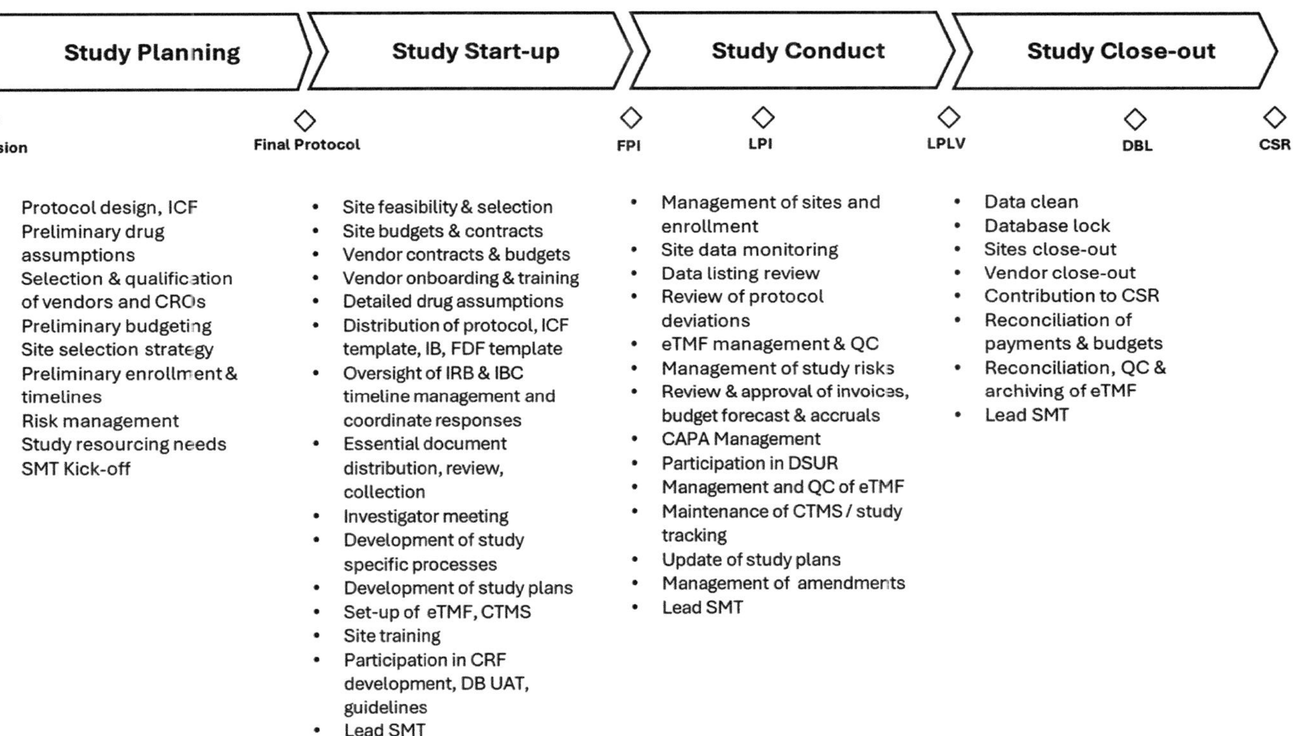

FIGURE 3.1
Clinical Operations Process.

Note: FPI = First Patient In; LPI = Last Patient In; LPLV = Last Patient Last Visit; DBL = Database Lock; CSR = Clinical Study Report; TMF = Trial Master File.

ensuring alignment and clarity among team members (Spilker 1991). These activities collectively form the foundation for subsequent phases of the clinical trial, setting the stage for efficient study start-up and successful study conduct (Friedman et al. 2010).

3.2.1.2 Study Start-Up

Clinical study start-up is a critical phase where meticulous planning and execution are imperative for the successful initiation of a clinical trial. During the study start-up phase, the Clinical Operations team transitions from planning to implementation, executing various tasks to prepare for the initiation of the clinical trial. The related activities in this phase are shown in Figure 3.1. This phase involves several key activities.

Firstly, site feasibility and selection are meticulously conducted to identify suitable sites based on patient demographics, infrastructure, and regulatory compliance (Friedman et al. 2010). Negotiations of site budgets and contracts are carried out to establish clear expectations and financial agreements. Similarly, vendor contracts and budgets are finalized, and vendors are onboarded and trained to ensure alignment with study protocols and standards (Spilker 1991).

Detailed drug assumptions are refined to guide drug-related activities effectively. The protocol, ICF, investigator brochure (IB) are submitted for Institutional Review Board (IRB) and Independent Ethics Committee (IEC) reviews and approvals, and essential documents are collected for regulatory compliance (ICH 2023).

Investigators are trained on study protocols and regulatory requirements to ensure they are well-prepared for their roles. Furthermore, study-specific processes are developed and documented to ensure consistency and adherence to protocols. Study plans are meticulously drafted to provide clear guidelines for study conduct (Friedman et al. 2010).

The eTMF and CTMS for the study are set up to facilitate efficient document management and study tracking. Site training is developed and conducted to educate site staff adequately. The Clinical Operations team actively participates in the development of case report forms (CRFs), user acceptance testing (UAT) of the database, and establishment of data management guidelines (Spilker 1991).

Finally, the Clinical Operations team leads the study management team, providing guidance, support, and oversight throughout the study start-up phase to ensure alignment with study objectives and timelines. These activities collectively ensure a robust foundation for the clinical trial, setting the stage for successful study execution and management.

3.2.1.3 Study Conduct

During the clinical study conduct phase, the Clinical Operations team undertakes various crucial responsibilities to ensure the smooth progress and integrity of the trial. This involves actively managing sites and enrollment, overseeing site data monitoring, and reviewing data listings to ensure accuracy and compliance with the protocol (Friedman et al. 2010). Additionally, the team diligently monitors and addresses protocol deviations while maintaining the electronic Trial Master File (eTMF) to ensure all essential documents are up-to-date and readily accessible (Spilker 1991).

Managing study risks is paramount, necessitating continuous risk assessment and the implementation of mitigation strategies. Financial oversight includes reviewing and approving invoices, completing budget forecasts, and managing accruals to maintain fiscal responsibility (FDA 2020d). Effective Corrective and Preventive Action (CAPA) management is implemented to address any issues promptly (ICH 2023).

Participation in the Development Safety Update Report (DSUR) is crucial for ongoing safety monitoring. The team also maintains and performs quality control on the eTMF CTMS for accurate study tracking (Friedman et al. 2010). Regular updates to study plans are made as necessary, and amendments to the protocol are managed efficiently to ensure compliance and data integrity throughout the study duration (Spilker 1991).

3.2.1.4 Study Close-Out

Clinical study close-out is a critical phase ensuring the integrity and completeness of research endeavors. It begins with meticulous scrutiny to ensure data cleanliness, meticulously validating each data point to guarantee accuracy and reliability (Friedman et al. 2010). Subsequently, the study database undergoes locking, safeguarding against further modifications and solidifying the dataset's integrity (Spilker 1991). Sites involved in the study are systematically closed out, ensuring compliance with protocols and regulatory standards. Concurrently, vendors integral to the study's execution are also closed out, their contributions thoroughly evaluated and accounted for (ICH 2023)

Moreover, close-out activities extend to the contribution to the Clinical Study Report (CSR), synthesizing findings for dissemination (ICH 2016). Financial aspects are meticulously managed, with payments and budgets reconciled to ensure transparency and accountability. Finally, the eTMF undergoes reconciliation, stringent quality control, and archival, preserving essential documentation for future reference and regulatory compliance. Each aspect of close-out is executed with precision, ensuring the study's conclusions are founded on a robust and comprehensive framework.

3.2.1.5 Managing Clinical Operations

Navigating the complexities of Clinical Operations requires meticulous planning, seamless execution, and effective communication. Establishing timelines, defining team roles, and setting study milestones are paramount.

A well-structured plan mitigates the risk of delays and ensures efficient resource allocation. Organizing the project team according to study requirements is essential for streamlined operations. Assigning roles such as project managers, study managers, CRAs, and quality control managers ensures effective coordination and communication throughout the trial. Effective stakeholder management, both internal and external, is indispensable. Keeping stakeholders informed and engaged facilitates informed decision-making and supports project objectives. Strategic communication plans tailored to different stakeholders. Continuous monitoring and follow-up with study sites are imperative throughout the trial duration. This involves assessing site performance, verifying data submissions, and ensuring protocol compliance to uphold trial integrity. Quality assurance activities are integral to clinical operations, encompassing site monitoring, data accuracy verification, and participant safety assurance. These measures safeguard the reliability and validity of trial outcomes.

Overall, Clinical Operations serves as the linchpin of successful clinical trials. Its multifaceted approach, encompassing planning, execution, and oversight, ensures trials are conducted rigorously and ethically.

3.2.2 Emerging Opportunities

Clinical trials serve as the cornerstone of evidence-based medicine, providing crucial insights into the safety, efficacy, and real-world utility of new therapies. However, conducting clinical trials is a complex and resource-intensive process that often requires significant time, effort, and investment. Numerous challenges exist in the current clinical trial process, including challenges in patient and site selection, slow enrollment due to complex protocols, inadequate site training, lack of awareness of the trial, and lack of interest in participating (Askin et al. 2023). Time-consuming manual processes in data collection, cleaning, and analysis further complicate the process. In fact, under-enrollment has been a significant barrier to trial success. Per Array (2023), nearly 80% of all trials fall short of meeting their initial enrollment deadlines, while 55% are prematurely terminated due to failure to achieve full enrollment. Lack of technological infrastructure to cope with the complexity of conducting clinical studies has contributed to unsuccessful trials (Harrer et al. 2019).

In recent years, the pharmaceutical industry has faced mounting pressure to accelerate the drug development process and bring new treatments to market more efficiently, particularly in response to emerging global health challenges such as the COVID-19 pandemic. Against this backdrop, enhancing clinical operations efficiency has emerged as a top priority for pharmaceutical companies, CROs, and other stakeholders involved in clinical research. By optimizing trial feasibility assessment, streamlining patient recruitment and retention efforts, and standardizing and automating data management and

	Trial Design	Trial Start-up	Trial Conduct	Trial Close-out
Advanced Data Analytics and AI Automation	Assess site performance using real-time monitoring	Mine EHRs and publicly available content, including trial databases and social media, to help match patients with trials by using NLP and ML	Assess site performance (e.g. enrollment and dropout rates) with real-time monitoring	Complete sections of final clinical trial report for submission by using NLP
	Assess site performance using real-time monitoring	Create drafts of investigator and site contracts and confidentiality agreements by smart automation	Analyze digital biomarkers on disease progression and other quality-of-life indicators	Data cleaning by ML methods
	Analyze and interpret unstructured and structured data from historical trials and scientific literature		Automate sharing of data across multiple systems	
AI-enhanced Mobile Applications, Wearables, Biosensors, and Connected Devices		Expedite recruitment and create a more representative study cohort through cloud-based applications	Enhance adherence through smartphone alerts and reminders	
		Simplify and accelerate the informed consent process using eConsent	eTracking of medication using smart pillboxes, and tools for visual confirmation of treatment compliance	
			eTracking of missed clinic visits and trigger non-adherence alerts	

FIGURE 3.2
AI and DHT-enabled applications in clinical trials. (Adapted from Taylor et al. 2020.)

analysis, organizations can overcome common bottlenecks and drive innovation in drug development.

Emerging technologies, such as AI and RWD integration, offer promising solutions to these challenges. AI can enhance patient matching, predict patient dropouts, and optimize site selection (Weissler et al. 2021). Integrating RWD into clinical trials can improve the relevance and applicability of study findings, though it presents challenges such as data standardization and interoperability. Digital health technologies, including mobile health apps and wearable devices, facilitate remote monitoring and data collection, thereby improving patient engagement and adherence (Harrer et al. 2019).

Furthermore, decentralized clinical trials (DCTs) are gaining traction as they allow for more flexible and patient-centric study designs. DCTs leverage telemedicine, home health visits, and digital data collection tools to reduce patient burden and improve access to trials, especially for populations that are traditionally underrepresented in clinical research (FDA 2023b). This approach not only enhances patient recruitment and retention but also ensures more diverse and representative study populations.

In the latest report by Taylor et al. (2020), a host of opportunities enabled by digital infrastructure and AI-based algorithms are provided as shown in Figure 3.2. In the following sections, we expound on selecting AI-driven solutions that are enhancing clinical trial operations.

3.3 Technology-Enabled Solutions to Clinical Operations

3.3.1 Trial Design

3.3.1.1 Patient Perspective

Clinical trials are intended to generate evidence supporting drug approval for unmet medical needs. As such, patients shall be at the front and center of these efforts. Developing a clinical study protocol from a patient-centric perspective involves designing trials that prioritize the needs, preferences, and experiences of participants. This approach begins with incorporating patient input into protocol development to ensure that the study addresses relevant health issues and outcomes that matter most to patients. Simplifying eligibility criteria is crucial to making trials more inclusive and reflective of the diverse patient population (Harrer et al. 2019). Patient-centric protocols also emphasize the minimization of patient burden by reducing the frequency of hospital visits, utilizing telemedicine, and incorporating home-based health assessments (FDA 2020d). Furthermore, clear and concise informed consent documents are essential to help patients understand the study's purpose, procedures, risks, and benefits, thereby fostering trust and willingness to participate (Emanuel et al. 2000). Employing digital health tools, such as mobile apps and wearable devices, can enhance patient engagement and adherence by providing real-time monitoring and support (Peng et al. 2020). By adopting a patient-centric approach, clinical trials can improve recruitment, retention, and overall patient satisfaction, ultimately leading to more robust and generalizable findings.

3.3.1.2 Optimizing Eligibility Criteria

One of the key considerations in clinical trial design is patient population and the feasibility of completing the study in the target population in a timely manner. Patient recruitment is one of the most time-intensive aspects of a clinical trial and may consume up to one-third of the study duration. Alarmingly, one in five trials fails to meet the requisite participant numbers, with nearly all trials surpassing anticipated recruitment timelines (Hutson 2024). Various methods have been suggested by researchers to optimize the eligibility criteria. One is to use eligibility criteria that require alternate and less invasive procedures to patient and clinical staff burden (Yang and Deepak 2023; Weissler et al. 2021). The others were centered on using AI technologies to streamline the optimization of eligibility criteria, reducing guesswork and manual labor. For example, at Stanford University, a team led by James Zou, a biomedical data scientist, pioneered Trial Pathfinder, a system designed to evaluate the impact of adjusting trial eligibility criteria on study efficiency and effectiveness (Liu et al. 2021). In a notable study, they applied Trial Pathfinder to drug trials targeting a specific type of lung cancer. The findings revealed that aligning criteria adjustments with Trial Pathfinder's recommendations could potentially double the pool of eligible patients without elevating the hazard ratio. Moreover, this system demonstrated efficacy across various cancer types, mitigating adverse outcomes by extending treatment eligibility to individuals with more severe conditions, who stand to benefit the most from the drugs. Notably, several pharmaceutical firms, including Roche, Genentech, and AstraZeneca, have embraced Trial Pathfinder. Recent advancements, such as AutoTrial developed by Sun's lab in Illinois, introduced a method for training large language models. This enables researchers to input trial descriptions, prompting the model to generate appropriate criterion ranges, such as body mass index thresholds, facilitating further automation and efficiency in trial design (Huston 2023).

3.3.2 Accelerating Trial Startup

Clinical trials are fundamental to the advancement of medical science, but their execution often faces challenges such as lengthy timelines, high costs, and logistical complexities. Accelerating trial startup is crucial for expediting the development of new therapies and bringing them to patients in need. In this section, we will explore innovative strategies and technologies aimed at streamlining the startup process, with a focus on feasibility assessment, smart automation for site contracts and confidential agreements, eConsent solutions, and DCTs.

3.3.2.1 Evidence-Based Feasibility Assessment

Feasibility assessment during trial planning is a process of evaluating the possibility for the successful conduct and delivery of a clinical program/trial in a particular geographical region characterized by project completion within timelines, targets and cost. There are three broad types of feasibilities (Rajadhyaksha 2010). At the program level, the feasibility concerns the entire program of multiple planned studies. On the other hand, study level feasibility aims to ensure that the trial is practical and the end goals achievable, within the intended setting, the resources required for delivery are available, and recruitment targets are realistic (Gloy et al. 2022). In contrast, feasibility at the site or investigator level is related to conducting the trial at a particular site. Establishing the feasibility of a clinical trial before starting the study is crucial in mitigating the risk of poor performance, insufficient recruitment, and an unacceptable high number of protocol violations (Butryn et al. 2016). As reported in the literature, one out of four RCTs are not completed as planned, owing to lower than projected recruitment (Bernardez-Pereira et al. 2014; Kasenda et al. 2014). Similar findings have also been reported regarding non-randomized clinical trials (Carlisle et al. 2015).

Traditionally, feasibility assessment relies on surveys of prospective investigators, past experience, and personal relationships within the clinical community. However, the robustness of this method is hampered by multiple issues, as much of the feedback is based on educated guesses and overestimation due to enthusiasm about the trial (Johnson 2015). Furthermore, the lack of access to the most current and comprehensive data makes it challenging to accurately assess patient availability at potential clinical sites. As a result, clinical trials frequently encounter under-enrollment issues, leading to delays and increased costs. Overly restrictive eligibility criteria can further exacerbate enrollment challenges by limiting the pool of eligible participants, particularly if these criteria do not align with the characteristics of the patient population.

In addition to enrollment challenges, another critical aspect of trial feasibility is ensuring that eligible patients are properly incentivized and aware of the trial opportunities available to them. Without effective strategies to address these issues, clinical trials may face significant hurdles that impede their progress and success.

Recognizing these challenges, pharmaceutical companies are increasingly turning to RWD sourced from EHRs to inform various aspects of clinical trial planning and execution. These data sources provide information such as diagnostic and procedure codes that enable the identification of eligible patients for the trial. By leveraging AI-enabled analytics, researchers can extract valuable insights from EHRs to support informed decision-making about trial feasibility.

One of the primary applications of RWD in clinical trial feasibility is the identification of high-enrolling sites. By analyzing historical patient data stored in EHRs, researchers can identify sites with a track record of successfully enrolling patients with similar demographic and clinical characteristics. This proactive approach allows sponsors to prioritize sites with a higher likelihood of meeting enrollment targets, thereby minimizing the risk of under-enrollment and associated delays. Furthermore, RWD can also be utilized to assess patient eligibility criteria and their impact on trial feasibility. AI-enabled analytics enable researchers to simulate the effects of modifying eligibility criteria on patient availability at potential sites (Clinical Trial Transformation Initiative (CTTI) 2024). By adjusting inclusion and exclusion criteria based on real-world patient data, sponsors can optimize trial protocols to better reflect the characteristics of the target patient population. This iterative process not only increases the likelihood of achieving enrollment targets within specified timelines but also minimizes the need for costly protocol amendments necessitated by overly restrictive eligibility criteria.

Several commercial entities, such as Quant Health and ConcertAI, provide support for optimizing inclusion and exclusion criteria, though peer-reviewed evidence regarding the effectiveness of their methods is not yet available. Several case studies and initiatives, such as those developed by the Clinical Trials Transformation Initiative (CTTI), have demonstrated the potential of RWD in enhancing trial feasibility and recruitment. These case studies highlight the successful application of RWD in identifying high-enrolling sites, refining eligibility criteria, and optimizing recruitment strategies based on real-world patient insights.

Moreover, the use of RWD for trial feasibility extends beyond site selection and patient recruitment to encompass various other aspects of clinical trial planning and execution. For example, researchers can leverage RWD to assess the feasibility of conducting decentralized or virtual trials, which offer potential advantages in terms of patient access, convenience, and retention (Harrer et al. 2019).

The integration of real-world data analytics into clinical trial planning and execution holds tremendous promise for enhancing trial feasibility and accelerating the drug development process. By leveraging insights from electronic health records (EHRs), sponsors can make more informed decisions regarding site selection, patient recruitment, and protocol optimization. This proactive approach not only increases the likelihood of successful trial outcomes but also contributes to the efficient allocation of resources and the timely delivery of new therapies to patients in need.

3.3.2.2 Machine Learning-Powered Site Selection

Effective site selection is essential for ensuring the success of clinical trials. Identifying high-performing sites with access to diverse patient populations and experienced research staff can be challenging. Machine learning (ML)-powered site selection solutions offer a data-driven approach to site identification and prioritization, enabling sponsors to identify sites with the highest likelihood of success based on objective criteria and historical performance data.

These solutions leverage advanced analytics techniques, including predictive modeling and clustering algorithms, to analyze a wide range of factors that impact site performance, such as patient demographics, disease prevalence, investigator experience, and historical recruitment rates. By identifying patterns and correlations in large datasets, ML algorithms can pinpoint sites that are best suited to meet the recruitment and retention goals of a given trial.

By streamlining the site selection process and focusing resources on sites with the greatest potential for success, ML-powered solutions can improve trial efficiency, reduce costs, and minimize the risk of under-enrollment. Additionally, by continuously monitoring site performance and adapting to changing conditions, these solutions help sponsors optimize their site selection strategies over time, further enhancing trial outcomes. The high operational burden in pivotal trials frequently prompts sponsors to curate a select list of preferred investigators, chosen for their expertise and track record from previous trials. Embracing AI tools offers an additional avenue to rank-order investigators, expediting site initiation and consequently enhancing recruitment efforts. This strategic integration of AI technology holds promise in streamlining the trial process, ultimately bolstering efficiency and positively influencing overall recruitment outcomes (Askin et al. 2023).

Overall, the integration of ML in site selection processes represents a significant advancement in clinical trial management. These technologies not only enhance the precision of site selection but also provide a scalable and adaptable approach to trial optimization, ensuring that clinical trials are conducted efficiently and effectively.

3.3.2.3 Smart Automation for Site Contract and Confidential Agreement

The process of negotiating and finalizing site contracts and confidential agreements with clinical trial sites can be time-consuming and resource intensive. However, integrating smart automation technologies can significantly expedite this process while ensuring compliance and transparency. One key aspect of smart automation is the use of contract management software equipped with AI capabilities. These platforms can streamline contract drafting, negotiation, and approval processes by leveraging natural language processing (NLP) algorithms to analyze contract terms and identify potential areas of contention or non-compliance (Infosys BPM n.d.; Ironclad n.d.). Additionally, these systems can automate routine tasks such as document routing and version control, freeing up valuable time for clinical trial personnel to focus on more strategic activities.

Furthermore, electronic signature solutions play a crucial role in accelerating the execution of site contracts and confidential agreements. By enabling stakeholders to sign documents electronically from anywhere in the world, these platforms eliminate the need for physical mailings and facilitate real-time collaboration and decision-making. Moreover, eSignature solutions often offer built-in audit trails and authentication mechanisms, ensuring the integrity and validity of signed documents.

Incorporating smart automation into the site contract and confidential agreement process not only accelerates the First Patient In (FPI) timeline but also enhances overall efficiency, compliance, and transparency. By leveraging AI-powered contract management software and electronic signature solutions, sponsors and clinical trial sites can expedite the initiation of trials while mitigating risks and ensuring regulatory compliance. This technological integration promises to streamline workflows, reduce administrative burdens, and enhance the overall quality and speed of clinical trial operations.

Overall, the adoption of smart automation technologies in clinical trials is a transformative step toward modernizing and optimizing the contract negotiation and finalization processes. By embracing these innovative tools, the industry can achieve greater efficiency, accuracy, and compliance, ultimately accelerating the delivery of new therapies to patients.

3.3.2.4 eConsent

Traditional paper-based informed consent processes have long been a bottleneck in clinical trial enrollment, often leading to delays in achieving the FPI milestone. However, the adoption of electronic consent (eConsent) solutions offers a promising avenue for expediting the consent process and enhancing patient engagement and comprehension.

Firstly, eConsent platforms leverage digital technologies to deliver interactive and multimedia-rich consent materials to study participants, enabling them to access, review, and sign consent forms electronically. These platforms often feature user-friendly interfaces, customizable content, and built-in comprehension assessments to ensure that participants fully understand the trial procedures, risks, and benefits before providing their consent (Grady et al. 2017).

Secondly, eConsent solutions offer several advantages over traditional paper-based approaches. For example, they enable real-time tracking of participant interactions with consent materials, allowing study coordinators to identify any areas of confusion or concern and address them promptly. This immediate feedback mechanism ensures that potential issues can be resolved quickly, enhancing participant understanding and satisfaction.

Thirdly, eConsent platforms can facilitate remote consent processes, allowing participants to provide informed consent from the comfort of their own homes or via telehealth consultations. This capability eliminates the need for in-person visits and reduces logistical barriers to enrollment, making it easier to include diverse and geographically dispersed populations in clinical trials.

Lastly, eConsent solutions can enhance data integrity and regulatory compliance by incorporating electronic signatures, audit trails, and version control mechanisms. By capturing electronic signatures and timestamps, these platforms provide verifiable evidence of participant consent and ensure adherence to regulatory requirements such as 21 CFR Part 11 (FDA 2020c). This digital traceability ensures that consent processes are transparent, secure, and compliant with legal standards.

Overall, eConsent represents a powerful tool for accelerating the FPI milestone in clinical trials while enhancing patient engagement, comprehension, and regulatory compliance. By leveraging digital technologies to modernize the consent process, sponsors and investigators can streamline trial enrollment, reduce administrative burdens, and ultimately bring new therapies to market more quickly and efficiently.

3.3.2.5 Patient Matching

Identifying and recruiting eligible patients for clinical trials is key to successful trial enrollment. However, it is a complex and time-consuming process, requiring dedicated and skilled clinical staff at the investigation site. Patient matching, the process of identifying individuals who meet the specific inclusion and exclusion criteria of a study, is often hindered by limited access to comprehensive patient data and inefficient screening methods.

AI-powered patient matching solutions offer a promising approach to streamlining the recruitment process and improving patient identification rates. These solutions leverage ML and NLP algorithms to analyze large volumes of patient data, including EHRs, medical histories, and demographic information, to identify individuals who are most likely to meet the eligibility criteria for a given clinical trial. By automating the screening process and identifying potential candidates more efficiently, AI-driven patient matching solutions enable sponsors and research sites to accelerate patient recruitment timelines and minimize the risk of under-enrollment. Moreover, by leveraging predictive analytics, these solutions can proactively identify and engage patients who may be eligible for future trials, thereby expanding the pool of potential participants and enhancing overall recruitment efforts.

Multiple clinical trial match systems have been developed for the purposes of either identifying the cohort of potentially eligible patients for a trial or for finding appropriate trials for a patient (Haddad et al. 2021; Zhang and Demner-Fushman 2017). Huston (2023) described an innovative system, called Criterion2Query, developed by researchers at Columbia University in New York City, that allows users to input inclusion and exclusion criteria in natural language or provide a trial's identification number. Subsequently, the system translates these criteria into a formal database query to identify matching candidates within patient databases. This work was further extended to aiding patients in their search for suitable trials. Initially, it utilizes Criteria2Query to extract criteria from trial descriptions. The subsequent phase generates pertinent questions to assist patients in refining their search. Additionally, TrialGPT, a collaborative effort between Sun's lab and the US National Institutes of Health, represents a method for leveraging a large language model to identify appropriate trials for individual patients (Huston 2023). This system evaluates whether a patient aligns with each criterion of a given trial, offering explanations for its decisions, and aggregates these assessments into a trial-level score. It then ranks multiple trials based on their suitability for the patient.

As pointed out by Askin et al. (2023), The efficacy of AI tools in recruitment hinges on the establishment of standardized language for eligibility criteria to facilitate system interoperability. To fulfill its purpose, the tool must possess the capability to comprehend and interpret input effectively. Therefore, a recommended approach involves combining structured data with insights derived from NLP of patient reports to augment information for eligibility screening. This fusion of structured and natural language data enhances the tool's capacity to accurately assess eligibility criteria, thereby optimizing recruitment processes.

3.3.2.6 Automating Trial Recommendation

In addition to identifying eligible patients for specific trials, AI plays a crucial role in automating the process of trial recommendation for individual patients. By analyzing patient data, including medical history, genetic profiles, and treatment preferences, AI algorithms can generate personalized trial recommendations tailored to each patient's unique characteristics and needs. Automated trial recommendation systems leverage NLP and predictive modeling techniques to sift through vast amounts of clinical trial data and match patients with relevant studies based on predefined criteria. These systems can consider factors such as disease type, stage, comorbidities, and geographic location to generate personalized recommendations that maximize the likelihood of patient participation and engagement. By streamlining the trial recommendation process and presenting patients with tailored options that align with their preferences and clinical profiles, AI-powered systems can significantly improve patient engagement and recruitment rates. Moreover, by continuously learning from patient interactions and feedback, these systems can refine their recommendations over time, further enhancing their effectiveness and relevance.

3.3.2.7 Decentralized Clinical Trials

DCTs mark a transformative shift in clinical research, aiming to expedite the FPI milestone and enhance patient accessibility and retention. Unlike conventional site-based trials, DCTs harness remote technologies such as wearables, mobile apps, and telemedicine platforms to facilitate participation across diverse geographic locations. These advancements reduce reliance on in-person visits, thereby mitigating logistical barriers to enrollment. For example, wearable devices monitor vital signs, activity levels, and medication adherence in real-time, offering timely insights into patient health and treatment outcomes. DCTs yield numerous advantages over traditional trials, including broader patient diversity, improved recruitment and retention rates, and decreased operational expenses. By broadening participant inclusion, DCTs accelerate the FPI milestone while enhancing the generalizability and external validity of study results. Moreover, they prioritize patient-centricity by providing flexibility and convenience through virtual visits and remote monitoring, reducing travel burdens and enhancing satisfaction and adherence.

However, the adoption of DCTs introduces distinctive challenges, particularly concerning data privacy, regulatory adherence, and technological infrastructure. Sponsors must uphold stringent privacy regulations such as GDPR and HIPAA and implement robust security measures to 2019 Successful implementation of DCTs hinges on meticulous planning and collaboration with regulatory bodies, investigators, and technology partners to ensure patient safety, uphold data integrity, and meet regulatory standards.

3.3.2.8 AI-Enabled EDC System for Clinical Trials

Traditional EDC systems, reliant on manual data entry and paper-based processes, have historically been associated with errors, delays, and inefficiencies that hinder trial progress and data quality. These legacy systems are often cumbersome to set up and maintain, limiting scalability and adaptability to evolving trial needs. However, with the integration of AI and advanced analytics, modern EDC systems are undergoing a transformation, empowering sponsors and research organizations to enhance data quality, streamline processes, and derive actionable insights to accelerate drug development.

AI-powered EDC platforms offer a suite of advanced functionalities including real-time data capture, automated data validation, and integrated analytics. These capabilities enable continuous monitoring of trial data, early detection of protocol deviations, and predictive insights into patient outcomes. Pharmaceutical companies and research institutions have leveraged AI-enabled EDC systems to achieve significant improvements in trial efficiency and outcomes.

3.3.3 DHT-Enhanced Study Conduct

3.3.3.1 eTracking of Medication

Electronic tracking (eTracking) of medication adherence has emerged as a pivotal advancement in clinical trials, addressing the challenges associated with traditional methods such as patient self-reporting and pill counts, which are often prone to bias and inaccuracies. In contrast, eTracking employs electronic sensors, smart packaging, and mobile health applications to monitor medication usage in real-time, offering objective insights into patient behavior and treatment adherence. AI and ML algorithms further enhance eTracking systems by analyzing adherence data, detecting patterns, and predicting adherence issues before they impact study outcomes. For instance, AI-driven video analysis can scrutinize patient medication intake, ensuring adherence to prescribed doses and alerting researchers to deviations promptly.

These technologies not only improve data accuracy but also enhance patient engagement through features like gamification and personalized incentives, motivating participants to adhere to their medication regimens and actively participate in the trial process. Research evidence supports the efficacy of eTracking solutions in clinical settings, demonstrating improved adherence rates and patient satisfaction compared to traditional monitoring approaches.

3.3.3.2 Smart Tools for Treatment Compliance

In addition to medication adherence, compliance with other aspects of the treatment protocol, such as dietary restrictions, lifestyle modifications, and follow-up appointments, is also crucial for the success of clinical trials. Smart tools powered by AI offer innovative solutions for enhancing treatment compliance and improving patient outcomes.

These tools leverage a combination of sensors, wearable devices, and mobile applications to monitor patient behavior and provide real-time feedback and support. For example, smart scales equipped with Bluetooth connectivity can automatically transmit weight measurements to a mobile app, allowing researchers to track changes in patient weight and identify potential deviations from the treatment plan.

By incorporating ML algorithms, smart tools can analyze patient data and identify patterns or trends that may indicate non-compliance or adverse events. For example, an AI-powered nutrition app can analyze dietary intake data and provide personalized recommendations for improving adherence to dietary restrictions.

Moreover, smart tools can facilitate remote monitoring and telemedicine consultations, enabling researchers to monitor patient progress and provide support and guidance from a distance. By leveraging telehealth technologies, researchers can overcome geographical barriers and improve access to care for patients who may have limited mobility or access to healthcare facilities.

AI-powered technologies offer innovative solutions for enhancing patient recruitment, adherence, and retention in clinical trials. By leveraging advanced analytics, ML algorithms, and digital health tools, researchers can overcome traditional barriers to success and improve the efficiency and effectiveness of their clinical operations. Moreover, by placing patients at the center of the trial process and providing personalized support and feedback, AI-powered solutions can enhance patient engagement and satisfaction, ultimately leading to better outcomes for patients, researchers, and sponsors alike.

3.3.3.3 Patient Engagement and Retention with AI

Patient engagement and retention in clinical trials remain significant challenges, with substantial dropout rates observed across various studies. Research indicates that nearly 40% of patients discontinue their prescribed medication within the first year of participation (Hutson 2024). To address these challenges, pharmaceutical companies, CROs, and research institutions are increasingly turning to RWD, AI, and advanced analytics to optimize clinical trial operations.

AI and ML are pivotal in enhancing patient engagement by reducing the burdens associated with participation. For instance, ML algorithms can passively collect data from routine clinical care to generate insights for investigational purposes, minimizing the need for additional invasive procedures or tests. Furthermore, AI applications like generative adversarial networks can analyze standard clinical slides to identify patients who may benefit from further imaging, thus optimizing patient management and resource allocation (Weissler et al. 2021).

AI-driven predictive algorithms are being developed to forecast patients at risk of discontinuing their participation in trials. These algorithms utilize historical data to identify early signs of patient disengagement, enabling proactive interventions from clinicians. Chatbots represent another innovative AI solution for improving patient engagement. Studies have shown that AI-powered chatbots can effectively address patient inquiries and concerns, providing timely and accurate responses comparable to those from healthcare professionals.

3.3.3.4 Enrollment, Recruitment Time, Study Duration Projection

3.3.3.4.1 Number of Enrolled Participants and Recruitment Time

Accurate projection of enrollment within predefined timeline during the design phase is of great importance to strategic planning and trial success. Many trials fall short of being completed within the specified enrollment timeframe, due to unforeseen events such as delays with site initiation and over-estimation of the enrollment rate. The over-estimation issue may stem from predominant reliance deterministic methods that fail to consider the range of uncertainties inherent in input data and the stochastic nature of recruitment fluctuations over time.

In recent years, various statistical methods have been proposed for modeling and predicting patient accrual (Liu et al. 2020). Bagiella and Heitjan (2001) utilized a homogeneous Poisson process to model recruitment. Anisimov and Fedorov (2007) enhanced this approach by employing a Poisson-Gamma mixture model to account for recruitment rate variation across multiple centers. The method proposed by Anisimov and Fedorov (2007) allows for prediction of both the number of recruited patients but also the recruitment time with statistical confidence. Applications of this model to real-life clinical trials showed it was fit for purpose. More recent methods developments included modeling time-dependent recruitment rate (Perperoglou et al. 2023), the use of non-homogeneous Poisson process while incorporating region-specific accrual in their framework (Deng et al. 2017). Most recently, Zhong et al. (2023) presented an approach based on generalized linear mixed-effects models and the use of non-homogeneous Poisson processes through a Bayesian framework to model the country initiation, site activation and subject enrollment. This model can be used to facilitate strategic planning at portfolio level.

3.3.3.4.2 Number of Events and Study Duration

When the primary endpoint is a time-to-event measure, such as death, the enrollment, enrollment time, and study duration are driven by the number of events as opposed to the number of participants to be enrolled. As it is not unusual that recruitment is slower than expected, in particular, for oncology trials, it is desirable to adjust enrollment strategy such as increasing the number of clinical sites to meet key timelines of the trial or extending enrollment or follow-up duration to meet the enrollment goal and ensure the study has sufficient power to meet the primary endpoint (Chang 2011). These adjustments become more feasible for trials that have pre-planned interim analyses. The results from the interim looks can be used to estimate key parameters needed to predict the number of events and study duration.

Consider an oncology study with a projected recruitment time T_R, follow-up time T_F, and pre-planned interim look at time T_I. It is assumed that the enrolment rate remains constant. Denote this rate as R, which represents the number of participants recruited per unit time. It is further assumed that the patient survival X_S and dropout X_D follow exponential distributions such that

$$X_S \sim E(\lambda_S)$$

$$X_D \sim E(\lambda_S).$$

Hence, the density functions of X_S and X_D are given as follows:

$$f_S(t) = \lambda_S \exp(-\lambda_S t)$$
$$f_D(t) = \lambda_D \exp(-\lambda_D t).$$

Let $N_S(T)$ and $N_D(T)$ be the numbers of deaths and dropouts at time T, respectively. Therefore (Chang 2011),

$$N_S(T) = \int_0^T \int_0^t R(r)\,dr f_S(t-r)\,dt.$$

Given $R(r) = R$ if $t \le T_I$ and $R(r) = 0$ if $t > 0$ and (**), the number of deaths can be calculated as

$$N_S(T) = \int_0^T \int_0^{\min(T_I,t)} R\,dr\,\lambda_S \exp\left[-\lambda_S(t-r)\right]dt.$$

or

$$N_S = \begin{cases} R\left(T - \dfrac{1}{\lambda_S} + \dfrac{1}{\lambda_S} e^{-\lambda_S T}\right) & \text{if } T \le T_I \\ R\left(T_I - \dfrac{1}{\lambda_S}\left(e^{-\lambda_S T_I - 1}\right) e^{-\lambda_S T}\right) & \text{if } T > T_I \end{cases} \tag{3.1}$$

Solving (3.1) for T, the time when there is the desired number of events is obtained as in (3.2):

$$T = \begin{cases} -\dfrac{1}{\lambda_S} \ln\left(\dfrac{\lambda_S N_S}{R} - T\lambda_S + 1\right) & \text{if } T \le T_I \\ -\dfrac{1}{\lambda_S} \ln\left(\left(T_I - \dfrac{N_S}{R}\right) e^{-\lambda_S T_I - 1}\right) & \text{if } T > T_I \end{cases} \tag{3.2}$$

Similarly, the number of dropouts can be estimated by

$$N_D = \begin{cases} R\left(T - \dfrac{1}{\lambda_D} + \dfrac{1}{\lambda_D} e^{-\lambda_D T}\right) & \text{if } T \le T_I \\ R\left(T_I - \dfrac{1}{\lambda_D}\left(e^{-\lambda_D T_I - 1}\right) e^{-\lambda_D T}\right) & \text{if } T > T_I \end{cases} \tag{3.3}$$

During the interim analysis at time T, the hazard rates λ_S and λ_D related to patient survival and dropout can be estimated based on the estimates $\widehat{N_S}$ and $\widehat{N_D}$ from the interim data, and equations (3.1) and (3.3). Denote these estimates as rates $\widehat{\lambda_S}$ and $\widehat{\lambda_D}$. Chang (2010) proposed to use simulations to estimate the additional number of events, N_{S1}, expected to be observed for the remainder part of the study (from time T to T_F).

Let K be the number of deaths observed by time T, and N_A the additional number of participants to be enrolled between T and the end of the study, $T_R + T_F$. The following procedure can be used to estimate the total projected number of deaths (Table 3.1).

Similarly, simulations can also be carried out to estimate the actual duration of the study (Table 3.2).

Projecting event occurrence using data from the early stage of a trial has been an area of active research. Various methods were proposed in the literature, including traditional statistical approaches and more recent advancements in ML models. More recently, ML models were explored to improve the accuracy of model prediction. For example, Fard et al. (2016) developed a Bayesian framework for improved performance in predicting event occurrence in clinical trials. These models can be updated using data from ongoing trials that become available, enhancing the prediction of remaining enrollment and time. In addition, to operationalize these approaches, advanced analytics and dashboards can be built to provide real-time insights and facilitate decision-making.

TABLE 3.1

Projection of Number of Events

Input	
Output	**Projected Number of Events at Study Completion T_I + T_F**
Step	**Description**
1	Set D = N
2	For i = 1 to N_A
	Generate enrollment time t_0i from U(T, TI)
	Generate survival time t from E(lambda)
	Generate censoring time r_i from E(omega)
	If r_i >= t_i, and t_0i + t_i <= T_I + T_F then
	End
3	Repeat 1–3 for m times
4	Calculate mean(D) and var(D)
5	Output mean(D) and var(D)

Source: Chang (2011).

TABLE 3.2

Projection of Trial Duration

Input	
Output	**Projected trial duration**
Step	**Description**
1	Set D = N
2	For i = 1 to N_A
	Generate enrollment time t_0i from U(T, TI+T_F)
	Generate survival time t from E(lambda)
	Generate censoring time r_i from E(omega)
	If r_i >= t_i, then A[D++]=t_0i+t_i
	Sort array A[] in ascending order
	Set T_E to be the last entry in A[]
	End
3	Repeat 1–3 for m times
4	calculate mean(T_E) and var(T_E)
5	Output mean(T_E) and var(T_E)

Source: Chang (2011).

3.3.3.5 Drug Supply in Clinical Trials

3.3.3.5.1 Importance of Drug Supply Optimization

Effective management of drug supply is essential for the success of clinical trials. Adequate drug supply ensures that study sites have access to the necessary medications and investigational products, enabling them to enroll and treat patients according to the study protocol. However, managing drug supply in clinical trials presents several challenges, including:

1. **Uncertain Demand**: Predicting the demand for investigational drugs in clinical trials can be challenging due to factors such as patient enrollment rates, protocol amendments, and site-specific variations in patient recruitment (Harrer et al. 2019).
2. **Supply Chain Complexity**: Clinical trials often involve multiple stakeholders, including drug manufacturers, distributors, and study sites, each with their own supply chain processes and timelines. Coordinating these activities and ensuring timely delivery of drugs to study sites can be complex and labor-intensive.
3. **Regulatory Compliance**: Regulatory agencies such as the FDA and EMA have stringent requirements for drug supply management in clinical trials, including traceability, accountability, and documentation of drug distribution activities. Ensuring compliance with these regulations adds another layer of complexity to drug supply management (FDA 2020b).

3.3.3.5.2 Predictive Analytics

Predictive analytics has emerged as a powerful tool in the optimization of drug supply management for clinical trials. By leveraging historical data, ML algorithms, and advanced statistical modeling techniques, predictive analytics enables sponsors and research organizations to forecast drug demand, anticipate supply chain disruptions, and optimize inventory levels. This approach ensures the timely and cost-effective delivery of investigational drugs to study sites and patients. For instance, predictive models can analyze patterns from past trials to predict future enrollment rates more accurately, allowing for better planning and reduced wastage. Additionally, predictive analytics can help identify potential bottlenecks in the supply chain, enabling preemptive measures to be taken to mitigate risks and ensure continuous drug availability.

3.3.4 Study Close-Out

The integration of AI, ML, and advanced analytics in clinical trial close-out activities significantly enhances efficiency and accuracy. AI-driven tools streamline data cleaning by automatically detecting and correcting discrepancies, reducing the risk of human error and expediting the process (Harrer et al. 2019). ML algorithms facilitate comprehensive data analysis by identifying patterns and insights that may be overlooked by traditional methods, thereby improving the robustness of the study conclusions (Gong et al. 2021). Additionally, AI applications aid in the writing of CSRs by generating drafts based on structured data inputs, which can then be reviewed and refined by researchers, thus saving substantial time and effort. For documents archiving, AI-powered systems ensure that all essential documents are accurately categorized and stored in eTMF, enhancing accessibility and compliance with regulatory requirements. These technological advancements not only streamline close-out activities but also enhance the overall quality and integrity of clinical trial data.

3.3.5 Implementation of Advanced Analytics

The implementation of advanced analytics in clinical operations is transforming the landscape of clinical trials by harnessing the power of predictive analytics to enhance efficiency, accuracy, and patient safety. Predictive analytics leverages vast datasets and sophisticated algorithms to forecast enrollment, predict drug supply needs, assess patient safety risks, impute missing data, and support risk-based monitoring (RBM). This section delves into these applications and illustrates their impact on clinical operations with detailed references.

3.3.5.1 Forecasting Enrollment

Accurate forecasting of patient enrollment is critical for the successful planning and execution of clinical trials. Predictive analytics can analyze historical enrollment data, site performance metrics, and demographic information to generate robust enrollment forecasts. This approach allows sponsors to identify potential recruitment bottlenecks and allocate resources more effectively. For instance, Gkioni et al. (2019) demonstrated the application of statistical models to predict patient accrual, significantly enhancing the reliability of enrollment projections. Similarly, Lan et al. (2019) developed models that incorporate time-dependent recruitment rates, further refining enrollment forecasts and enabling proactive management of recruitment strategies.

3.3.5.2 Predicting Drug Supply Demand

Predictive analytics also plays a crucial role in managing drug supply chains for clinical trials. By analyzing trial design parameters, patient enrollment rates, and historical usage patterns, predictive models can forecast drug demand, anticipate supply chain bottlenecks, and optimize inventory levels to meet study requirements. This ensures that sufficient drug supplies are available at all trial sites, minimizing the risk of stockouts or overstocking. A study by Heitjan et al. (2015) highlighted the benefits of using Poisson-Gamma mixture models to predict drug supply needs, demonstrating significant improvements in supply chain efficiency. Predictive analytics algorithms can also be utilized to assess the impact of various risk factors, such as manufacturing delays, regulatory changes, and geopolitical events, on drug supply availability in clinical trials. By quantifying the probability and severity of supply chain disruptions, these algorithms enable sponsors to proactively identify and mitigate potential risks, ensuring continuity of drug supply and minimizing the impact on study timelines and outcomes.

3.3.5.3 Predictive Patient Safety Risk

Ensuring patient safety is paramount in clinical trials. Predictive analytics can identify patients at higher risk of adverse events by analyzing clinical data, patient demographics, and genetic information. This allows for early interventions and personalized monitoring plans. For example, Weissler et al. (2021) showed how ML algorithms could predict adverse drug reactions, enabling researchers to mitigate risks proactively. Additionally, AI-driven models have been used to analyze EHRs and identify patients with potential safety risks, improving the overall safety profile of clinical trials (Yang and Khatry 2023).

3.3.5.4 Imputing Missing Data

Missing data is a common challenge in clinical trials that can compromise the validity of study results. Advanced analytics techniques, such as multiple imputation and ML algorithms, can accurately estimate missing values, ensuring the integrity and completeness of the dataset. Zhong et al. (2023) discussed the application of generalized linear mixed-effects models for imputing missing data, which improved the robustness of clinical trial analyses. These methods not only enhance data quality but also reduce the need for data cleaning and manual imputation efforts.

3.3.5.5 Supporting Risk-Based Monitoring

RBM is a strategic approach that focuses on the most critical data and processes in clinical trials. Predictive analytics can identify high-risk sites and patient populations by analyzing real-time data from multiple sources, such as EHRs, site performance metrics, and patient-reported outcomes. This allows sponsors to prioritize monitoring efforts and allocate resources more effectively. Ali and Luqman (2023) demonstrated how predictive models could support RBM by identifying sites with higher risks of data quality issues, enabling targeted interventions and reducing overall monitoring costs.

3.3.6 Considerations and Challenges

While predictive analytics offers significant potential benefits for enhancing clinical operation efficiency, several considerations and challenges must be addressed to maximize its effectiveness and impact.

3.3.6.1 Data Quality and Availability

Predictive analytics relies on accurate and reliable data to generate meaningful insights and predictions. Ensuring data quality and availability, particularly in the context of clinical trials where data may be fragmented or incomplete, requires robust data governance processes and integration with existing systems such as electronic data capture (EDC) and clinical trial management systems (CTMS). Data standardization and harmonization efforts are critical for overcoming issues related to data inconsistency and fragmentation (Beck et al. 2018).

3.3.6.2 Model Validation and Calibration

Predictive analytics models require validation and calibration to ensure their accuracy and reliability for their intended purposes. Stakeholders must conduct rigorous testing and validation exercises using historical data and real-world scenarios to assess the model's performance and identify potential sources of bias or error. Techniques such as cross-validation and sensitivity analysis are essential for assessing model robustness and generalizability (Maleki et al. 2020).

3.3.6.3 Regulatory Compliance

Predictive analytics solutions for drug supply optimization and patient outcomes must comply with regulatory requirements and industry standards, including Good Manufacturing Practice (GMP) guidelines and Good Clinical Practice (GCP) regulations. Ensuring compliance with these regulations requires robust validation and documentation of predictive models, as well as adherence to data privacy and security requirements. Implementing compliance frameworks and audit trails can help ensure that predictive models meet regulatory standards.

3.3.6.4 Future Directions and Opportunities

Looking ahead, predictive analytics holds tremendous promise for transforming drug supply management in clinical trials. By leveraging emerging technologies such as AI, ML, and big data analytics, stakeholders can unlock new insights, optimize supply chain processes, and improve patient outcomes in ways that were previously not possible. Some of the key future directions and opportunities for predictive analytics in drug supply management include:

1. **Real-Time Predictions:** Advances in AI and ML algorithms enable real-time predictions of drug demand, supply chain performance, and patient outcomes, allowing stakeholders to make proactive decisions and respond quickly to emerging trends and events.
2. **Personalized Supply Chains:** Predictive analytics can enable personalized supply chain solutions tailored to the specific requirements of individual clinical trials, including patient demographics, disease characteristics, and geographic location. This approach can significantly enhance the efficiency and effectiveness of clinical operations.
3. **Integration with Blockchain:** Integration with blockchain technology enables secure and transparent data sharing across the drug supply chain, enhancing traceability, accountability, and integrity of drug distribution activities. Blockchain can provide immutable records of drug movement, ensuring compliance and reducing the risk of counterfeit drugs.

3.4 Case Studies

In this section, we use three case examples to illustrate how predictive modeling based on AI, ML, or advanced statistical methods can be used to increase the efficiency of clinical trials. The first study by Beaulieu et al. (2021) sought to use a predictive ML modeling without the use of vase line vital capability measures for use in clinical trials during the COVID-19 pandemic. The second example concerns prediction of trial enrollment rates based on study design features (Bieganek et al. 2022) while the last describes a statistical modeling approach to project drug supply demand of a pivotal oncology trial (Yang et al. 2018).

3.4.1 Design of Eligibility Criteria for ALS Clinical Trials

3.4.1.1 Background

Amyotrophic lateral sclerosis (ALS) is a dire form of motor neuron disease. It manifests through the gradual breakdown of nerve cells in both the spinal cord and brain. Frequently referred to as Lou Gehrig's disease, in honor of the renowned baseball player who succumbed to it, ALS is one of the most devastating afflictions impacting nerve and muscle functionality. The heterogenicity of ALS patient population complicates the design of clinical trials. Researchers face challenges in achieving a homogeneous study population to enhance the chances of demonstrating therapeutic efficacy (van Eljik et al. 2019). To address this, eligibility criteria are employed to enroll participants who are more likely to adhere to the protocol or derive benefit from the treatment, often necessitating the exclusion of individuals with prolonged disease durations or those unlikely to survive the follow-up period. Consequently, a considerable number of patients may be excluded, raising concerns about the trial's generalizability and applicability to a broader population. An analysis of 38 ALS trials by van Eljik et al. (2019) revealed that conventional eligibility criteria excluded close to 60% of ALS patients and did not effectively mitigate the variability in survival rates.

In the past two decades, the majority of ALS clinical trials have mandated a minimum baseline percent predicted vital capacity (BL%VC) for patient inclusion. Typically, these trials have set BL%VC inclusion thresholds ranging from 50% to 80%. Studies have demonstrated that BL%VC serves as a prognostic indicator for survival, and the exclusion of patients with low BL%VC aims to minimize the likelihood of participants not surviving the duration of the study.

However, routine spirometry can be impractical in ALS clinical trials under certain special circumstances. For example, one notable impact of the global COVID-19 pandemic was on the collection of %VC from ALS patients due to safety concerns. Andrews et al. (2020) observed that 69% of 61 surveyed ALS clinics faced difficulties in conducting spirometry, even within clinical settings. Additionally, the surge in telemedicine has exacerbated the challenge of obtaining routine spirometry, a trend likely to persist beyond the pandemic's resolution, although its intensity may vary. Telemedicine offers a pragmatic solution, particularly beneficial for individuals with ALS, for whom traveling to appointments can be arduous. Remote assessment of respiratory function for both patient care and inclusion in clinical trials is actively being researched (De Marchi et al. 2020; Capozzo et al. 2020; Vasta et al. 2021; Govindarajan et al. 2020). Consequently, the frequency of VC measurements may decrease compared to pre-pandemic standards, both in the short and potentially long term.

Beaulieu et al. (2021) developed a ML survival model without the use of baseline vital capability measures (VC-Free). When compared to a multivariate model that includes vital capacity (VCI), the performance of the model was marginally reduced. It was also shown that both VC-Free and VC outperformed a univariate reference model (UNI) that predicted survival using only the baseline value of %CV.

3.4.1.2 Training and Validation Data

As described in Beaulieu et al. (2021), The data utilized for training predictive models were sourced from the PRO-ACT Database, comprising records from over 10,700 ALS patients involved in 23 phase II/III trials. This extensive database encompasses more than 10 million longitudinally collected data points, incorporating demographic, laboratory, and medical data, as well as survival and family histories (Zach et al. 2015). To ensure robust model training, a subset of the PRO-ACT database was selected to minimize missing data among baseline predictors and guarantee the availability of follow-up data for assessing outcomes, specifically the time from baseline to tracheostomy or death.

For model validation, data from 188 placebo patients enrolled in the VITALITY-ALS trial were employed as an external test dataset, representing a contemporary clinical trial population (Shefner et al. 2019). Additionally, data from 630 patients treated at the Emory University ALS clinic between 2008 and 2015 served as another external validation dataset, reflecting a clinic population (Traxinger et al. 2013). All patients within these external datasets underwent multiple visits starting from either their baseline trial visit or initial clinic appointment, with comprehensive data typically encompassing the full ALS Functional Rating Scale-Revised (ALSFRS-R) questionnaire, vital capacity (VC), and vital signs.

3.4.1.3 Predictive Features

The modeling building utilized a diverse set of baseline measures collected in both clinical trial and clinic settings (Table 3.3). Amyotrophic Lateral Sclerosis Functional Rating Scale – Revised (ALSFRS-R) score was included in the full set feature as it was previously report that ALSFRS-R was strong prognostic for ALS disease progression. Slopes for all ALSFRS-R scores and percent vital capacity (%VC) were calculated from symptom onset to baseline, assuming

TABLE 3.3
Features Used in Model Development

Source	Feature	VC-Free	VCI	UNI
Demographics	Age (years)	✓	✓	
	Sex	✓	✓	
	Height	✓	✓	
	Weight	✓	✓	
	BMI (kg/m)	✓	✓	
Disease Characteristics	Time since symptom onset	✓	✓	
	Time since disease diagnosis	✓	✓	
	Site of symptom onset	✓	✓	
	Riluzole use	✓	✓	
Vitals	SBP	✓	✓	
	DBP	✓	✓	
	Pulse	✓	✓	
	Months since symptom onset	✓	✓	
ALSFRS-R Scores		✓	✓	
Vital Capacity	%VC		✓	✓

a state of "full health" wherein all ALSFRS-R question scores were 4 and percent VC was 100%, beginning the day before symptom onset. All the features were used for training the VCI model while %VC was excluded in the development of the VC-Free model. The UNI model was created to provide an additional frame of reference in assessing the performance of the VC-Free model. This model predicts patient survival, utilizing a single predictor %VC at baseline. The approach mirrors the current strategy adopted to enhance survival enrichment in ALS clinical trials, which involves establishing an inclusion criterion based on a minimum baseline %VC threshold (Beaulieu et al. 2021).

3.4.1.4 Model Development

3.4.1.4.1 Model Training

3.4.1.4.1.1 COX PROPORTIONAL MODEL

In survival analysis, two functions are of particular interest: survival function $S(t)$ and hazard function $h(t)$. The survival function is the probability for a patient to survive beyond time t, and the hazard function is defined as the instantaneous death rate at item t conditional on survival beyond t. The two functions are mathematically related to each other through the following equation:

$$h(t) = \frac{S'(t)}{S(t)}.$$

Therefore a hypothesis of the survival function can often be translated into a hypothesis of the hazard rate. The Cox's proportional hazards model is a popular survival analysis method. In the model, the hazard of a patient is assumed to be proportional to the baseline hazard of the patients of the same disease. Mathematically, the model can be expressed as

$$h(t) = h_0(t)\ \exp(\beta X)$$

where $h(t)$ is the hazard function, $h_0(t)$ is the baseline hazard, $\beta = (\beta_1, \beta_2, \ldots, \beta_m)$ are coefficients, and $X = (x_1, x_2, \ldots, x_m)$ are explanatory variables or predictors. The coefficients β can be estimated by optimizing the log partial likelihood function:

$$\sum_{i=1} \beta \mathrm{X}_{\mathrm{i}} - \log \sum_{j:t_j \geq t_i} \exp(\beta X_j)$$

where the summation is over the set of subjects k where the event has not occurred before time t_i (including subject i itself).

3.4.1.4.2 *Gradient Boosting Algorithm*

For each of the two models, VCI and VC-Free, instead of optimizing the loss function in (xx), a gradient boosting model was trained through maximize the loss function, using the training dataset,

$$\sum_{i=1} f(X_i) - \log \sum_{j:t_j \geq t_i} f(\beta X_j)$$

where $f(X)$ is the output from the gradient boosting model.

In contrast, the UNI model utilized a Cox proportional hazards model fitted with a single predictor. Each model generated a predicted log hazard, enabling the ranking of patients by their relative predicted risk of death. This log-hazard was further transformed using the baseline cumulative hazard to provide predicted survival probabilities over time.

3.4.1.4.3 *Model Validation*

The performance of each model was evaluated for discrimination and calibration through one year (Steyrberg et al. 2014). The former was assessed, using the concordance C-statistic (Harrell et al. 1996) while the latter was based on two measures, calibration-in-the-large and calibration slope (Crowson et al. 2016).

The C-statistic, also known as the concordance statistic or the area under the Receiver Operating Characteristic (ROC) curve, is a measure of the goodness of fit for binary outcomes in a logistic regression model. In clinical studies, the C-statistic provides the probability that a randomly selected patient who has experienced an event (such as a disease or condition) will have a higher risk score than a patient who has not experienced the event. Its value ranges from 0 to 1, with a higher value indicating better discrimination between patients who experience the event and those who do not and a value close to 0.5 indicating poor discrimination. In contrast, calibration-in-the-large serves as a fundamental gauge of alignment between observed and predicted risks. It is calculated by determining the disparity between the mean observed risk and the mean predicted risk. The calibration slope is computed by conducting a regression of the observed outcome against the predicted probabilities. This estimated slope offers insight into the agreement between the observed and predicted probabilities, with a value close to 1 indicating good calibration (Crowson et al. 2016).

For each of the above three models, internal 10-fold cross-validation was conducted using the PRO-ACT training set. Additionally, external model validation was carried out using the VITALITY-ALS clinical trial and Emory clinic datasets. These internal and external validations were intended to assess the reproducibility and generalizability of the models. Since BL%VC is a known prognostic variable, patients with missing BL%VC were excluded from the model training and validation to ensure fair comparisons of the models, resulting in inclusion of 5,251, 188, and 581 participants from PRO-ACT, VITALITY-ALS, and Emory Clinic, respectively. It is also worth noting that because of controlled clinical trial settings, participants in the PRO-ACT and VITALITY-ALS were enrolled in the trials based on a set of eligibility criteria. In contrast, the participants in the Emory Clinic dataset are more variable, and reflective of the real-world ALS population.

Additional validations of the VCI and VC-Free models were also carried out, using simulated datasets. Specifically, random samples of size 250 were drawn from the Emory clinic dataset. For each sample, patients were classed into model-positive and model-negative groups, based on if the predicted probability of surviving at one year exceeds 50% or not. The former represents the enriched population for a trial, whereas the latter being excluded from the trial. The simulations were iterated 100 times. The cutoff probability of 50% was chosen, in light of the fact that recent ALS trials have utilized BL%VC > 50% as an inclusion criterion, which exactly corresponds to a 50% predicted probability of one-year survival from the UNI model (refer to Figure 3.3).

Key metrics such as actual mortality within one-year, median survival, and the proportion of screened patients included versus excluded, as determined by predictions from each model, were estimated, based on the data across all simulation runs. Kaplan–Meier curves were generated for patients who were either included or excluded from the trial, based on the predictions from each model (UNI, VCI, VC-Free).

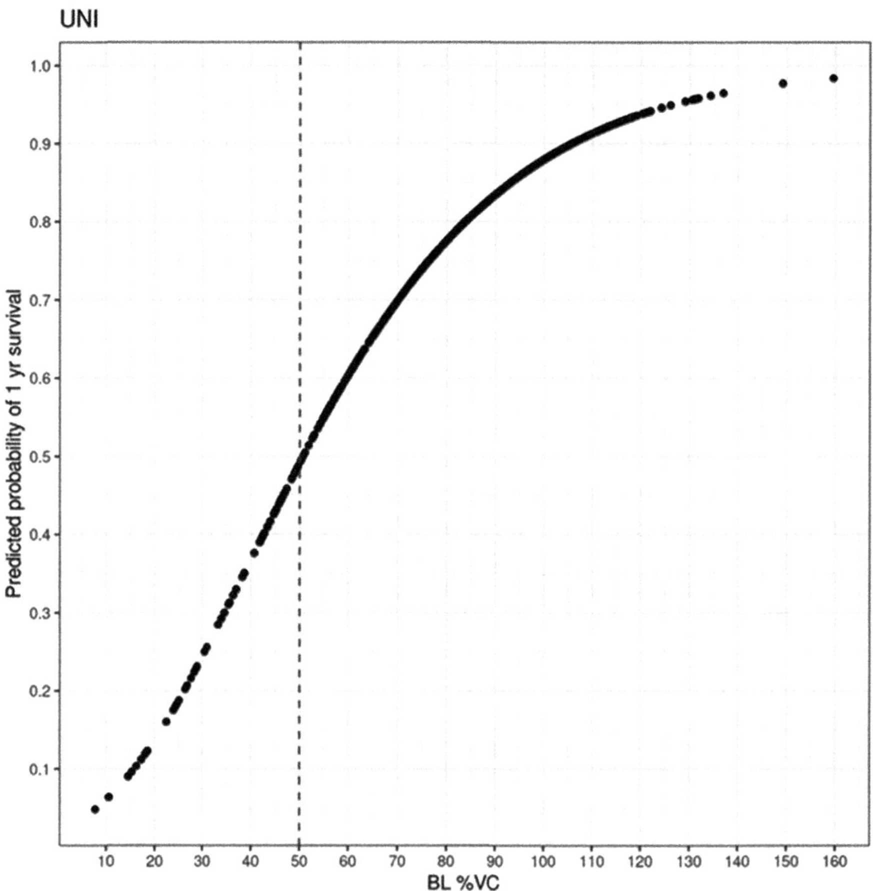

FIGURE 3.3
BL%VC vs. predicted probability of one-year survival from UNI model. A vertical red line is plotted at BL 50%VC, a common ALS clinical trial inclusion criteria cutoff. (Adapted from Beaulieu et al. 2021.)

3.4.1.5 Results

Table 3.4 provides a descriptive summary of patient characteristics across the training and validation datasets from PRO-ACT, VITAITY-ALS trials, and Emory Clinic. While most of the patient characteristics are comparable across the three groups, broader ranges and increased variability were observed in baseline characteristics with the dataset from Emory Clinical, due to a more diverse patient cohort from Emory Clinic compared to the clinical trial datasets. Of note, the median follow-up duration for patients from the Emory clinic was 15.9 months, whereas it stood at 12.5 months for both the PRO-ACT and VITALITY-ALS placebo arm cohorts. Within the PRO-ACT internal validation subset, 1,017 patients (19.4%) succumbed to the condition within one

TABLE 3.4

Patient Characteristics

	PRO-ACT (*N* = 5,251)	**VITALITY-ALS (*N* = 188)**	**Emory Clinic (*N* = 581)**
Sex, *n* (%)			
Female	1,987 (37.8)	65 (34.6)	228 (39.2)
Male	3,264 (62.2)	123 (65.4)	353 (60.8)
Age (years)			
n	5,251	188	581
Mean (SD)	55.6 (11.57)	55.9 (10.63)	59.8 (12.24)
Median	56.4	57	60
Min, Max	18, 84	30, 86	22, 91
BMI (kg/m)			
n	5,238	188	527
Mean (SD)	25.8 (4.6)	27.3 (4.34)	26.6 (5.46)
Median	25.3	26.2	25.8
Min, Max	14.4, 53.3	19.7, 39.5	15.6, 69.7
Bulbar onset, *n* (%)			
No	4,068 (77.5)	156 (83)	430 (74)
Yes	1,183 (22.5)	32 (17)	151 (26)
Limb onset, *n* (%)			
No	1,361 (25.9)	32 (17)	169 (29.1)
Yes	3,890 (74.1)	156 (83)	412 (70.9)
Riluzole use, *n* (%)			
No	1,076 (20.5)	47 (25)	252 (43.4)
Yes	2,629 (50.1)	141 (75)	329 (56.6)
Missing	1,546 (29.4)	--	--
Months since symptom onset			
n	5,251	188	580
Mean (SD)	22 (14.19)	21.5 (16.19)	26.5 (38.22)
Median	18.4	18.3	13.5
Min, Max	0.5, 287.2	2.1, 123.9	0.3, 414
Months since disease diagnosis			
n	3,097	188	546
Mean (SD)	10.1 (11.13)	8.1 (6.03)	8.2 (27.96)
Median	6.9	6.8	0
Min, Max	0, 242.2	0, 23.5	â^'6.9, 418.4
Baseline ALSFRS-R score			
n	5,248	188	557
Mean (SD)	37.1 (6.04)	38.3 (5.05)	35.9 (7.45)
Median	38	39	37
Min, Max	13, 48	22, 47	10, 48

(*continued*)

TABLE 3.4 (Continued)

Patient Characteristics

	PRO-ACT (*N* = 5,251)	VITALITY-ALS (*N* = 188)	Emory Clinic (*N* = 581)
Baseline % expected vital capacity			
n	5,251	188	581
Mean (SD)	83.5 (19.19)	87.4 (15.04)	81 (24.39)
Median	83.5	87.4	84
Min, Max	0, 286.7	40.2, 132.9	7.7, 159.5
Follow-up time (months)			
n	5,251	188	581
Mean (SD)	13 (6.69)	11.3 (2.48)	20.1 (15.72)
Median	12.5	12.5	15.9
Min, Max	0.7, 122.9	2.1, 14.3	0.2, 75.1
Survival through one year, *n* (%)			
Censored	4,234 (80.6)	171 (91)	520 (89.5)
Dead	1,017 (19.4)	17 (9)	61 (10.5)

Source: Beaulieu et al. (2021).

year. Similarly, among the VITALITY-ALS participants, 17 individuals (9.0%) met this outcome within the same timeframe. In comparison, 61 patients (10.5%) from the Emory clinic dataset experienced mortality within one year.

Model performance was evaluated both in terms of discriminant power characterized by a C-statistics, and agreement between the observed and predicted values measured by a calibration stope, and calibration-in-the-large as previously described. The results are summarized in Table 3.5.

From the table, for the internal validation and VITALITY-ALS datasets, the C-statistics of the VCI and VC-Free models are all above 0.76, showing good discriminant performance, with the VCI model being slightly better than the VCI-Free model (PRO-ACT All Folds, C-Statistics (95% CI): VCI 0.80 (0.79, 0.81) vs. VC-Free 0.77 (0.75, 0.78); VITALITY-ALS, C-Statistics (95% CI): VCI 0.79 (0.70, 0.88) vs. VC-Free 0.76 (0.66, 0.86)). When applying the two models to the Emory Clinic dataset, VC-Free shows a marginal improvement (Emory Clinic, C-Statistics (95% CI): VCI 0.78 (0.73, 0.83) vs. VC-Free 0.79 (0.74, 0.85). Both models clearly outperform the UNI model across the three datasets.

Interestingly as regard to performance measured by the calibration slope, judged by the absolute deviation from 1, VC-Free has slightly improved performance than VCI for the PROC-CT and Emory datasets (PRO-ACT All Folds, Calibration Slope (95% CI): VC-Free 0.91 (0.97, 102) vs. VCI 0.93 (0.88, 0.97); Emory Clinic, Calibration Slope 1.17 (0.90, 1.46) vs. VCI 0.82 (0.0.61, 1.05)), but a little worser performance than VCI for the VITALITY-ALS dataset (VITALITY-ALS, Calibration Slope (95% CI): VC-Free 1.16 (0.58, 1.76) vs. VCI 1.12 (0.62, 1.63)). Consistent with performance measured by the C-statistics,

TABLE 3.5

Summary of Model Performance

Data Source	Validation Cohort	C Statistics			Calibration Slope			Calibration-in-the-Large		
		VCI	VC-Free	UNI	VCI	VC-Free	UNI	VCI	VC-Free	UNI
Internal	PRO-ACT, Fold 1	0.80	0.78	0.68	1.03	0.98	1.05	0.16	0.02	−0.04
		(0.76, 0.84)	(0.74, 0.82)	(0.63, 0.74)	(0.86, 1.20)	(0.79, 1.15)	(0.72, 1.36)	(−0.03, 0.34)	(−0.07, 0.21)	(−0.25, 0.16)
	PRO-ACT, Fold 2	0.82	0.77	0.72	1.12 (	1.14	1.12	−0.18	−0.23	−0.25
		(0.78, 0.86)	(0.72, 0.81)	(0.67, 0.67)	0.93, 1.12)	(0.90, 1.37)	(0.84, 1.38)	(−0.40, 0.02)	(−0.46, −0.02)	(−0.47, −0.05)
	PRO-ACT, Fold 3	0.82	0.76	0.75	0.86	0.93	1.44	−0.02	0.09	0.26
		(0.78, 0.85)	(0.72, 0.80)	(0.71, 0.79)	(0.72, 0.99)	(0.75, 1.12)	(1.17, 1.71)	(−0.21, 0.16)	(−0.10, 0.27)	(0.07, 0.43)
	PRO-ACT, Fold 4	0.79	0.80	0.71	0.80	1.23	1.34	−0.22	0.24	−0.01
		(0.75, 0.83)	(0.76, 0.84)	(0.66, 0.76)	(0.65, 0.94)	(1.03, 1.42)	(1.01, 1.67)	(−0.43, −0.03)	(0.05, 0.42)	(−0.20, 0.18)
	PRO-ACT, Fold 5	0.80	0.76	0.74	1.00	0.95	1.20	0.17	0.03	0.10
		(0.77, 0.84)	(0.72, 0.84)	(0.70, 0.79)	(0.84, 1.16)	(0.77, 1.13))	(0.93, 1.45)	(−0.02, 0.36)	(−0.16, 0.22)	(−0.10, 0.29)
	PRO-ACT, Fold 6	0.81	0.74	0.72	0.92	0.92,	1.40	−0.27	−0.01	0.03
		(0.77, 0.85)	(0.70, 0.79)	(0.76, 0.77)	(0.76, 1.08)	0.71, 1.12)	(1.10, 1.69)	(−0.49, −0.08)	(−0.22, 0.19)	(−0.17, 0.21)
	PRO-ACT, Fold 7	0.79	0.77	0.76	0.89	0.92	1.35	0.00	0.02	−0.18
		(0.74, 0.83)	(0.73, 0.81)	(0.71, 0.80)	(0.71, 1.04)	(0.74, 1.08)	(1.06, 1.64)	(−0.19, 0.19)	(−0.18, 0.20)	(−0.39, 0.02)
	PRO-ACT, Fold 8	0.81	0.76	0.76	1.01	1.02	1.42	−0.04	−0.06	−0.03
		(0.77, 0.85)	(0.72, 0.81)	(0.72, 0.81)	(0.85, 1.04)	(0.83, 1.20)	(1.16, 1.67)	(−0.25, 0.15)	(−0.27, 0.13)	(−0.23, 0.17)
	PRO-ACT, Fold 9	0.79	0.77	0.64	0.86	0.87	1.05	−0.19	0.09	−0.05
		(0.75, 0.83)	(0.73, 0.81)	(0.59, 0.70)	(0.70, 1.01)	(0.70, 1.03)	(0.70, 1.40)	(−0.39, 0.01)	(−0.10, 0.27)	(−0.25, 0.14)
	PRO-ACT, Fold 10	0.84	0.76	0.70	0.91	0.86	1.15	−0.12	0.03	−0.03
		(0.81, 0.87)	(0.72, 0.80)	(0.65, 0.75)	(0.78, 1.04)	(0.67, 1.04)	(0.87, 1.43)	(−0.31, 0.07)	(−0.17, 0.22)	(−0.23, 0.15)
	PRO-ACT,	0.80	0.77	0.72	0.93	0.97	1.24	−0.07	0.03	−0.02
	All Folds	(0.79, 0.81)	(0.75, 0.78)	(0.70, 0.73)	(0.88, 0.97)	(0.91, 1.02)	(1.15, 1.33)	(−0.13, −0.01)	(−0.04, 0.09)	(−0.08, 0.04)
External	VITALITY-ALS	0.79	0.76	0.59	1.12	1.16	0.65	−0.48	−0.50	−0.71
		(0.70, 0.88)	(0.66, 0.86)	(0.43, 0.74)	(0.62, 1.63)	(0.58, 1.76)	(−0.34, 1.67)	(−1.00, −0.04)	(−1.02, −0.06)	(−1.23, −0.27)
	Emory Clinic	0.78	0.79	0.63	0.82	1.17	0.43	−1.09	−0.80	−0.95
		(0.73, 0.83)	(0.74, 0.85)	(0.57, 0.70)	(0.61, 1.05)	(0.90, 1.46)	(0.14, 0.71)	(−1.35, −0.85)	(−1.06, −0.56)	(−1.22, −0.71)

Source: Beaulieu et al. (2021).

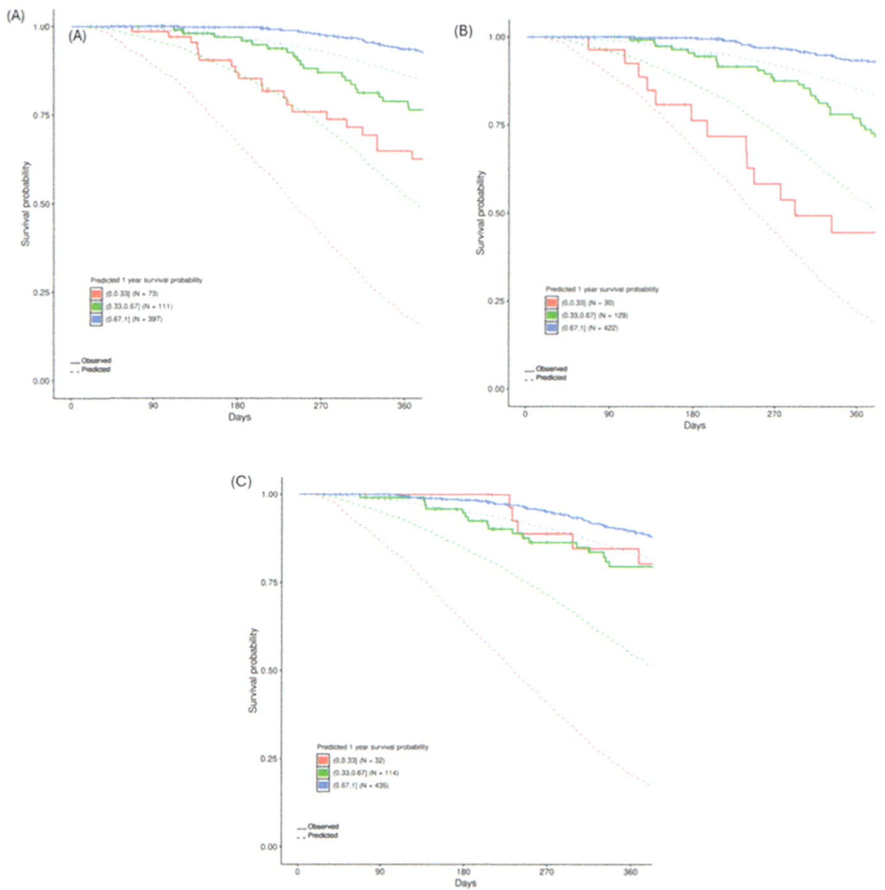

FIGURE 3.4
Visual translation of discrimination and calibration from all Emory clinic patients. (A) VCI model, (B) VC-free model, and (C) UNI model.

both the VC-Free and VCI models perform well evidenced by their respective calibration slopes being close to 1 and are better than the UNI model.

Finally, Calibration-in-the-large (with 95% confidence intervals) yielded the following results: VCI: 0.07 (−0.13, −0.01), VC-Free: 0.03 (−0.04, 0.09), UNI: −0.02 (−0.08, 0.04). Calibration on external datasets exhibited a decreased performance when compared to internal validation across all three models. Notably, the VC-Free model demonstrated superior calibration concerning the external Emory Clinic data in contrast to both the VCI and UNI models. This distinction is visually evident in Figure 3.4B for the VC-Free model, where actual Kaplan–Meier curves are compared to mean predicted curves across various prediction-defined risk strata. This comparison highlights the superior performance of the VC-Free model relative to the VCI (Figure 3.4A) and UNI models (Figure 3.4C).

The performance of the three models were further assessed by comparing model-positive and model-negative groups, with respect to the measures of strata size, death through one year, and median survival. The results are presented in Figure 3.5 and Table 3.6. The Kaplan–Meier curves in Figure 3.5 depict actual one-year survival for the entire Emory population, segmented by the 50% predicted survival cutoff applied to patient predictions from each model. These curves distinctly reveal a greater separation in survival outcomes when strata are delineated using the VCI and VC-Free models compared to the UNI model.

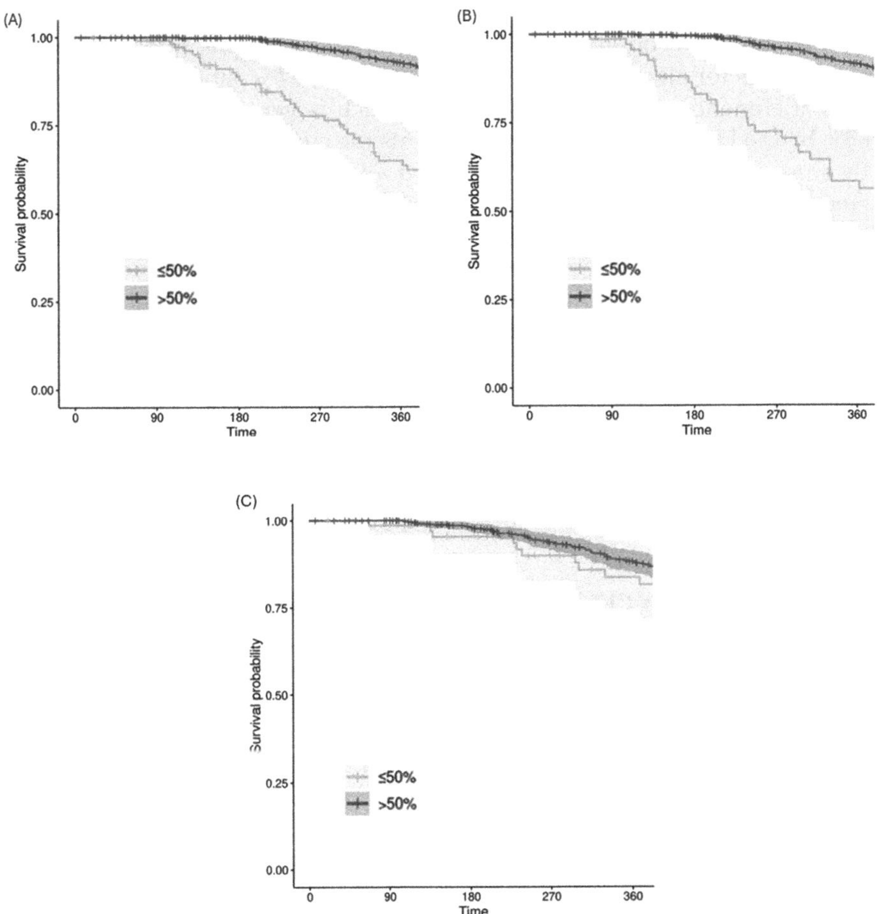

FIGURE 3.5
Kaplan–Meier curves plotted for patients with 50% (excluded) or >50% (included) predicted probability of survival at one year based on predictions from each model: (A) VCI, (B) VC-Free, and (C) UNI. (Adapted from Beaulieu et al. 2021.)

TABLE 3.6

Results of 100 Simulations of $N = 250$ Sampled from Emory Clinic Data

Data Source	Strata Size		Death through One Year		Median Survival (Days)	
	VCI	VC-Free	VCI	VC-Free	VCI	VC-Free
VCI	80 (76, 83)	20 (17, 24)	6 (4, 8)	27 (18, 34)	896 (769, 1,021)	454 (393, 542)
VCI-Free	87 (85, 90)	13 (10, 15)	7 (4, 10)	31 (22, 42)	843 (740, 984)	430 (330, 520)
UNI	88 (84, 90)	12 (01, 16)	10 (7, 13)	13 (3, 20)	782 (686, 935)	741 (578, 917)

Strata size, death through one year, and median survival of each of model between the included (>50% predicted one-year survival) and excluded (≤50% predicted one-year survival) groups were quantified using 100 simulations for all three models. From Table 3.6, 87% patients were selected for enrichment by the VC-Free model, which is very comparable with the 88% by the UNI method, which represents the current practice in ALS trials. By comparison, the VCI model included a lower percentage of patients for enrichment (80%). On average, 10% of the patients deemed included by the UNI model experienced mortality within one year, indicating a slight reduction from the 13% observed in the excluded group. Moreover, these included patients exhibited only a marginal increase in median survival (mean = 782 days) compared to the excluded group (mean = 741 days). In contrast, both the VCI and VC-Free models identified cohorts for inclusion that were notably enriched for survival, with a lower percentage of deaths through one year (VCI: 6%, VC-Free: 7%) and substantially higher median survival times (VCI: 895 days, VC-Free: 843 days) compared to the excluded population (VCI: deaths through one year: 27%, median survival: 454 days; VC-Free: deaths through one year: 31%, median survival: 430 days).

3.4.2 Prediction of Clinical Trial Enrollment Rates

3.4.2.1 Background

Under-enrollment stands as a significant factor contributing to the failures of numerous clinical trials. An essential aspect of trial feasibility analysis involves evaluating whether the trial can adhere to pre-established timelines for completion. Accurately forecasting the enrollment timeline during the planning phase holds paramount importance for both corporate strategic planning and trial success. Traditional approaches rely on statistical models like Poisson-Gamma to characterize the enrollment process and forecast enrollment rates. However, these models heavily depend on the estimated recruitment rate, a piece of information typically unavailable before trial initiation. Consequently, such methods offer limited utility for assessing enrollment before the trial launch. Alternative endeavors have focused on identifying factors influencing enrollment rates, yet many of these investigations examined only a handful of predictor variables, potentially yielding suboptimal predictive models.

Moreover, these studies often draw data from narrow trial populations, such as single health systems or specific diseases, limiting the generalizability of identified factors and their quantitative relationships with enrollment rates across a broader spectrum of nationally conducted clinical trials.

Addressing these limitations, Bieganek et al. (2022) delved into ML for clinical trial enrollment predictions, leveraging available information prior to trial initiation. Utilizing a dataset comprising 46,724 trials from ClinicalTrials.gov, the authors aimed to develop generalizable models across different institutions and organizations, incorporating various features related to study design and recruitment institutions. Moreover, they sought to construct predictive models independent of estimated enrollment rates, ensuring robust performance. Their approach involved applying a range of state-of-the-art supervised learning methods for enrollment rate prediction and comparing the information content among models constructed with different feature types to identify the most informative predictors for enrollment rate prediction.

3.4.2.2 Data

ClinicalTrials.gov is a web-based repository maintained by the National Library of Medicine (NLM), containing information on both publicly and privately supported clinical studies across a wide spectrum of diseases and conditions. Details regarding specific clinical studies are provided and regularly updated by the sponsor or principal investigator of each study and are available for download. They are made available for download for a variety of research purposes. Bieganek et al. (2020) focused on completed interventional clinical trials conducted within the United States. Despite the repository containing data on over 350,000 clinical trials conducted worldwide over the past two decades, the filtration process narrowed the scope to approximately 56,000 studies. Further refinement involved filtering for studies where enrollment status was marked as "Actual" rather than "Estimated," and including only those studies with a minimum duration of 5 days. Consequently, the final dataset comprised 46,724 studies. The distribution of total trials per year is illustrated in Figure 3.6.

3.4.2.3 Enrollment Rate

For each study, enrollment rate r is defined as

$$r = \frac{\text{total number of enrollment}}{\text{Study duration}}$$

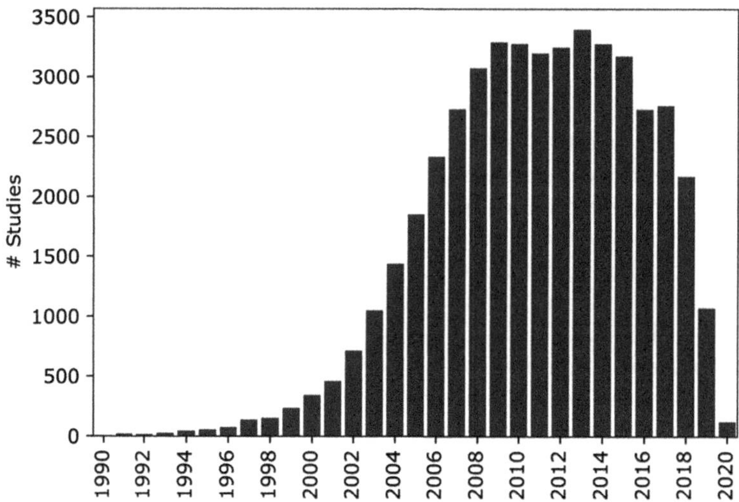

FIGURE 3.6
Number of US based clinical trials per year. (Adapted from Bieganek et al. 2022.)

TABLE 3.7
Stratification of Enrollment Rate

Enrollment Rate	Definition
Low	$r \leq 25$
Medium	$25 < r \leq 100$
High	$r > 100$

where the quantities both the numerator and denominator were extracted from the ClinicalTrials.gov website.

In essence, r represents the number of enrolled participants per year. To facilitate enrollment rate prediction, the clinical trial enrollment rate was stratified into three categories: low, medium, and high in Table 3.7.

Enrollment rates of 25 and 100 corresponded to the 48th and 77th percentiles of the enrollment rate distribution. A time-dependent drift in the distribution of class membership is observed (Figure 3.7). Some of the drift observed in the more recent years can be ascribed to a selection bias stemming from the exclusion of trials initiated in recent years that have not yet concluded.

3.4.2.4 Features Selection and Data Pre-Processing

Since the primary objective was predicting clinical trial enrollment rates prior to study initiation, Bieganek et al. (2022) selected candidate predictors or features available beforehand. These predictors, listed in Table 3.8, were

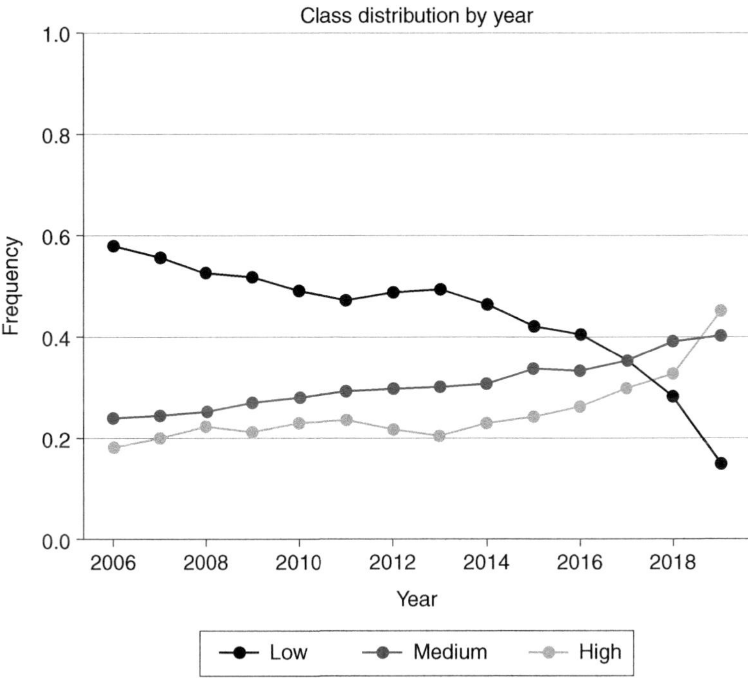

FIGURE 3.7
Distribution of enrollment rates over time. The curves at the top, in the middle, and at the bottom in the plot correspond to the Low, Medium, and High enrollment rate classes, respectively. (Adapted from Bieganek et al. 2022.)

TABLE 3.8
Feature Sets Used for Information Content Analysis

Feature	Complete	Population	Study Center	Study Design	MeSH
Study Phase	x			x	
Funding Agency Class	x			x	
Randomized Allocation	x			x	
Intervention Model	x			x	
Intervention Type	x			x	
Eligibility Gender	x			x	
Accepts Healthy Volunteers	x			x	
Eligibility Minimum Age	x			x	
Eligibility Maximum Age	x			x	
Study Center Cout	x		x		
Study Population	x	x	x		
Institute Core	x		x		
MeSH Terms	x				x

Source: Bieganek et al. (2022).

derived from ClinicalTrials.gov XML data. In addition, the feature list also includes study population and institution score. The size of population was sourced from the U.S. Census Bureau (2019), totaling the populations of metropolitan areas surrounding the site(s) while the institution score, quantifying performance of an institution, was derived based on the Nature Index scores for 2018 calendar year. Recognizing that the Nature Index metrics were only available for the top 500 U.S. institutions, a score of 0 was assigned to those not in the top 500 list. Although ideally, institution scores should be calculated before a study's commencement, the annual Nature Index scores began in 2016. Despite this, it was assumed that the institution score is largely time-independent, thus incorporating the 2018 score into the dataset without significant bias.

It is worth noting that Medical Subject Headings (MeSH) terms were included as features due to their informative nature regarding the trial's research topic and availability from ClinicalTrials.gov. Within ClinicalTrials.gov, each trial is tagged with keywords describing its scope. The ClinicalTrials.gov team assigns two sets of MeSH terms to each trial: one for the conditions studied and another for the interventions used. These MeSH keywords are included in the XML file downloadable for each trial.

3.4.2.5 Model

3.4.2.5.1 Model Training

To constrict the predictive model, Bieganek et al. (2022) explored various multi-class classification algorithms, included multinomial logistic regression, k-nearest neighbors (KNN), multinomial elastic net, support vector machine (SVM), and random forest. Additionally, a dummy classifier was included as a reference when assessing the performance of the above models. This dummy classifier makes predictions according to the prior distribution of the target variable within the training set.

To examine the predictive performance of different types of features in the dataset, these models were trained on various sets of features and their predictive performances compared to a model trained with all features. As listed in Table 3.8, these feature sets include: (1) Population: The population from which participants can be recruited; (2) Study Center: Population, facility count, and institution score; (3) Study Design: Information related to the design of the clinical trial; (4) MeSH: Characteristics regarding the medical domain of the trial; and (5) Complete: A combination of items (1)–(4).

The hyperparameters of each model, wherever applicable, were optimized, through a grid search, based on the training data within the inner loop of the nested time-series cross-validation. Table 3.9 presents the models along with hyperparameters and grid for turning the models. The logistic regression and the dummy model do not have hyperparameters to tune.

TABLE 3.9

Hyperparameters and Values for Grid Search for Model Tuning

Model	Parameter	Value
KNN	# of Neighbors	5, 10, 20, 50, 100
Elastic Net	C	1×10^{-4}, 7.7×10^{-4}, 6.0×10^{-3}, 4.6×10^{-2}, 0.36, 2.78, 21.5, 167, 1,290, 1,000
	L1-ratio	.1, .2, .5, .7, .9, .95, .00, 1
Random Forest	# of Trees	200, 400, 600, 800
	Min Sample Per Leave	1, 2, 5, 10
SVM	Degree	1, 2, 3, 4
	C	10^{-4}, 10^{-3}, 10^{-2}, 10^{-1}, 1, 10, 100, 1,000, 10,000
Logistic Regression	NA	
Dummy Model	NA	

Source: Bieganek et al. (2022).

All the models were trained with and without the MeSH terms. For the former, the training dataset consisted of all 4,624 observed MeSH terms whereas for the latter, the training data comprised only the most frequently occurring MeSH terms, defined as having terms appearing in at least 200 clinical trials were considered. This resulted in 112 included terms. The common MeSH terms in the training dataset are Diabetes Mellitus," "Depression," "Breast Neoplasms," "Depressive Disorder," and "Syndrome." Overall, 53% of the trials in the training dataset contain at least one of the 112 frequently occurring MeSH terms.

3.4.2.5.2 Model Validation

Typically, random samples are drawn from a dataset for model training and validation purposes. The intent is to ensure observations are independently and identically distributed to ensure the validity of model performance evaluation. However, time-series data are not independently and identically distributed. This makes the standard cross-validation approach inappropriate. To address issue, Bieganek et al. (2022) implemented a nested time-series cross-validation design (Cawley and Talbot 2010; Bergmeir and Benítez 2012; Cerqueira et al. 2020), which is a variant of the traditional cross-validation approach, using early data in the time-series to train and optimize the model (inner cross-validation) and later data from the series to evaluate the performance of hyperparameter-tuned models (outer cross-validation).

Specifically, 14 validation datasets were used for the outer cross-validation. Each dataset consisted of a single year of data spanning from 2006 to 2019. Data collected from years preceding the validation dataset were employed to train models and select the most effective hyperparameters through a cross-validation hyperparameter grid search process. The data underwent a process for scaling and missing data imputation. For details, refer to Bieganek et al. (2022).

3.4.2.5.3 Performance Metrics

Because of the predicted enrollment rates of the models falling into low, medium, and high categories, metrics related to multi-class classifications were used to evaluate model performance. These include the multi-class area under the receiver operating characteristic curve (AUC), accuracy, recall, and precision. The formulas for computing the performance metrics are presented in Table 3.10. The calculations of these metrics utilize a confusion matric, C, such that the C_{ij} element of the confusion matrix is the count of the observations for which the predicted class is c_i and the true class is c_j.

It is of interest to assess the influence of various predictive features as related to patient population, study center, study design, and MeSH terms on the performance of the predictive models. To that end, the models were trained on different feature sets and their predictive performances were compared with the model trained with all features. Table 3.8 describes the features in each of the sets. Of note, feature sets incorporating MeSH terms (feature sets (4) and (5)) were either utilizing all 4,624 observed MeSH terms or solely focusing on the most frequently observed MeSH terms.

3.4.2.5.4 Selection Bias and Domain Adaptation

Classifier learning methods often assume that the training data are randomly selected examples from the same distribution as the test examples, where the learned model is intended to make predictions. In many real-life

TABLE 3.10

Performance Metrics: AUC, Recall, Precision, Accuracy

Metrics	Definition	
AUC	$AUC = \frac{2}{K(K-1)}\sum_{i=1}^{K}\sum_{j\neq i}^{K} AUC(c_i, c_j)$	where $AUC(c_i, c_j)$ is the two-class AUC with c_i and c_j being the positive class and negative class, respectively.
Recall	$\frac{1}{K}\sum_{j=1}^{K}\frac{C_{jj}}{m_j}$	where $m_j = \sum_{i=1}^{K} C_{ij}$ is the total number of samples with true class c_j.
Precision	$\frac{1}{K}\sum_{i=1}^{K}\frac{C_{ii}}{n_i}$	where $n_i = \sum_{j=1}^{K} C_{ij}$ is the total number of observations with predicted class c_i.
Accuracy	$\frac{1}{N}\sum_{i=1}^{K} C_{ii}$	

applications, however, this assumption is violated. It is valuable to utilize bias correction in the model building when sample selection bias is known to exist. Since only completed clinical trials were included in the investigation by Bieganek et al. (2022), it is evident from Figure 3.5, there was clearly a decreasing trend in number of completed studies in the years close to the data cutoff year 2020. This inevitably introduced a selection bias that was in favor of selecting studies of shorter durations in the years close to 2020, thus resulting in a difference in the training data and the validation data. To counter this problem, Bieganek et al. (2022), implemented domain adaptation to correct the difference in sampling distribution. This bias correction method filtered out the training samples with duration longer than τ – the maximum study duration in the validation set.

3.4.2.6 Results

Unless otherwise specified, the results presented below are for models trained using all MeSH terms.

3.4.2.6.1 Predictive Performance of Models

The ML models presented in Table 3.11 were trained across different feature sets outlined in Table 3.8.

The performance of model trained with all features and on the 14 validation sets from 2006 to 2019 is illustrated in Figure 3.8. It is apparent that all of these models outperformed the dummy classifier, and comparable performance was observed across the methods with respect to the four performance metrics. Among all the models, SVM exhibited the best overall performance from 2006 to 2019, with an AUC (SD) of 0.81 (0.02), recall (SD) of 0.58 (0.02), precision of 0.59 (0.04), and accuracy (SD) of 0.62 (0.05). Following closely, the random forest model ranked second, performing only slightly worse than SVM, with an AUC (SD) of 0.78 (0.03), recall (SD) of 0.57 (0.02), precision (SD) of 0.58 (0.04), and accuracy (SD) of 0.62 (0.05).

Additionally, Bieganek et al. (2022) examined model performance on a dataset including all features and most frequent MeSH terms. For this evaluation, the random forest classifier emerged as the best-performing model among all the methods that completed performance assessment across all experimental settings. It is also interesting to point out that the random forest maintained similar performance compared to when all MeSH terms were included, with an average AUC from 2006 to 2019 of 0.78 ± 0.02 (mean $\pm$ std).

3.4.2.6.2 Information Content in Different Types of Features

To evaluate the relative information content in different feature types, predictive models were developed using the feature sets in Table 3.8. For illustration, the results of the random forest classifier are used to examine the impact of different feature sets (Figure 3.9). Among the feature sets examined, the study

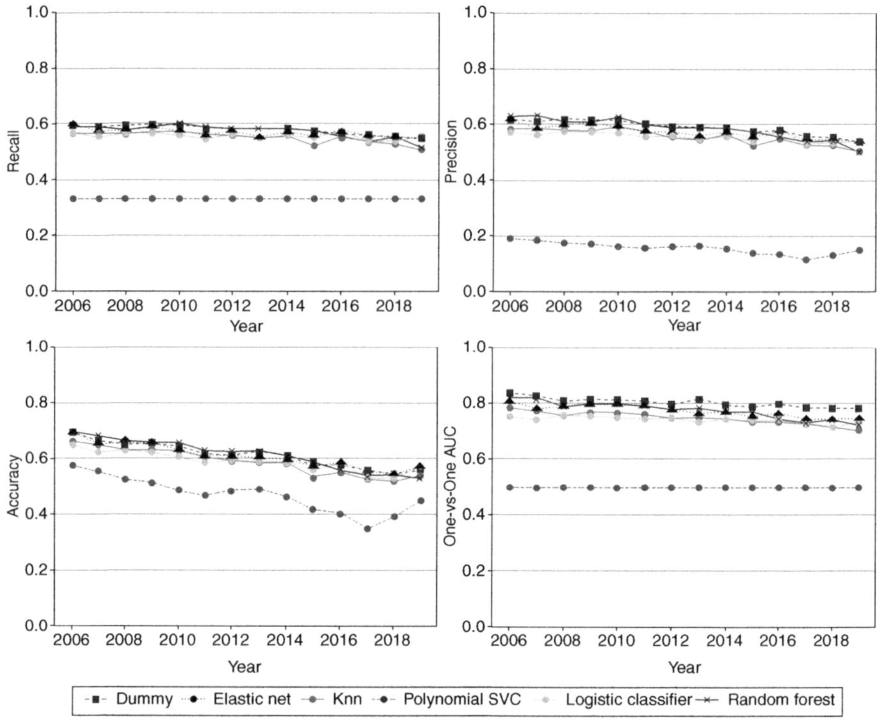

FIGURE 3.8
Predictive performance of models constructed with various classification methods using the complete set of 4,636 features. (Adapted from Bieganek et al. 2022.)

design-related feature set demonstrated the highest predictive performance, yielding an AUC of 0.76 ± 0.02 across all years. Conversely, the remaining feature sets exhibited weak to moderate predictive capabilities regarding enrollment rate individually. Specifically, models based solely on the population of the recruitment region, study center characteristics, and MeSH terms yielded AUCs of 0.59 ± 0.01, 0.62 ± 0.01, and 0.67 ± 0.02, respectively. Furthermore, combining these features with those of study design only marginally improved model performance. Specifically, the AUC for the complete set of features reached 0.78 ± 0.03, compared to 0.76 ± 0.02 for the model with study design features alone.

3.4.2.6.3 Selection Bias and Domain Adaption

As reported by Bieganek et al. (2022), interestingly, when models were trained on the full dataset, incorporating all MeSH terms, the impact of domain adaptation on model performance was negligible. Figure 3.10 depicts the performance metrics of the random forest model between 2006 and 2020 with and without domain adaptation when models were on a dataset excluding MeSH terms appearing fewer than 200 times. Two observations can be made from the plots. Firstly, the impact of domain adaptation on predictive performance

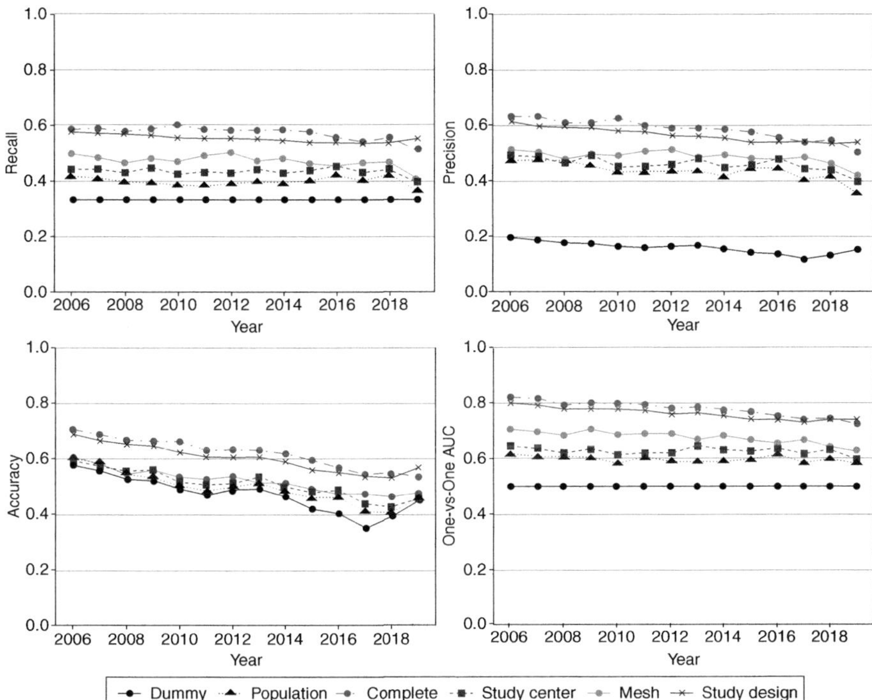

FIGURE 3.9
Predictive performance of random forest models with different feature subsets. (Adapted from Bieganek et al. 2022.)

for earlier years was minimal, due to little to none distribution shift between the training and validation data observations can be made. In contrast, domain adaptation aligned the performance for recent years more closely with that of earlier years. Specifically, the average AUC across 2016 to 2019 was 0.76 ± 0.01 with domain adaptation, compared to 0.73 ± 0.01 without it.

3.4.3 Forecasting Drug Supply Demand

In this section, we present a case example demonstrating how predictive analytics can be used to forecast the demand for reference product in a biosimilar clinical study (Yang et al. 2018).

3.4.3.1 Background

A continuous and adequate drug supply is a key component of conducting a successful clinical trial. For biosimilar studies, drug supply can be more challenging as these studies require sourcing reference products from the open market. Specifically, for each reference product, there are usually several

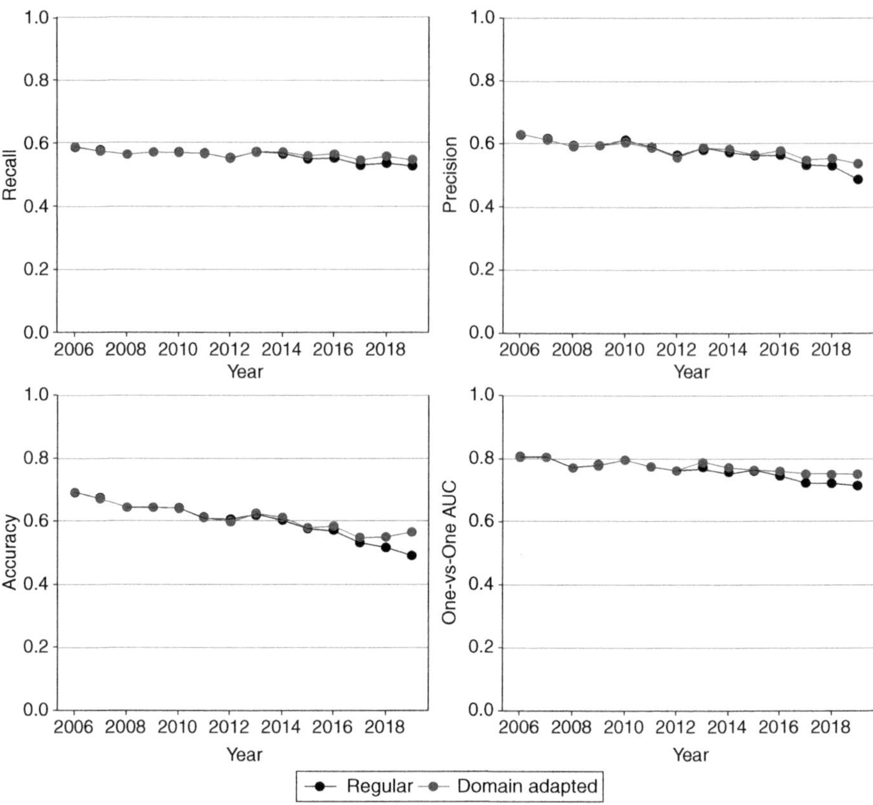

FIGURE 3.10
Predictive performance of random forest models with and without domain adaptation on dataset with reduced MeSH features. (Adapted from Bieganek et al. 2022.)

biosimilars in clinical development. Therefore, sourcing the reference product can be very competitive. In addition, since biological products are often intended for repeated or long-term use, large quantities of the reference drug are needed over an extended period. Furthermore, regulatory authorities for certain regions require use of the reference product in the clinical studies which is specifically produced for that region. This limits the number of procurement sources. Advanced planning based on accurate prediction of the demand is key to making the interchangeability study less costly and complete according to the plan. It is also advantageous to break down drug supply by calendar time so that the procurement of the drug can be planned accordingly.

3.4.3.2 Modeling Approach

In an oncology study, drug supply can be estimated in terms of the expected total number of cycles of treatment needed for each month for the entire duration of the study. When this estimate is available, the amount of drug can be calculated with the input of other parameters such as dose levels, e.g., 15 mg/kg, and average weight of the patients in the study. A key factor in the estimation of the number of cycles of treatment for a given month is the conditional probability for a patient to stay on the treatment for the month. This can be accomplished through modeling the time to treatment discontinuation, T. Assume T follows an exponential distribution $P[T > t] = e^{-\lambda t}$. Let t_0 be the time at the beginning of the month, and t ($> t_0$) a future time and $d = t - t_0$. It can be derived that the conditional probability for a patient to receive treatment at times from t_0 to t can be derived

$$Y(t) = P(T > t | T > t_0) = \frac{\text{Prob}(T > t \text{ and } T > t_0)}{\text{Prob}(T > t_0)} = \frac{\text{Prob}(T > t)}{\text{Prob}(T > t_0)} = \frac{e^{-\lambda t}}{e^{\lambda t_0}} = e^{-\lambda d} \quad (3.4)$$

From (3.4), the total number of cycles needed from t_0 to t is calculated as

$$\frac{d}{\text{cycle duration}} \times (N_a + N_{dr}/2), \quad (3.5)$$

where N_{dr} is the number of dropouts from t_0 to t and N_a is the number of active subjects at time t:

$$N_{dr} = \text{No. of active subjects at } t_0 \times [1 - Y(t)]$$

$$N_a = \text{No. of randomized subjects at } t - \text{No. of total dropouts at } t.$$

Note that we assume that the distribution of dropout is uniform in the time interval $[t_0, t]$. Therefore, on average, these dropout patients only have half of the planned cycles of treatment. It is also important to point out that in order to estimate the number of active subjects at the beginning of a time interval, the recruitment rate needs to be specified.

For the study of interest, 80 subjects were planned to be randomized to the study within 6 months with the first subject randomized in December 2017. The rate of recruitment was projected to be constant and drugs were administered in each 14-day cycle. Drug supplies were planned to be provided until January 2020 when the study was expected to be complete. The median duration of treatment time was 190 days (about 6.24 months) for the reference product. Based on the exponential model presented above, the model parameter was estimated by

$$\lambda = -\frac{\ln(0.5)}{6.24} = 0.111$$

From (3.4)

$$Y(t) = e^{-0.111d}.$$

Based on (3.5), the number of treatment dropouts and cycles by calendar time were estimated iteratively, and the results presented in Table 3.11.

From the table, there were 62 active subjects by 30-June-18. The probability of subjects who were still active by 31-July-18 was 0.893. Therefore, the number of dropouts from 30-June-18 to 31-July-18 was calculated as $62 \times (1 - 0.893) = 7$. Since, by 31-July-18, 80 subjects were randomized, and the total number of dropouts was 24. Therefore, the number of active subjects by 31-July-18 was 80 – 24 = 56. From (3.5), the total number of cycles until

TABLE 3.11

Estimate of Number of Cycles of Treatment Needed by Calendar Time

Calendar Time	d (mons)	$Y(t)$	No. of Randomized Subjects	No. of Drop-out	Active	Total Dropouts	No. of Cycles
31-December-17			1	0	1	0	
31-January-18	1	0.893	13	0	13	0	29
28-February-18	1	0.903	26	1	25	1	51
31-March-18	1	0.893	39	3	35	4	80
30-April-18	1	0.896	52	4	44	8	99
31-May-18	1	0.893	65	5	53	12	122
30-June-18	1	0.896	80	5	62	18	139
31-July-18	1	0.893	80	7	56	24	130
31-August-18	1	0.893	80	6	50	30	116
30-September-18	1	0.896	80	5	44	36	101
31-October-18	1	0.893	80	5	40	40	93
30-November-18	1	0.896	80	4	36	44	81
31-December-18	1	0.893	80	4	32	48	75
31-January-19	1	0.893	80	3	28	52	67
28-February-19	1	0.903	80	3	26	54	54
31-March-19	1	0.893	80	3	23	57	54
30-April-19	1	0.896	80	2	21	59	46
31-May-19	1	0.893	80	2	18	62	43
30-June-19	1	0.896	80	2	16	64	37
31-July-19	1	0.893	80	2	15	65	34
31-August-19	1	0.893	80	2	13	67	31
30-September-19	1	0.896	80	1	12	68	27
31-October-19	1	0.893	80	1	10	70	25
30-November-19	1	0.896	80	1	9	71	21
31-December-19	1	0.893	80	1	8	72	20
31-January-20	1	0.893	80	1	7	73	18
Total							1,591

Source: Yang et al. (2018).

31-January-20 was 1,591 in this case. This provided sufficient information to allow for the estimation of the total number of kits/vials needed at certain time or total. In practice, a certain percent of overage, for example, 15–20%, is usually included to ensure sufficient drug supplies. If needed, the post-study drug supplies can be also estimated.

3.5 Concluding Remarks

In summary, the integration of advanced technologies such as AI, predictive analytics, and ML holds immense promise for transforming clinical trial operations. By embracing innovation and leveraging data-driven insights, stakeholders can enhance efficiency, improve patient outcomes, and accelerate the development of life-saving therapies. AI and predictive analytics can optimize various aspects of clinical trials, from patient recruitment to drug supply management and patient safety monitoring. These technologies enable real-time predictions of patient enrollment rates, allowing for proactive adjustments to recruitment strategies and ensuring timely trial completion. Additionally, predictive models can forecast drug supply needs, preventing shortages or overproduction, thereby reducing costs and ensuring that trials run smoothly.

ML algorithms play a crucial role in identifying and mitigating patient safety risks. By analyzing vast amounts of data from EHRs and other sources, these algorithms can detect patterns indicative of adverse events, enabling early intervention and enhancing patient safety. Furthermore, ML can impute missing data, which is a common issue in clinical trials, ensuring the integrity and completeness of the dataset. This capability supports RBM by highlighting potential issues that warrant closer scrutiny, thereby optimizing the allocation of monitoring resources and improving overall trial quality.

Despite these benefits, the adoption of advanced technologies in clinical trials is not without challenges. Data privacy is a paramount concern, as the handling of sensitive patient information must comply with stringent regulations such as GDPR and HIPAA. Ensuring regulatory compliance requires robust validation and documentation of predictive models, along with adherence to data privacy and security standards. Model validation and calibration are essential to ensure that predictive analytics tools are accurate, reliable, and free from bias. Rigorous testing using historical data and real-world scenarios is necessary to validate these models and build trust among stakeholders.

Looking forward, the integration of AI and predictive analytics with emerging technologies such as blockchain can further enhance the transparency, traceability, and integrity of clinical trials. Blockchain can facilitate secure data sharing across the supply chain, ensuring accountability and preventing data tampering. Personalized supply chains tailored to the specific needs of individual trials, real-time predictions of supply chain performance, and patient outcomes represent exciting future directions. By addressing current challenges and leveraging these innovations, the clinical trial industry can realize the full potential of advanced technologies, ultimately delivering more efficient, effective, and patient-centric research.

4

Quality by Design for Clinical Trials

4.1 Introduction

Clinical trials are intended to evaluate the safety and efficacy of medical interventions by way of generating patient data. Reliable evidence from clinical trials serves as the basis for key decision-making throughout the lifecycle of clinical development. As such, the integrity and quality of the trial design and data not only influence clinical development strategy, but also regulatory decisions. However, ensuring the integrity and quality of clinical trials is also a costly, complex, multifaceted endeavor, fraught with challenges because of cultural barriers, technological limitations, and evolving regulatory requirements. In this chapter, we discuss the shift from the current retrospective and reactive quality management of clinical trials to an integrated, proactive, risk-based, and cost-efficient model based on Quality by Design (QbD) principles. As we discuss the intricacies of leveraging innovative methods inspired by this change, we aim to elucidate their potential to revamp traditional clinical trial design, planning, and quality oversight, thereby enhancing the integrity and reliability of clinical trial data. Several use cases are provided to highlight the transformative role of artificial intelligence (AI), machine learning (ML), and advanced analytics in ensuring participant protection and reliability of clinical trial data.

4.2 Clinical Evidence Generation

Drug developers design clinical trials to answer specific research questions related to a medical product. Each trial follows a specific protocol that describes study objectives, patient population, therapeutic intervention, procedures, and study visits at which patient outcomes are measured. It also

DOI: 10.1201/9781003226086-4

includes design considerations such as sample size, randomization, and statistical methods to ensure unbiasedness in data collection and analysis.

As described in Chapter 2, upon finalization, essential information is extracted from the study protocol to build various trial management systems. Of note is the Electronic Data Capture (EDC) system for collecting and managing patient data. The EDC is managed by clinical data management (CDM) – a function with specially trained staff, tools, and processes responsible for the collection and management of clinical data. The goal of CDM is to ensure that clinical trial data are accurate, reliable, and fit for purpose; that is, the data are of the required quality for addressing study objectives. Like most aspects of clinical research, there are regulations and guidance documents that stipulate specific requirements for EDC and clinical data (FDA 2003, 2013a; ICH 2016; EMA 2023a). In addition to the regulatory guidelines, there are also industry best practices, established by organizations such as the Society for Clinical Data Management (SCDM) (SCDM 2013).These regulations and guidance documents, along with industry best practices, provide a robust framework for ensuring the effective and compliant use of EDC systems in clinical trials.

The process of clinical data generation is depicted in Figure 4.1. It begins with protocol design. Once the protocol is the data manager identifies the data items to be collected, and frequency of collection with respect to the visit schedule. The CDM team then crafts Case Report Forms (CRFs). The CRFs clearly indicate the type of data to be entered. In addition, specification of

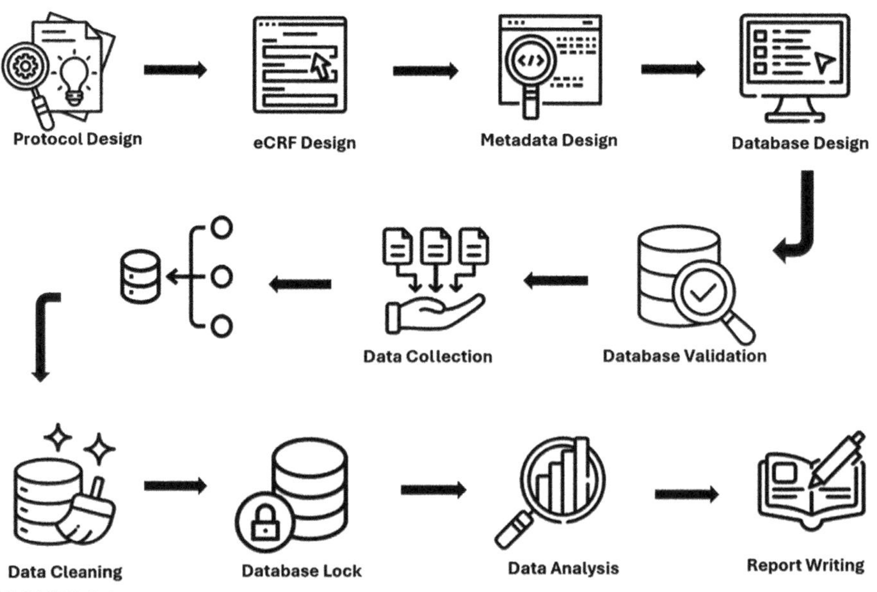

FIGURE 4.1
Process of clinical data generation.

metadata, which are information or metrics about study conduct as related to data collection, will be developed. Subsequently, a database is built to facilitate the downstream CDM activities. In accordance with regulations, the database is tested to ensure it is fit for its intended purpose, before being put into production. During the study, patient data are collected at clinical trial sites by the site investigator or supporting staff and entered in the CRFs in the database. Data cleaning is a critical step in ensuring high-quality data. After the data is declared clean, the database is locked to prevent any changes to the data. The final steps involve analyzing data, interpreting, and reporting findings. Researchers employ statistical techniques to draw conclusions about the products or procedures under study. Findings are documented in a clinical study report (CSR) and submitted to regulatory authorities for review, ensuring transparency and accountability in the research process.

4.3 Quality Assurance and Control

4.3.1 Current Method

The current method for ensuring clinical data quality is primarily focused on protocol development, training clinical staff at site for data collection, routine quality checks, and ongoing monitoring of clinical conduct and data generation.

4.3.1.1 Protocol Development

The current approach to clinical data quality begins with the development of a protocol that clearly defines the study's objectives, endpoints, design, methodology, inclusion and exclusion criteria, and other critical elements. This process entails collaboration among various stakeholders, including the study sponsor, investigators, and regulatory authorities. The protocol undergoes meticulous review to ensure compliance with regulatory standards and scientific rigor (ICH 2016). However, little attention has been given to the impact on downstream clinical activities such as patient recruitment.

4.3.1.2 Training

Ensuring data quality also necessitates thorough training of investigators and site personnel. Training includes both the use of the database and key CDM documents such as Data Management Plan (DMP) and CRF Completion Guidelines that outline the database design, data entry, and tracking guidelines, quality control procedures, guidelines for serious adverse event (SAE) reconciliation, discrepancy management, data extraction and transfer, and database lock. In addition, training is also conducted to ensure consistent implementation of trial procedures across all study sites and the generation of reliable data, protocol compliance, and adherence to regulatory requirements.

4.3.1.3 Data Quality Assurance

Data collection begins with the enrollment of study participants and continues throughout the duration of the clinical trial. During the study conduct, several measures are implemented to warrant the accuracy, completeness, and reliability of clinical trial data. This includes double-entry of each data point, routine data cleaning, through both programmatic checks, also known as edit checks and manual review to resolve discrepancy (SCDM 2013). In addition, drugs and medical terms are coded, using dictionaries or classification systems like the Medical Dictionary for Regulatory Activities (MedDRA), the International Classification of Diseases (ICD), and Current Procedural Terminology (CPT) to ensure data consistency, and on an on-going basis SAE reconciliation is being carried out to ensure that all SAEs recorded in the clinical trial database are accurately matched with those reported in the safety database.

4.3.1.4 Monitoring and Auditing

The current approach also heavily relies on on-site monitoring and auditing to ensure data integrity and quality. Monitoring is typically carried out by study teams or independent clinical research organizations (CROs). During the on-site visits, data are reviewed against source documents to identify any discrepancies. Often, 100% source data verification (SDV) is carried out. In contrast, auditing entails a systematic review of the clinical trial by regulators to ensure adherence to protocol and regulatory requirements, often conducted by regulatory authorities.

4.3.2 Limitations

In clinical trials, there has been a long-held belief that more data and oversight would lead to higher quality of the trial and better outcomes. In practice, this is realized through collecting data on every aspect of the trial, frequent on-site visits, and 100% SDV. This approach has become so ingrained that deviating from it is seen as risky (Dietrich et al. 2018). However, mounting evidence suggests that an obsessive focus on the collection of large volume of data and the accuracy of every data point, regardless of its significance, contributes minimally, if at all, to trial integrity and data quality. Meanwhile, it incurs substantial expenses and efforts. Stakeholders within the clinical trial community have voiced concerns about the current model for conducting clinical trials, citing its high costs, lack of sustainability, and potential hindrance to generating reliable evidence (Meeker-O'Connell, Glessner, Behm et al. 2016; Meeker-O'Connell, Sam et al. 2016; Collin and Macmmahan 2001; Cardiac Arrhythmia Suppression Trial (CAST) Investigators 1989).

Over the past 20 years, there has been a significant surge in both the quantity and complexity of clinical trials. These developments introduce fresh hurdles to overseeing clinical trials, especially due to the widening spectrum of clinical investigator expertise, variations in site infrastructure, diverse treatment options, and evolving standards of healthcare. Additionally, the geographical spread of trials poses its own set of challenges. In the meanwhile, the expense of bringing a prescription drug to market approval continues to escalate, with recent estimates pegging it at $2.6 billion – a staggering 145% surge after adjusting for inflation from 2003 to 2014 (DiMasi et al. 2016). Roughly two-thirds of the total drug R&D expenditures are tied to the clinical phases of development, with 70% stemming from Phase II/III clinical trials. Since the financial burden of drug development ultimately falls on patients and society, it becomes crucial to manage the costs of drug research and development (R&D) while ensuring the integrity of the data.

Consequently, there's growing interest in tailored approaches that consider how both trial design and trial conduct impacts quality. Implementing such approaches could alleviate unnecessary burdens on trial participation and conduct, thereby enhancing the efficiency of drug development and expediting the delivery of new products to patients.

4.4 Regulatory Developments

Recognizing the need for modernizing the approach to clinical trial integrity and quality, in recent years, regulatory authorities such as the ICH and FDA have published several guidelines that recommend addressing the clinical trial quality issue from a holistic quality risk management perspective.

On October 6, 2021, the ICH formally finalized and adopted the *ICH guideline E8 (R1) General considerations for Clinical Studies* (ICH 2021a). The guideline stresses the importance of adopting the QbD approach to clinical research, by focusing on clinical trial factors that are critical to quality (CtQ) to ensure the protection of study participants and the generation of reliable and meaningful results, and management of risks to those factors using a risk-proportionate approach. It also emphasizes the establishment of an appropriate framework for the identification and review of CtQ factors at the time of trial design and planning.

The FDA (2013b) guidance *Oversight of Clinical Investigations – A Risk-Based Approach to Monitoring* provides recommendations for implementing risk-based monitoring strategies in clinical trials. The guidance recommends a systematic assessment of risks associated with clinical trials, considering factors such as study design, patient population, investigational product, and data collection methods. The guidance also emphasizes the importance of centralized monitoring, coupled with more targeted on-site monitoring in enhancing the efficiency and effectiveness of monitoring activities (Sakamoto and Buyse 2016). Subsequently, in April 2023, FDA published the guidance *Risk-Based Approach to Monitoring of Clinical Investigations Questions and Answers* (FDA 2023b), expanding on its previous RBM guideline by providing additional information to facilitate the implementation of RBM.

In 2016, in a major revision to the ICH E6(R1) guideline: *Good Clinical Practice (GCP)*, the ICH published ICH E6(R2), *Good Clinical Practice: Integrated Addendum to ICH E6(R1)* (ICH 2016), which was later broadly adopted by various Health Authorities. The guideline recommends a risk-based approach to quality management at both the protocol and system levels. It also introduces Quality Tolerance Limits (QTLs) to monitor CtQ factors and track trial progress toward endpoints and elevates the implementation of QTLs in clinical trials as part of quality management system (QMS).

ICH E6(R3) *Good Clinical Practice (GCP)* (ICH 2023) builds on key concepts outlined in ICH E8(R1) *General Considerations for Clinical Studies* (ICH 2021a). This includes fostering a quality culture and proactively designing quality into clinical trials and drug development planning, identifying factors critical to trial quality, and engaging stakeholders, as appropriate, using a proportionate risk-based approach. Specifically, it recommends QbD should be implemented to identify data and processes that are critical to ensure trial quality and the risks that threaten the integrity of those factors and ultimately the reliability of the trial results. Clinical trial processes and risk mitigation strategies implemented to support the conduct of the trial should be proportionate to the importance of the data being collected and the risks to trial participant safety and data reliability. Trial designs should be operationally feasible and avoid unnecessary complexities.

Together these regulatory documents have ushered in a risk-based and lifecycle approach to clinical trials. Including the approach as part of the QMS provides greater visibility across clinical trial functions.

4.5 Quality by Design for Clinical Trials

4.5.1 Quality by Design Principles

The current approach to clinical trial integrity and quality is reactive, inefficient, and costly. To address these limitations, a holistic approach centered on error prevention is essential. This process begins at the protocol planning stage and continues throughout the lifecycle of a trial, effectively building quality into both the scientific and operational aspects of the trial. For example, when designing a trial, careful consideration should be given to integrating simplified yet well-justified eligibility criteria, flexible visit schedules, diverse data capture methods, and streamlined collection of pertinent clinical events with centralized review of critical safety and efficacy information (EMA 2013; Meeker-O'Connell, Glessner et al. 2016).

Equally important is enhancing quality oversight by focusing on data and other aspects of the trial that are critical to the safety of participants and the reliability of the trial results (ICH 2016). This holistic, more focused approach aligns with the principles of QbD.

The notion of QbD was initially introduced by Juran JM in the early 1990s. Broadly, it characterizes quality as the absence of significant errors. In the context of a clinical trial, this could entail errors in study execution or inaccuracies in data collection and/or reporting that impact the predetermined study endpoints, thus undermining study validity or compromising patient rights or safety (Juran 1992; SCDM 2013).

QbD is a systematic approach to pharmaceutical development, focusing on building quality into the fabric of clinical trial conduct from the outset. QbD entails identifying CtQ factors and risks to key processes and data of the clinical trial, and subsequently designing the trial and processes to consistently yield outcomes aligning with predefined quality standards. The QbD principles underpin a risk-based QMS, where risk to quality is identified and the level and method of risk control are commensurate with the level of risk (FDA 2023c).

4.5.2 Industry Initiatives

In response to the changing regulatory landscape, several quality initiatives have been launched. The Clinical Trials Transformation Initiative (CTTI) expanded the QbD framework into a comprehensive, actionable resource for clinical trial development and execution. This encompasses a series of core recommendations alongside a practical toolkit designed to facilitate the planning and implementation of a successful QbD framework (Meeker-O'Connell, Glessner et al. 2016).

TransCelerate devised a methodology that transitions monitoring procedures away from an overly focused approach on SDV toward a comprehensive, risk-driven monitoring strategy (TransCelerate 2020). This shift in monitoring philosophy utilizes centralized and off-site mechanisms to holistically monitor critical study parameters, while incorporating adaptive on-site monitoring to bolster site operations, ensure subject safety, and maintain data quality. TransCelerate has also developed a framework to guide the industry in implementing QTLs (Bhagat et al. 2021). QTLs are part of that risk-based approach and serve as an added control for risks to factors that are CtQ. Similarly, AVOCA Quality Consortium (AVOCA 2020) developed a process for establishing CtQ factors and setting QTLs for parameters critical to data integrity and patient safety.

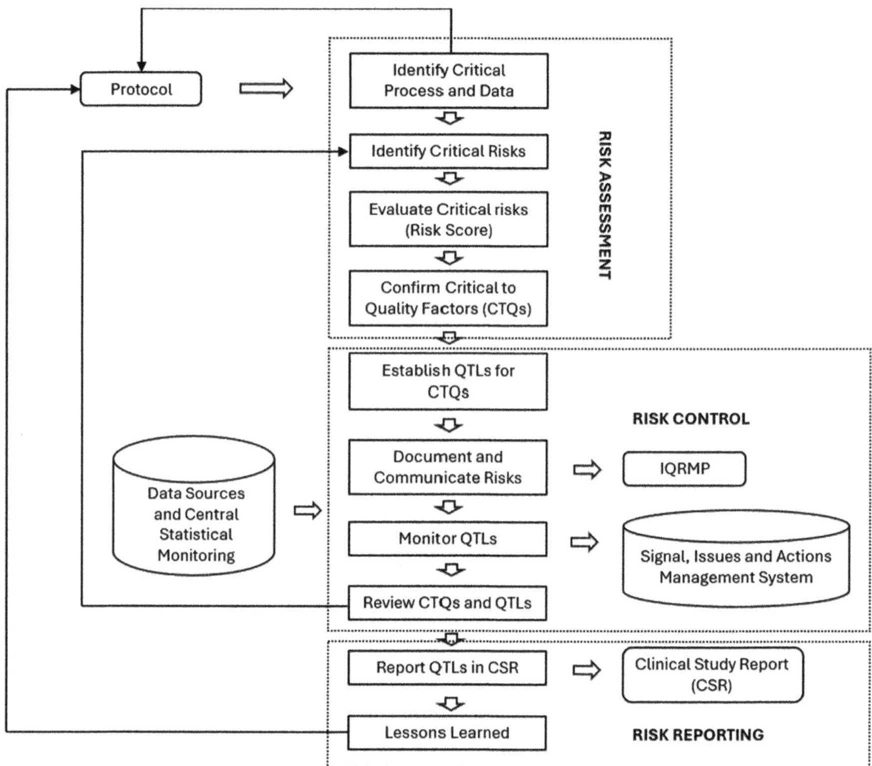

FIGURE 4.2
Process of implementing QbD for clinical trials as part of quality risk management. (Adapted from AVOCA 2020.)

To support a shift from reactive to proactive, and to prevent risks from becoming issues that undermine clinical activities, TransCelerate published a concept paper outlining its vision of a Clinical Quality Management System (CQMS) tailored to clinical development and the unique requirements of clinical development stakeholders. This CQMS is designed to enable the consistent and efficient delivery of reliable data that can be used by an organization, its partners, health authorities, healthcare providers, and patients to make informed decisions.

As depicted in Figure 4.2, QbD permeates every stage of a trial, commencing with protocol design and the identification of CtQ factors and risks to processes and data which can impact patient safety and the reliability of trial results (Bhagat et al. 2021). It extends 2021 through risk control during trial execution and culminates in analyzing and reporting the impact of significant deviations in the CSR (ICH 2021a; FDA 2023c).

4.6 Implementation of QbD

Implementation of the QbD approach to clinical trials starts at protocol conception and continues throughout the lifecycle of the study. By applying a workflow similar to the Plan-Do-Check-Act (PDCA) schema (Figure 4.3), the entire clinical trial process is set for continuous improvement (Landray et al. 2012; Dietrich et al. 2018).

At the "Plan" phase, the study protocol should be carefully designed with broad input from experts and stakeholders. In addition, key risks to human subject protection, successful execution of the study, and the reliability of results should be identified, and the plan for risk control developed. The "Do" aspect is centered on the implementation of the risk management strategies developed in the "Plan" phase. It encompasses training of site personnel about various aspects of the trial and operations, establishing infrastructures, processes, and tools to meet protocol requirements, and tailor procedures and operations to enhance quality. The "Check" is to assess if the trial is in a state of control and determine if remediations such as more frequent onsite visits are needed to address emerging issues. Finally, the information gathered throughout the study should be reviewed. The findings may lead to the modification of risk assessment and changes to all relevant components

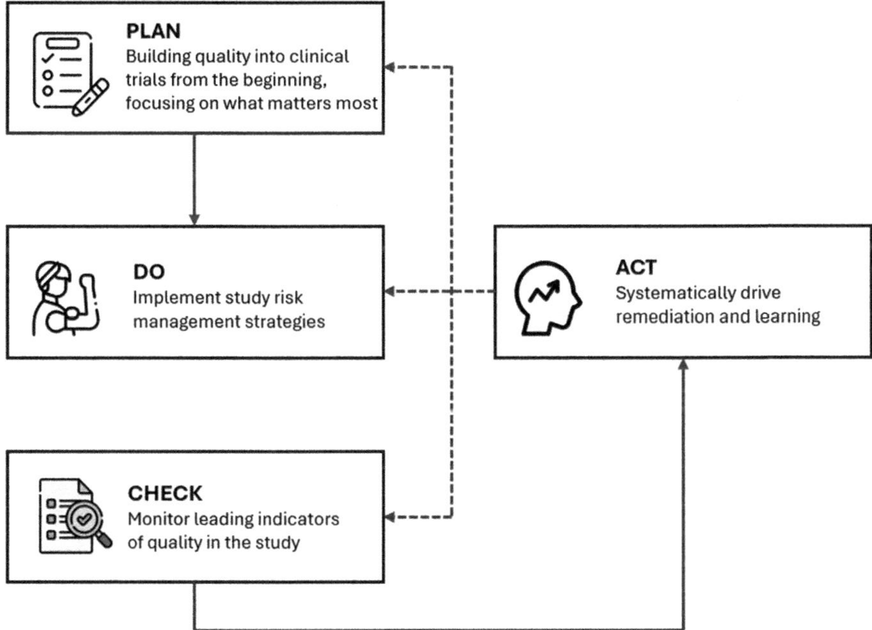

FIGURE 4.3
QbD implementation framework: Plan, Do, Check, Act. (Adapted from Dietrich et al. 2018.)

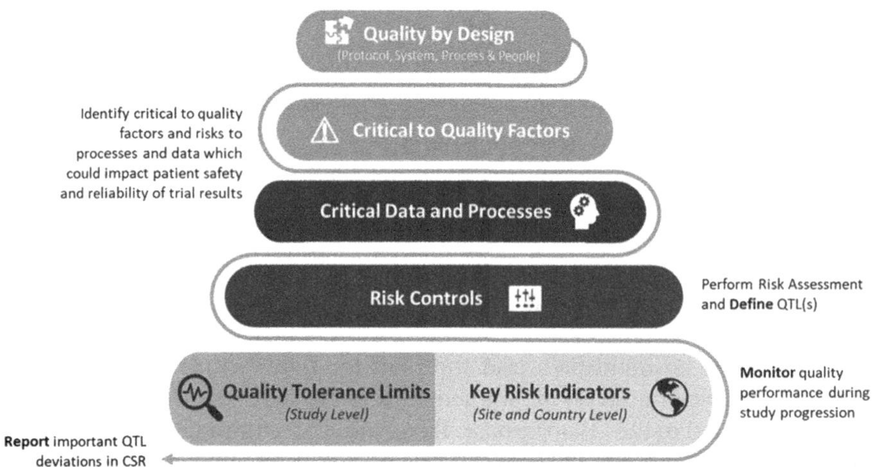

FIGURE 4.4
Risk-based quality management components. (Adapted from Bhagat et al. 2021.)

of the PDCA cycle, including the study protocol, test procedures, and monitoring plan.

As shown in Figure 4.4, implementation of QbD in clinical trials involves several key components, including the identification of CtQ factors, the establishment of QTLs, quality performance monitoring through QTLs and key risk indicators (KRIs), and reporting important QTL deviation in CRS. In the following sections, we discuss those concepts in detail.

4.6.1 Critical to Quality Factors

Per ICH E8(R1),

> critical to quality factors are attributes of a study whose integrity is fundamental to the protection of study subjects, the reliability and interpretability of the study results, and the decisions made based on the study results. These quality factors are considered to be critical because, if their integrity were to be undermined by errors of design or conduct, the reliability or ethics of decision making would also be undermined.
>
> (ICH 2021a)

CtQ factors encompass elements with the potential to influence participant protection and/or the reliability of trial outcomes, such as:

- Primary objectives
- Safety objectives
- Patient eligibility
- Investigational product exposure

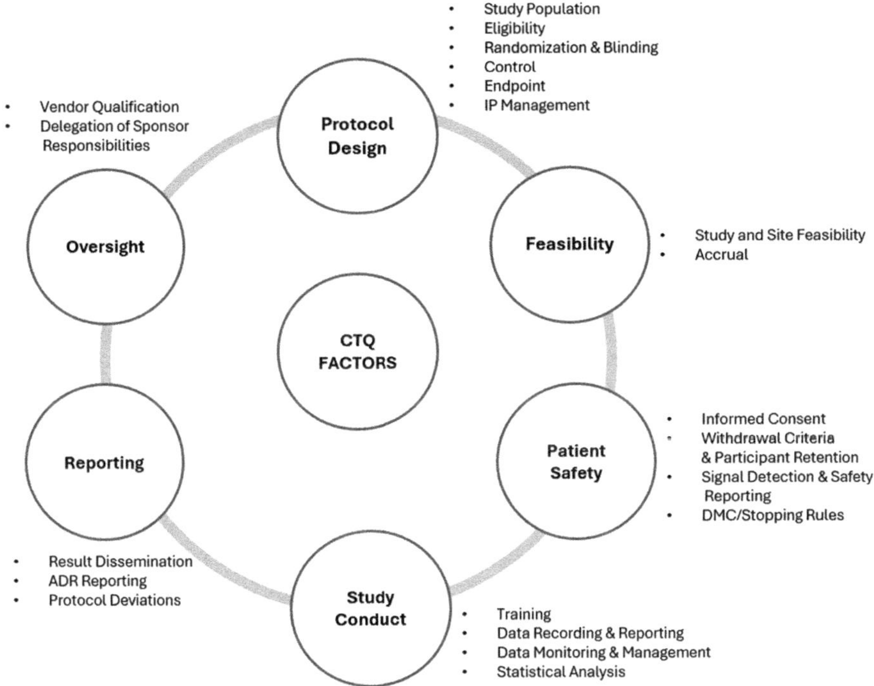

FIGURE 4.5
CtQ factors throughout the lifecycle of a clinical trial.

During protocol design, careful consideration of CtQ factors is fundamental to building quality into trials. Concurrently, clinical trial sponsors should establish appropriate risk management strategies to safeguard trial participants and the integrity of trial outcomes This involves implementing controls such as QTLs and KRIs, which are discussed in detail in the following sections.

Figure 4.5 outlines potential CtQ factors related to protocol design, study feasibility, patient safety, study conduct, result reporting, and vendor management. It is important to note that the clinical team should identify and select CtQ factors that match the needs of the particular trial and its stakeholders.

4.6.2 Quality Tolerance Limit

4.6.2.1 QTL Process

4.6.2.1.1 Definition

A QTL is a level, point, or value associated with a parameter that should trigger an evaluation if a deviation is detected to determine if there is a possible systematic issue (i.e., trend has occurred). A QTL represents the maximum risk associated with a CtQ factor that is deemed to be inconsequential (ICH 2021a). It is imperative that CtQs and associated QTLs are defined during trial planning and design and incorporated in the protocol.

4.6.2.1.2 *Setting Threshold of QTL*

Establishing thresholds of QTLs for a study requires cross-functional efforts, and input from subject matter experts including study clinician and biostatistician. These levels should be proportionate to the risks associated with their corresponding CtQ factors. As such, a holistic approach that considers various information is recommended. Bhagat et al. (2021) suggested that the process should consider the following information: (1) Trial-level risk management plan (including controls); (2) Number of participants; (3) Number of sites; (4) Trial Duration – adequate duration of the trial is a consideration to implementing the QTL process and implementing any remedial actions as a part of the QTL process; (5) Recruitment rate; (6) Trial Design (e.g., dose escalating cohorts because of the small number of participants in each cohort); and (7) Trial population.

QTL thresholds may be established, based on external historical data and internal data. However, it is important to curate the historical data to identify and remove erroneous data points and systematic bias. In the absence of well-established levels, statistical methods can be used to determine the QTL threshold as discussed in Section 4.8.

While QTLs are trial specific, QTLs that are in a current QTL library or were used in previous trials can be used as a starting point. A sample QTA library with parameters and thresholds is provided by Bhagat et al. (2021).

4.6.2.1.3 *Monitoring*

The CtQs will be consistently monitored against the predefined QTLs throughout the trial's duration. Moreover, any indication of a trend suggesting a potential deviation from the predefined QTL should prompt an evaluation to determine if corrective action is necessary to prevent such deviation. To provide study teams with early opportunities to mitigate risks to participant safety or the reliability of trial results and prevent QTL deviations, early action thresholds or alert limits could be determined for QTL parameters. This additional threshold empowers study teams to intervene before a significant deviation from a QTL occurs, if deemed necessary.

Consistent with the PDCA principles, due to the systematic nature of findings based on QTLs, some of the risk mitigation methods would benefit future studies by identifying risks and associated remediating methods to control them proactively (Bhagat et al. 2021).

4.6.2.1.4 *Reporting*

After study closeout, a QTL summary report is generated, which includes a list of deviations from the defined QTLs and actions take to address those issues. The findings are included in the CSR for regulatory submissions.

4.6.2.2 Relation between QTL and KRI

QTLs and KRIs serve as vital tools in managing risks identified early in the clinical development process. Both are defined and monitored to oversee factors CtQ throughout the trial's execution. Occasionally, KPIs and QTLs may encompass the same parameter such as under-reporting of serious safety events. However, they differ in their scope. Specifically, KRIs are typically assessed at the site level to guide site monitoring activities, whereas QTLs provide a higher-level overview of overall trial quality. For example, the timeliness to complete CRFs of the primary endpoints require oversight to ensure key timelines are not impacted but may not carry the same significance for human subject protection or the reliability of trial results as QTLs. In a risk-based approach, some KRIs, such as the above-mentioned timeliness in CRF completion, may not be suitable as QTL parameters (Bhagat et al. 2021).

AVOCA (2019) categorizes risks of clinical trials into three categories, standard risks, heightened risks, and critical risks (Figure 4.6). The standard risks are those inherent in the conduct of clinical trials and monitored through established clinical operation processes aligned with GCPs and regulation.

It is important to distinguish QTL from KRI (Bhagat et al. 2021). As a subset of the standard risks, key risks are those that may affect (1) participant protection; (2) reliability of trial results; and (3) other parameters as related to study performance, data quality, organizational reputation, etc. (AVOCA 2020)). In contrast, critical risks are a subset of key risks that may negatively impact

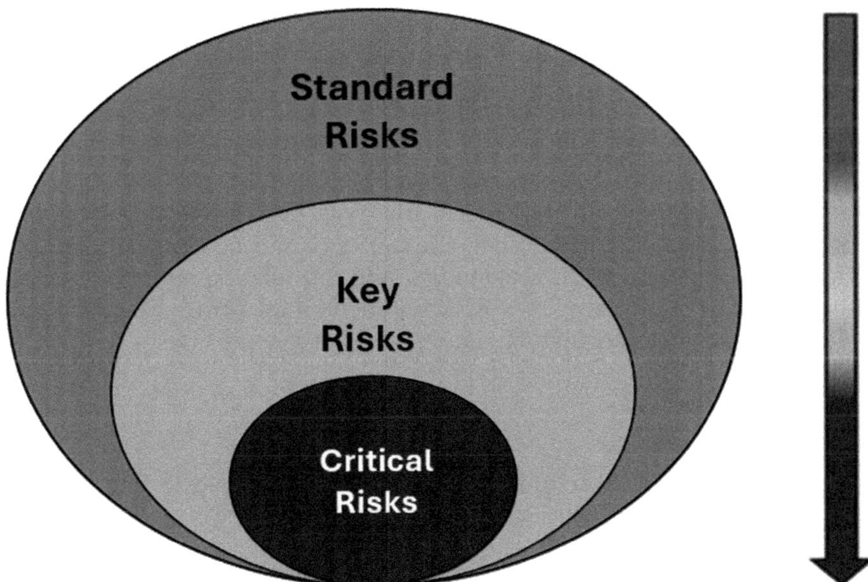

FIGURE 4.6
Various types of risks of clinical trials. (Adapted from AVOCA 2020.)

TABLE 4.1

Components of Risk-Based Monitoring

Component	Description
Risk Assessment	Conduct a comprehensive risk assessment at the outset of clinical trials to identify potential risks to data quality and patient safety.
Risk-Based Monitoring Plan	Based on the risk assessment, develop a tailored monitoring plan that prioritizes monitoring activities according to the level of risk.
Centralized Monitoring	Implement centralized approaches of reviewing aggregate data from an ongoing trial. Using visual analytics to detect unusual patterns at patient/site level, predict potential issues, remediate risk.
Trigger-Based Monitoring	Implement trigger-based monitoring mechanisms to identify deviations from expected data patterns and signal potential issues for further investigation, reducing levels of SDV.
Quality Oversight	Ensure the integrity and reliability of clinical trial data, including the use of data analytics and risk mitigation strategies.

trial integrity such as participant protection and quality such as reliability of trial results. They are typically established and monitored at trial level.

4.6.3 Risk-Based Monitoring

RBM is a critical component of clinical QMS, enabling direct monitoring resources toward areas of heightened risk. The RBM principles can be utilized for monitoring clinical trials both at the trial and site/country level. Effective application of RMB can reduce the overall burden of clinical monitoring while ensuring participant safety and quality data. Table 4.1 outlines several important components of the RBM method:

4.6.3.1 Risk Assessment

The risk-based monitoring of clinical trial data begins with the identification of KRIs, assessing their relative importance and occurrence frequency across various phases of clinical studies, and formulating strategies for risk mitigation. For example, a list of KRIs outlined on the British Medicines and Healthcare Products Regulatory Agency website encompasses:

- Recruitment rates
- Screen failure rates
- Timeliness of CRF submission/completion compared to actual patient progress in the trial
- Query rates
- Time taken to resolve queries versus the number of active queries (at the site level)

- Reporting of SAEs
- Instances of missed or delayed visits/data
- Participant withdrawals/dropouts
- Protocol/GCP non-compliance instances documented/reported
- eCRF audit trail data regarding completion times in relation to visits or expected timelines

While KRIs offer effectiveness to a certain extent, their implementation is not without complexity. They necessitate predefined parameters, programming, testing, and validation. Additionally, determining thresholds for abnormal analytical results requires statistical analysis of historical data and adaptation to the unique characteristics of each trial (Beauregard et al. 2015).

KRIs are chosen based on their detectability, likelihood of occurrence, and criticality in impacting quality. Risk factors selected for RBM should have associated data periodically available from EDC. KRIs are ranked according to their relative risk, considering their probability of occurrence and potential impact on data quality, subject safety, or trial integrity. Table 4.2 illustrates a numerical system for categorizing KRI risk levels as low, medium, or high across different study phases, including start-up, execution, and close-out. Indeed, the relative risk ranking of KRIs may evolve as sites progress through trial phases, as the likelihood of different risks occurring can vary with experience, data acquisition rate, or treatment exposure (Beauregard et al. 2015).

TABLE 4.2

An Example of Risk-Ranking of Key Risk Indicators

	Risk			Rank		
Key Risk Indicator	**Participant Safety**	**Data Quality**	**Trial Integrity**	**Startup**	**Conduct**	**Closeout**
Enrollment Rate				L	M	L
Screen Failures				M	M	L
Withdrawal				L	M	L
Out of Range Visit Rate				L	M	M
Missing Dose Rate				N/E	H	H
Missing Endpoint Data				H	H	M
Overdue Visit Entry				M	M	H
Time to Data Entry				M	H	M
Query Rate				H	H	H
Time to Query Resolution				L	H	M
Error Rate				H	H	H
Deviate Rate				M	H	H
Adverse Event Rate				L	H	M

Source: Beauregard et al. (2015).

4.6.3.2 Monitoring Plan

Following the identification of KRIs, a risk-based monitoring plan is developed, taking into account study complexity, endpoints, heterogeneity of patient population, geography, experience of clinical investigator, and sponsor's prior experience with the CI. Additionally, the capability of the EDC used for the trial in providing timely quality metrics report, relative safety of the investigational product, stage of the study, and quantity of data need to be factored in the development of the monitoring plan.

4.6.3.3 Centralized Monitoring

4.6.3.3.1 Methods

Recent rise in the adoption of electronic systems and records, along with advancements in statistical methods and advanced analytics present opportunities for incorporating centralized monitoring as a part of a risk-based approach to monitoring. Using aggregated study data and metadata in real-time across site, this approach enable cross-site comparison and identify quality issues, such as delayed data entry, earlier than when relying on on-site monitoring alone. Sites with multiple issues detected through centralized monitoring, such as delays in assessments or missing assessments, may also provide an early signal that the sponsor should promptly determine whether there is a need for a site visit and corrective actions to minimize the likelihood of similar issues occurring during the remainder of the clinical investigation (FDA 2023b).

Centralized monitoring processes offer enhanced capabilities to complement and potentially reduce the necessity for on-site monitoring, aiding in distinguishing between reliable and potentially unreliable data. Reviewing accumulated data from centralized monitoring, which may include statistical analyses, serves several purposes: (1) Identifying missing or inconsistent data, outliers, unexpected lack of variability, and protocol deviations; (2) Examining data trends such as range, consistency, and variability within and across sites; (3) Evaluating for systematic or significant errors in data collection and reporting, either at individual sites or across multiple sites, as well as detecting potential data manipulation or integrity issues; (4) Analyzing site characteristics and performance metrics; and (5) Facilitating the selection of sites and/or processes for targeted on-site monitoring.

4.6.3.3.2 Workflow

Although the particulars may vary significantly among studies, a risk-oriented centralized monitoring plan commonly workflow. Firstly, data from diverse sources are automatically extracted and transferred to a central repository. Secondly, the data are analyzed, summarized, and displayed on a central dashboard. Tailored to provide visual cues for high-risk sites, the dashboard displays risks along with KRIs and/or QTLs. Different color codes can be utilized to signify different levels of risks. When a site shows a QTL or KRI level is breached, an investigation plan is devised, and root cause analysis carried out. Lastly, beyond the visualization of risks, the workflow often utilizes advanced analytics to predict potential risks.

4.6.3.4 Trigger-Based Monitoring

Trigger-based monitoring in clinical trials is an advanced approach that leverages real-time data analytics to enhance the efficiency and effectiveness of monitoring activities. This method involves the implementation of predefined triggers or thresholds to identify deviations from expected data patterns, signaling potential issues that require further investigation. By focusing on specific indicators, trigger-based monitoring allows for the timely detection of anomalies, such as unusual patient responses, data inconsistencies, or protocol deviations. This proactive approach significantly reduces the need for extensive SDV, as it targets only the areas where irregularities are detected, rather than applying blanket checks across all data. Consequently, trigger-based monitoring not only streamlines the monitoring process but also allocates resources more efficiently, ensuring that critical issues are addressed promptly. This method enhances data integrity, participant safety, and overall trial quality, while also reducing the operational burden and costs associated with traditional monitoring methods.

4.6.3.5 Quality Oversight

Significant issues should be thoroughly evaluated in a timely manner at the appropriate levels (for example, sponsor, clinical sites) as described in the monitoring plan. A root cause analysis followed by appropriate corrective and preventive actions should be undertaken promptly to reduce the impact of the identified issue on the rights, safety, and welfare of participants in the clinical investigation and/or the integrity of the data. Additionally, the risk assessment and monitoring plan should be reviewed and revised, as needed, to help ensure the risk of recurrence is decreased, or if possible, eliminated. In instances in which corrective actions modify study processes, the protocol and/or associated investigational plans should be amended to reflect changed processes.

Related systemic issues should be identified and resolved promptly to help ensure that investigation quality, including the rights, safety, and welfare of investigation participants and data integrity, is maintained. Examples of preventive and corrective actions that may be warranted include but are not limited to (1) improved training for the clinical investigator and site staff; (2) halting enrollment at a clinical site pending resolution of identified issues; (3) clarifying or revising the protocol and/or other related investigational plans and documents; and/or (4) modifying vendor service agreements to ensure adequate trial support.

Lastly, significant issues identified through monitoring and oversight activities and the actions to be taken should be documented and communicated to the appropriate parties, which may include, but are not limited to (1) sponsor management; (2) sponsor teams; (3) clinical sites; (4) institutional review boards; (5) other relevant parties (for example, DMCs and relevant contract research organizations); and (6) applicable regulatory agencies, including FDA, when appropriate.

4.6.3.6 Challenges and Considerations

Risk-based monitoring represents a paradigm shift in clinical trial monitoring practices, offering a more targeted, efficient, and cost-effective approach to ensuring data quality and patient safety. The FDA guidance document "Oversight of Clinical Investigations: A Risk-Based Approach to Monitoring" and ICH E6(R2) (ICH 2016) provide valuable frameworks and recommendations for implementing risk-based monitoring strategies. By conducting comprehensive risk assessments, developing tailored monitoring plans, leveraging technology and data analytics, and ensuring regulatory compliance, sponsors can enhance the efficiency and effectiveness of monitoring activities while maintaining compliance with regulatory requirements. However, effective implementation of the method presents a unique set of challenges as well.

4.6.3.6.1 Cultural Shift and Stakeholder Engagement

Implementing risk-based monitoring may require a cultural shift within organizations accustomed to traditional monitoring practices.

Stakeholder engagement, including investigators, study coordinators, and regulatory authorities, is crucial for fostering acceptance and adoption of risk-based monitoring approaches.

4.6.3.6.2 Resource Allocation and Training

Adequate resources, including personnel, technology, and training, are essential for successful implementation of risk-based monitoring strategies.

Training programs should be provided to clinical trial staff to ensure proficiency in risk assessment methodologies, monitoring techniques, and regulatory requirements.

4.7 Statistical Methods for Risk Monitoring

4.7.1 Control Charts

During the conduct of a clinical trial, CtQ and key risk factors are monitored as data are being collected, and risk factors estimated. This continued vigilance is meant to provide assurance that the trial is in a state of control. Over the years, various statistical methods have been developed and used for trending analysis. Control charts are perhaps the simplest and most effective graphical tools for such an analysis to detect issues related to special cause variations. There are many types of control charts that can be used for routine monitoring of analytical method performance. These include Shewhart individual control chart, also known as individual/moving range (I-MR) control chart, exponentially weighted moving average (EWMA) chart, cumulative sum (CUSUM) chart, and J-chart (or zero control chart). A typical control chart consists of a centerline, lower and upper warning and control limits, also known as alert and action limits.

The centerline and limits can be established based on historical data collected as previously discussed. The chart provides a visual means for identifying out of trend observations, unusual shifts and variability indicative of potential performance issues. The data are assumed to follow a statistical distribution such as normal distribution. This allows for the quantification of the magnitude of trend or shift that qualifies as a rare event. From example, under the normality assumption of the data, the probability for nine points in a row to be greater than the population mean is 0.12%. Several well-established rules have been suggested to detect out-of-control results, based on control charts. Notable are the rules developed by WECO (Montgomery 2013). Table 4.3 lists the Nelson Rules.

TABLE 4.3

Nelson Rules Applied to a Control Chart for Detecting an Unusual Shift or Trend

Rule	Description	Indication
1	One point exceeds 3 standard deviations from the center line	One sample is out of control
2	Nine or more points in a row on the same side of the center line	There is a mean shift in performance
3	Six or more points in a row continuously increasing or decreasing	A trend exists
4	Fourteen or more points in a row, alternating in direction	There is a negative correlation between neighboring points
5	Two out of 3 data points on the same side and more than 2 standard deviations away from the centerline	A possible increase in process/assay variability
6	Four out of 5 points on the same side and more than 1 standard deviation away from the centerline	A possible increase in process/assay variability
7	Fifteen points in a row within 1 standard deviation of the centerline	A possible decrease in process/assay variability
8	Eight points in a row on both sides of the centerline but none within 1 standard deviation of the centerline	Non-random sample

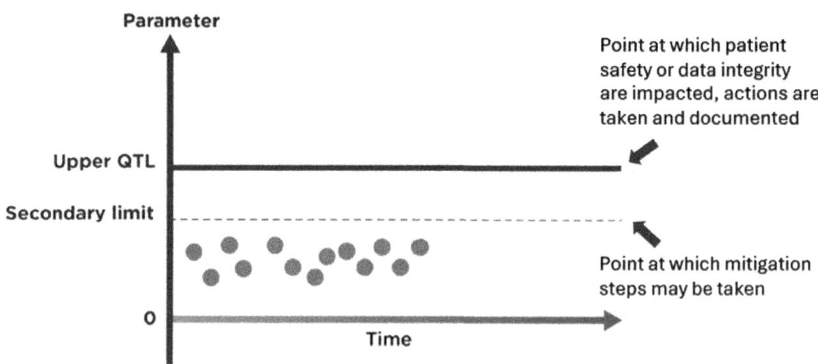

FIGURE 4.7
QTL and secondary limit for monitoring CtQ factors. (Adapted from AVOCA 2020.)

In the context of monitoring CtQ factors, QTL is the upper control limit (UCL). Excursion beyond this limit will trigger a formal investigation. To mitigate the risk of QTL deviation, an early action threshold, also known as secondary limits, can be specified to render early opportunities to detect the adverse trend (Figure 4.7).

4.7.2 Alert and Action Limits

In the following, we introduce a few control charts commonly used for the control of CtQ and key risk factors. To simplify discussion, we use alert and action limits to represent the secondary limit and the limit for either QTL parameters or key risk factors, respectively. An alert limit is a threshold that, when exceeded, indicates that a risk may have drifted from its expected level. Alert limit excursions may also lead to additional monitoring with higher frequency and more intense scrutiny of the data and other CtQ or key risk factors. Alert level excursions constitute a warning and do not necessarily require corrective action. By contract, an action limit is a level that, when exceeded, indicates that the risk is probably drifting from its acceptable level. Exceeding an action limit triggers an immediate investigation to identify the root cause and take corrective measures to bring the risk to its tolerable range.

4.7.3 Classical Frequentist Methods

4.7.3.1 Normal Distribution

Historically it was a common practice to use the mean ± 2 or 3 standard deviations (SDs) as alert and action limits, respectively (Wilson 1997). When the risk factor data, such as time taken to resolve queries, are normally distributed, these limits have a probabilistic interpretation. That is, they cover fixed proportions of all possible observations. When the data are not normally distributed, transformation of the data may help "normalize" the distribution, and the alert and action limits can be set, using the transformed data, and back-transformed to the limits on the original scale.

4.7.3.2 Non-Parameter Intervals

Alternatively, non-parametric methods can be used to set alert and action limits. These methods do not rely on an underlying assumption of the distribution of the data. Rather, they make use of the empirical distribution of the data. For example, the alert and action levels can be set as the 95th and 99th percentiles of the empirical distribution, estimated as follows:

$$F(x) = \frac{\#\text{ of observations} \leq x}{N},$$

where $x \geq 0$ and N is the total number of observations in the dataset. The alert and action levels based on the empirical distribution are observed microbial counts that cover pre-specified proportions of the historical data.

A more rigorous approach is to use the non-parametric tolerance interval suggested by Conover (1999). Consider $X_1, \ldots, X_n$ to be a random sample of microbial counts and $X_{(1)}, \ldots, X_{(n)}$ to be the ranked values. Let q and $1-\alpha$ be the percentage of population covered by the tolerance interval and the confidence level, respectively. Further, we let m be an integer such that

$$q = \frac{4n - 2(m-1) - \chi^2_{2m}(1-\alpha)}{4n - 2(m-1) + \chi^2_{2m}(1-\alpha)},$$

where $\chi^2_{2m}(1-\alpha)$ is the $(1-\alpha)$ percentile of a chi-squared distribution with $2m$ degrees of freedom. Thus, with probability of at least $(1-\alpha)$ the 1-sided tolerance limit $X^{(n-m)}$ covers *100q%* of the population (Conover 1999).

However, there are several drawbacks to the non-parametric methods. First of all, as pointed out by several researchers (Christensen, et al. 2003; Yang et al. 2013), the reliability of these non-parametric approaches depends heavily on the amount of data available. In the absence of large datasets, these approaches produce alert and action levels that far exceed the nominal significance levels, thus artificially creating wider ranges. Secondly, they run the risk of selecting as the alert/action limits the outlying data points caused by out-of-control conditions. This again results in inflated estimates of alert and action limits. In addition, there is also no optimality criterion, such as the shortest interval width, for selecting these intervals. As a result, there is a risk that the chosen intervals may be much larger than what is needed to cover a certain percentage of the future observations. Therefore, recent efforts have been centered on setting alert and action levels using parametric model-based methods.

4.7.3.3 Binomial

Many risk factors, such as recruitment, screen failure, and queries, are measured by percentage or rate. The rate, p, is an unknown parameter. Consider screen failure. Let X denote the number of patients (out of n) who failed screening. Under the assumption that the patients' screening was conducted independently, X follows a binomial distribution Binomial (n, p). That is, the probability for X to be equal to the value k $(= 0, 1, \ldots, n)$ is given by

$$P[X = k] = \binom{n}{k} p^y (1-p)^{n-k}. \tag{4.1}$$

The above distribution can be used to make inference about the response rate p.

4.7.3.4 Beta-Binomial to Capture Heterogeneity

When the historical data used to set up the alert and action limits were from multiple sites, a hierarchical model that accounts for variation due to side and variability of data collected at different time period within a clinical site is more suitable for describing the data. The issue can be addressed by assuming that p follows a beta-distribution. That is,

$$\pi(p) = \frac{1}{B(\alpha,\beta)} p^{\alpha-1}(1-p)^{\beta-1}, \quad \alpha > 0, \ \beta > 0, \tag{4.2}$$

where $B(\alpha,\beta)$ is the normalization constant.
From (4.1) and (4.2), the marginal distribution of X is the beta-binomial

$$P[Y = y] = \binom{n}{y} \frac{B(y+\alpha, n-y+\beta)}{B(\alpha,\beta)}.$$

Let $1-\gamma$ be the confidence level of an UCL and $(\hat{\mu},\hat{k})$ be the maximum likelihood estimate (MLEs) of the distribution parameters. UCL can be chosen such that

$$\hat{\text{UCL}}_{\text{Exact}}^{\text{NB}} = \min\left\{n : \sum_{x=0}^{n} \hat{g}(x \mid \hat{p}, \hat{\mu}, \hat{k}) \geq 1-\alpha\right\}.$$

4.7.3.5 Poisson Distribution

The number of immediately reportable events reported late is a QLT parameter. Setting the alert and action limits for this parameter is desirable. The Poisson distribution has been widely applied to model count data. If X follows a Poisson distribution, the probability for X to be equal to a number x is given by (Haight 1967):

$$f(x|\lambda) = \frac{\lambda^x e^{-\lambda}}{x!},$$

where λ is the mean count in a unit sample. One unique characteristic of the Poisson distribution is that the mean of the distribution is equal to the variance. The parameter λ can be estimated using maximum likelihood. Let $\hat{\lambda}$ be such an estimator.

In general, a one-sided UCL can be obtained, using either of the following two methods, one of which is based on normal approximation and the other is deemed to be exact. Using the approximate method, the 1-sided UCL is given by

$$\hat{\text{UCL}}^{\text{P}}_{\text{Approximate}} = \hat{\lambda} + z_{1-\alpha}\sqrt{\hat{\lambda}/n},$$

and the exact 1-sided UCL by

$$\hat{\text{UCL}}^{\text{P}}_{\text{Exact}} = \min\left\{n : \sum_{x=0}^{n} f(x \mid \hat{\lambda}) \geq 1-\alpha\right\},$$

where $z_{1-\alpha}$ is the upper $(1-\alpha)\times 100\text{th}$ percentile of the standard normal distribution.

However, when applied to environmental data, the Poisson model is often inadequate as the actual variability in the data is often larger than that expected under the Poisson assumption. This phenomenon is often referred to as overdispersion. For count data of immediately reportable events reported late, several factors may contribute to overdispersion. Chief among these is that measurements are taken at different times and various investigator sites. Conceivably the mean count may differ from time to time and site to site due to various factors. Although at each fixed time or location, the number of under-reported reportable events may exhibit behavior that can be described through a Poisson distribution, the collective data as a whole may not follow any Poisson distribution. Blindly applying a single-parameter Poisson model to such data consisting of several subpopulations may underestimate the variability of the data, leading to lower alert and action limits, and thus causing more frequent false alarms of microbial excursion.

4.7.3.6 Negative Binomial

The negative binomial distribution is a variant of Poisson distribution that has increased flexibility to address the issue of overdispersion. It can be viewed as either a generalization of the geometric distribution or a mixture of Poisson distributions (Hoffman 2004). Hoffman (2004) and Christensen et al. (2003) suggest using the negative binomial distribution to correct for the effect of overdispersion. Assume X follows a negative binomial distribution NB(μ, κ), with density function given by

$$g(x_i \mid \mu, k) = \frac{\Gamma\left(1/k + x_i\right)(k\mu)^{x_i}}{x_i!\Gamma\left(1/k\right)(1+k\mu)^{1/k+x_i}},$$

where $\mu > 0$ and $k > 0$. The distribution of X has a mean μ and variance $\mu(1+k\mu)$. It is evident that the variance is greater than mean, whereas they are equal for the Poisson distribution. As a consequence, the negative binomial distribution is useful for modeling overdispersed data.

Hoffman (2004) suggested two methods to set alert and action limits using the negative binomial distribution. The first is based on the cumulative density function (CDF) of the negative binomial distribution with the parameters being estimated by their MLEs. The second method determines the limits based on a chi-square approximation to the negative binomial distribution. Both methods are briefly described in the following.

Let $1-\alpha$ be the confidence level for a one-sided UCL and $\left(\hat{\mu}, \hat{k}\right)$ be the MLEs of the distribution parameters. The UCL is determined such that

$$\hat{\text{UCL}}_{\text{Exact}}^{\text{NB}} = \min\left\{n : \sum_{x=0}^{n} \hat{g}(x \mid \hat{\mu}, \hat{k}) \geq 1-\alpha\right\}.$$

Because of the discrete nature of the data and approximation of the true model parameters, the above limit might not be truly exact (Hoffman 2004).

The second method by Hoffman (2004) uses the following approximation suggested by Guenther (1972):

$$P\left[X \leq \text{UCL}\right] = P\left[\chi_v^2 \leq \frac{2\text{UCL}+1}{1+\mu k}\right] = 1-\alpha,$$

where χ_v^2 is a random variable following a chi-squared distribution with v degrees of freedom $v = 2\hat{\mu}/\left(1+\hat{k}\hat{\mu}\right)$. Solving the above equation with parameters (μ, k) being replaced by their MLEs gives rise to an estimate of the UCL:

$$\text{UCL}_{\text{Approximate}}^{\text{NB}} = \left[\chi_v^2(1-\alpha)\left(1+\hat{\mu}\hat{k}\right)-1\right]/2$$

where $\chi^2_v(1-\alpha)$ is the $100(1-\alpha)^{th}$ percentile of the chi-squared distribution with v degrees of freedom.

4.7.4 Bayesian Approach

As previously discussed, setting up QTL not only relies on historical data but also opinions of KOLs and subject matter experts. In addition, as new data either from the current study or other sources become available, QTL may be updated. In contrast with Frequentist methods, Bayesian This is made possible by the Bayes' rule, which is briefly described in the following.

4.7.4.1 Bayes Theorem and Posterior Distribution

Let θ, y, and $\tilde{y}$ denote the parameter of interest, observable data, and future data, respectively. Bayesian methods begin by describing θ and y through a prior distribution for the parameters $\pi(\theta)$ and a conditional probability function for the data $\pi(y|\theta)$. The latter is also called the likelihood function after y is observed and is often depicted as $L(\theta|y)$. The frequentist believes that all information concerning the unknown parameter . is contained in the likelihood $\pi(y|\theta)$ after the experiment is complete and consequently bases all statistical inference about θ solely upon $\pi(y|\theta)$. In the Bayesian approach, the prior information is updated through the Bayes' Theorem to give rise to a posterior distribution:

$$\pi(\theta|y) = \frac{\pi(\theta)\,\pi(y|\theta)}{\pi(y)}$$

where $\pi(y) = \int \pi(\theta)\,\pi(y|\theta)\,d\theta$.

The posterior distribution has two key components, $\pi(\theta)$ and $\pi(y|\theta)$. The former, which is referred to as prior distribution of θ, represents the information of the parameters before the current data are observed. The latter is the likelihood to observe the current data, conditioned on $\boldsymbol{\theta}$. Together, they represent the totality of knowledge of the parameter up to the current data. The Bayes' theorem also provides a means for updating the posterior distribution as new data become available, using the current posterior as the prior for the future posterior distribution. This is very useful for studies, such as dose-finding trials (see Chapter 5), in which observations are sequentially obtained and estimates are continuously updated based on new information.

4.7.4.2 Inference about Parameters

Inference regarding θ can be made based on the posterior distribution. For example, either the mode, mean, or median of the distribution may be used as an estimate of the parameter. Moreover, an interval estimator, (a, b), can also be constructed for a univariate parameter θ such that

$$P\left[a \le \theta \le b \mid y\right] = 1 - \alpha.$$

The interval (a, b) is often referred to as $(1-\alpha)\times 100\%$ Bayesian credible interval. Unlike the frequentist confidence interval, the interval in (2.5) has a probabilistic interpretation; it covers the parameter with a probability of $1-\alpha$.

4.7.4.3 Bayesian Methods for Alert Action Limits

4.7.4.3.1 Normal Distribution

Suppose

$$X_i \sim N\left(\mu, \sigma^2\right) \; (i = 1, 2, \ldots, n)$$

are independent historical data collected from analytical testing. Assume that priors

$$\mu \mid \sigma^2 \sim N\left(\mu_0, \frac{\sigma^2}{\kappa_0}\right)$$

$$\sigma^2 \sim \text{inverse-Gamma}\left(\frac{\gamma_0}{2}, \frac{\gamma_0 \sigma_0^2}{2}\right),$$

where μ_0, κ_0, σ_0^2, and γ_0 are the hyperparameters, the predictive probability distribution of $\tilde{X}$ can be derived (Colosimo and Castillo 2007).

The joint prior is

$$\left(\mu, \sigma^2\right) \sim \text{NIG}\left(\mu_0, \kappa_0, \gamma_0, \sigma_0^2\right).$$

It can be shown that the joint posterior distribution is

$$\left(\mu, \sigma^2\right) \mid \bar{X}, s^2 \sim \text{NIG}\left(\mu_n, \kappa_n, \gamma_n, \sigma_n^2\right),$$

where

$$\mu_n = \frac{\kappa_0}{\kappa_0 + n}\mu_0 + \frac{n}{\kappa_0 + n}\bar{X},$$

$$\kappa_n = \kappa_0 + n, \; \gamma_n = \gamma_0 + n,$$

$$\gamma_n \sigma_n^2 = \gamma_0 \sigma_0^2 + (n-1)s^2 + \frac{\kappa_0}{\kappa_0 + n}\left(\bar{X} - \mu_0\right)^2.$$

Further, the marginal posterior distributions are

$$\sigma^2 \mid \bar{X}, s^2 \sim \text{inverse-Gamma}\left(\frac{\gamma_n}{2}, \frac{\gamma_n \sigma_n^2}{2}\right)$$

$$\mu \mid \bar{X}, s^2 \sim t\left(\mu_n, \frac{\sigma_n^2}{\kappa_n}, \gamma_n\right).$$

Hence, the alert and action limits can be obtained as

$$\text{Alert Limit} = \mu \mid \bar{X}, s^2 \sim t_{\alpha_1}\left(\mu_n, \frac{\sigma_n^2}{\kappa_n}, \gamma_n\right),$$

$$\text{Action Limit} = \mu \mid \bar{X}, s^2 \sim t_{\alpha_2}\left(\mu_n, \frac{\sigma_n^2}{\kappa_n}, \gamma_n\right),$$

where $t_{\alpha_1}\left(\mu_n, \frac{\sigma_n^2}{\kappa_n}, \gamma_n\right)$ and $t_{\alpha_2}\left(\mu_n, \frac{\sigma_n^2}{\kappa_n}, \gamma_n\right)$ are the $(1-\alpha_1)\times 100^{\text{th}}$ and $(1-\alpha_2)\times 100\text{th}$ percentiles of a non-central t distribution $t\left(\mu_n, \frac{\sigma_n^2}{\kappa_n}, \gamma_n\right)$, respectively, with $0 < \alpha_2 < \alpha_1 < 1$.

The values of α_1 and α_2 are chosen per expert opinions.

While the Gaussian-data conjugate prior provides a simple solution, it is possible that historical information is better represented by a non-conjugate prior. In such a case, the posterior distribution for (μ, σ^2) cannot be written in a closed form and so, generally, sampling of (μ, σ^2) allows estimation of their distributions.

4.7.4.3.2 Binomial

Consider the previous example. It is assumed that the probability p has a beta prior distribution of $\text{Beta}(\alpha_1, \alpha_2)$.That is,

$$\pi(p) \propto p^{(\alpha_1 - 1)}(1-p)^{\alpha_2 - 1}, \tag{4.3}$$

where the symbol "$\propto$" means "proportional to."

From (4.1) and (4.3), it follows that the posterior is also a beta distribution

$$\pi(p \mid y) \propto p^y (1-p)^{n-y} \times p^{(\alpha_1 - 1)}(1-p)^{\alpha_2 - 1} = p^{\alpha_1 + y - 1}(1-p)^{\alpha_2 + n - y - 1}. \tag{4.4}$$

From (4.4), it can be shown that

$$p \mid y \sim \text{Beta}\left(\alpha_1 + y, \alpha_2 + n - y\right).$$

Similar to the normal case, the alert and actions limits can be chosen to be the $\left(1-\alpha_1\right)\times 100\text{th}$ and $\left(1-\alpha_2\right)\times 100\text{th}$ percentiles of $Beta\left(\alpha_1 + y, \alpha_2 + n - y\right)$, respectively.

Poisson

We assume that the λ of the Poisson distribution in Section 4.7.3.5 has a prior distribution of gamma(α,β). Hence,

$$f\left(\lambda\right) \propto \frac{\lambda^{\alpha-1} e^{-\lambda/\beta}}{\beta^{\alpha}}.$$

It can be shown that the posterior distribution of $\theta = \left(p, \lambda\right)$ is given by

$$f\left(\theta | x_1, \ldots, x_n\right) \propto \lambda^{\sum_{i=1}^{n} x_i + \alpha - 1} e^{-\lambda(n+1/\beta)}.$$

This implies that the posterior distribution of $\theta \mid x_1, \ldots, x_n$ follows a gamma distribution $gamma\left(\sum_{i=1}^{n} x_i + \alpha, n + 1/\beta\right)$. Therefore, the alert and actions limits can be calculated to be $\left(1-\alpha_1\right)\times 100\text{th}$ and $\left(1-\alpha_2\right)\times 100\text{th}$ percentiles of gamma$\left(\sum_{i=1}^{n} x_i + \alpha,\, n + 1/\beta\right)$. Therefore gamma $\left(\sum_{i=1}^{n} x_i + \alpha, n + 1/\beta\right)$, respectively.

4.7.4.4 Inference of Future Observations

During the trial conduct, CtQ and key risk factors are continuously monitored as new data emerge. While alert and action limits are useful in discerning the state of control of the trial, it is of interest to predict the future state as related to excursion beyond the alert or action limit given the data cumulated at each time point. In other words, based on the current data $\boldsymbol{y}$ what is the likelihood for the CtQ or key risk factor measure, $\tilde{\boldsymbol{y}}$, in the next reporting period to exceed the alert or action limit. This question can be relatively easily addressed within the Bayesian framework. Specifically, the posterior predictive distribution is obtained by

$$\pi\left(\tilde{\boldsymbol{y}} \mid \boldsymbol{y}\right) = \int \pi\left(\tilde{\boldsymbol{y}} \mid \theta\right) \pi(\theta | \boldsymbol{y}) d\theta$$

In essence, $\pi\left(\tilde{\boldsymbol{y}} | \boldsymbol{y}\right)$ is the sampling distribution of the future observations $\tilde{\boldsymbol{y}}$ weighted over the updated distribution of the parameter θ. Based on this distribution, prediction regarding the behavior of $\tilde{y}$ can be readily made. For example, both point and interval estimates of $\tilde{y}$ can be obtained from the distribution in (2.8). It is worth noting that inference based on this distribution is different from that of the classical frequentist using the condition distribution $\pi\left(\tilde{y}\,|\,\hat{\theta}\right)$, where $\hat{\theta}$ is an estimate of the parameter. As noted by Colosimo and del Castillo (2007), the classical prediction interval often does not consider the uncertainty in the parameter θ. There are many applications of predictive distribution. It is particularly useful for multi-stage clinical trials in defining stopping rules (Lee and Liu 2008), power calculation (Spiegelhalter et al. 2004), and model validation and diagnosis (Berry and Stangl 1996).

Consider the previous example of time taken resolve data queries. The posterior predictive distribution of $\tilde{X}$ can be obtained through the following integration:

$$p\left(\tilde{X}\,|\,\boldsymbol{X}\right) = \int\!\!\int p\left(\tilde{X}\,|\,\mu, \sigma^2\right) p(\mu, \sigma^2\,|\,\boldsymbol{X})\, d\mu\, d\sigma^2,$$

which yields

$$\tilde{X}\,|\,\boldsymbol{X} \sim t\left(\mu_n, \frac{\sigma_n^2\left(\kappa_n + 1\right)}{\kappa_n}, \gamma_n\right)$$

Thus, the probability $P(\tilde{X} > q_0\,|\,\boldsymbol{X})$ can be estimated either through numerical or Monte Carlo integration. The threshold q_0 is either the alert or action limit. If this probability is higher, the timeliness of query resolution is of concern.

For an illustration, consider a clinical site that has been monitored for time to query resolution. Typically, the site is expected to resolve queries within business days. Knowing some additional effort regarding the same data issue may be needed after the first round of query, the sponsor set up alert limit at 12 days. Suppose that data from the previous 25 reporting periods gave rise to the mean query resolution time of $\bar{X} = 9.9$ days with a variance of $s = 1.49$ days. Furthermore, it is assumed that in the current reporting period, data suggests that the mean query data is an $X_{current} = 11.7$ days. Even though the observed time for query resolution is within the alert limit of 12 days, the sponsor desired to understand how likely the time for query resolution is trending up in the next reporting period.

With hyperparameters $\mu_0 = 10$, $\sigma_0^2 = 3.2$, $\kappa_0 = 2$, and $\gamma_0 = 3$, the posterior predictive probability $P(\tilde{X} > X_{Current}\,|\,X)$ can be directly calculated through the T distribution about 3%. The result indicates that it is highly unlikely that query resolution at the site would be problematic.

4.7.4.5 Selection of Priors

The key component of Bayesian inference is in the selection of prior distributions. Since the prior reflects expert opinion and depends on personal judgment, it can be subjective in nature. Because most controversies surrounding Bayesian analysis are centered on the selection of priors, care should be given in prior selection. Before data collection, one must assess the relative accuracy and weight of the prior information (van de Schoot, Kaplan et al. 2013).

Three major categories of prior distributions are conjugate priors, non-conjugate priors, and non-informative priors. A conjugate prior yields a posterior that belongs to the same family of the prior distribution. There are two important advantages in the use of a conjugate prior. First, it is mathematically and computationally simple, particularly in sequential studies (Colosimo and del Castillo 2007). Second, in many cases, the prior may directly be viewed as additional data. For example, a $\text{Beta}(\alpha, \beta)$ prior for an experiment with binary response discussed previously can be considered as data from an early experiment which had α successes and β failures. In a similar vein, Morita et al. (2008) formulated the effective sample size that is contributed by any prior distribution, whether it is conjugate or not. Unfortunately, conjugate priors can sometimes be too restrictive to reflect the expert knowledge and other prior beliefs.

In contrast, a non-conjugate prior can vary in complexity and so one may always find a non-conjugate prior that possesses the flexibility to characterize historical knowledge. The extra complexity often comes at the price of no closed form for (2.4). Although use of non-conjugate priors previously posed significant computational challenges, with the advances in Markov chain Monte Carlo (MCMC) methods, the issue has been attenuated.

A key subtype of prior distributions is called non-informative prior, which is well suited for cases in which little knowledge about the parameters is available. For example, a flat prior that is uniform over the entire real line is noninformative in the sense that the parameter value is believed to be equally likely to assume any value. Over the past decades, a variety of methods for deriving non-informative priors have been suggested (Berger 1985). Jeffrey (1961) proposed a general method that constructs non-informative priors based on the so-called invariant principle, which states that posterior inferences based on either the original prior $\pi(\theta)$ or the prior $\pi(g(\theta))$, with $g(\theta)$ being a one-to-one transformation, results in the same conclusion. Under such constraint, Jeffreys showed that a prior $\pi(\theta)$ meeting the invariant principle satisfies

$$\pi(\theta) \propto \sqrt{\det[I(\theta)]}$$

where "det" stands for determinant and $I(\theta)$ is the Fisher's information matrix given by

$$I(\theta) = -E_\theta\left[\frac{d^2\log\left[\pi(y|\theta)\right]}{d^2\theta}\right]$$

As an example, when the response $Y \sim N(\mu, \sigma^2)$, the Jeffreys non-informative prior of $\theta = (\mu, \sigma^2)$ is given by

$$\pi(\theta) \propto \frac{1}{\sigma^3}.$$

4.7.4.6 Bayesian Computation

As noted by Robert (2013), it has long been a bane of the Bayesian approach that the solutions it proposes are intellectually attractive but inapplicable in practice. The difficulty in implementing Bayesian solutions had long hindered the adoption of these methods; however, the advances of Bayesian computational methods in the past two decades have spurred interest in Bayesian applications in various areas including drug research and development. In this section, we discuss two computational methods, namely, Monte Carlo simulation and MCMC.

4.8 Applications of AI and Machine Learning in Data Quality

In recent years, AI and ML have made significant inroads in ensuring clinical trial participant safety and data quality. Virtually all aspects of clinical trial conduct involve applications of these technologies to minimize manual work, human errors, and enhance efficiency. In this section, we discuss various usages of AI and ML tools.

4.8.1 Automation of Data Collection, Review, and Cleaning

4.8.1.1 Automation

Clinical data review, cleaning, and query are essential processes in clinical trials that ensure the integrity, accuracy, and reliability of medical data. However, these tasks often involve labor-intensive manual efforts, which can be time-consuming, error-prone, and costly. With the advent of AI and ML technologies, there is an opportunity to automate these processes, making them more efficient, accurate, and scalable. Application of ML, particularly natural language processing (NLP), coupled with data standardization, can facilitate the automation of these processes. For example, NLP techniques have shown to be effective for extracting information from unstructured clinical text data, such as physician notes, radiology reports, and discharge summaries while named entity recognition (NER) algorithms can identify and classify entities such as diseases, medications, and procedures mentioned in clinical notes, facilitating automated data review. The AI-enabled data automation reduces time, costs, and the potential for errors linked with manual data extraction, whether in prospective trials or retrospective reviews. Although this application necessitates addressing variable data structures and sources, it has demonstrated early success in fields such as cancer (Rajkomar et al. 2018), epilepsy (Choi et al. 2017), and depression (Beam and Kohane 2018), among others (Wang, Casalino et al. 2018).

4.8.1.2 Automated Query Generation and Resolution

High-quality clinical data is paramount for accurate data analysis and robust clinical decision-making. However, clinical datasets often contain errors, missing values, and inconsistencies that require thorough cleaning and pre-processing. AI and ML techniques can automate these tasks, reducing the burden on clinical staff and improving data quality. In recent years, clinical trial data has undergone standardization and structuring through initiatives like CDISC and streamlined processes implemented by pharmaceutical companies. This standardized format enhances data compatibility across trials, and integration of these data for training ML models for data queries. Supervised learning algorithms such as Support Vector Machines (SVM) and Random Forests can be trained on labeled datasets to identify abnormal patterns in clinical data. Unsupervised learning techniques like clustering algorithms (e.g., K-means, DBSCAN) can group similar data points together, helping in the detection of outliers and anomalies. ML algorithms such as k-Nearest Neighbors (k-NN) and Multiple Imputation by Chained Equations (MICE) can predict missing values based on the relationships between variables in the dataset. Deep learning models like Recurrent Neural Networks (RNNs) and Long Short-Term Memory (LSTM) networks can learn temporal dependencies in sequential clinical data and impute missing values accordingly. Robust statistical methods and ML algorithms can identify outliers in clinical datasets that may arise due to measurement errors or data entry mistakes. Outliers can be corrected or removed using techniques such as winsorization, trimming, or robust regression. Saama (2024) described an AI-assisted data review process, though the techniques have not been described in peer-reviewed literature. Kikawa and Nakajima (2020) also described a pilot study, exploring ML algorithms for automating clinical data review. AI and ML offer automated solutions to streamline data review processes by employing algorithms to detect patterns, outliers, and deviations from expected norms. These innovative approaches challenge traditional methods and have the potential to reshape the future of clinical data cleaning.

4.8.1.3 AI for Risk-Based Monitoring

AI and ML provide an opportunity to power risk-based monitoring. Several technologies have emerged to directly monitor patient adherence to treatment protocols by remotely tracking dosing. Traditionally, clinicians would rely on direct observation of dosing, especially in challenging populations like individuals with substance use disorder or schizophrenia (Koesmahargyo et al. 2020). However, innovative methods now include the utilization of "smart" pills that emit a signal upon ingestion and smartphone-based technologies that use computer vision through video to remotely confirm patients ingesting medication (Koesmahargyo et al. 2020). These data sources provide continuous monitoring of each dosing event, enabling dynamic and individual-level risk prediction rather than static population-level forecasts. Dynamic risk prediction facilitates the efficient allocation of known effective treatment resources and communications. Given the widespread and varied nature of adherence issues in clinical treatment and limited resources for patient care management, optimizing efficacy in adherence is imperative. A dynamic prediction system focused on medication adherence in clinical research was reported by (Koesmahargyo et al. 2020). Several other tools were also developed, using video capture devices with built-in ML algorithms to confirm medication intake (Askin, Burkhalter et al. 2023). More broadly, by harnessing ML algorithms, stakeholders in clinical trials can identify high-risk areas, predict potential anomalies, and allocate resources efficiently for targeted interventions, including more frequent on-site visits and heightened SDV. This strategy not only optimizes monitoring procedures but also strengthens data integrity by proactively identifying discrepancies and deviations. Various AI-based methods have been explored to predict non-adherence to treatment protocols. For example, Ménard Barmaz et al. 2019) developed a ML-based flagging system for predicting sites for under-reporting adverse events (AEs). Other opportunities extend to trial participants' safety oversight, using near-real-time data from wearable devices and sensors and AI algorithms (Weissler et al. 2021). Faster access to actionable insights about clinical trial participants is advantageous, particularly for those with life-threatening or debilitating conditions. This is consistent with the new patient-centric clinical development paradigm.

AI and ML technologies offer transformative opportunities to streamline manual clinical data review, cleaning, and query processes in healthcare. By automating these tasks, AI-driven systems can enhance efficiency, accuracy, and scalability while reducing the burden on healthcare professionals. However, successful implementation requires addressing challenges related to data privacy, interpretability, fairness, integration, and ongoing validation.

With careful consideration of these factors, AI has the potential to revolutionize the way clinical data is managed and analyzed, ultimately improving patient outcomes and advancing medical research.

4.9 Case Examples

In this section, we present three applications of a Bayesian method for monitoring treatment discontinuation (Yang and Novick 2019), and ML model for predicting under-reporting of AEs (Ménard et al. 2019), and AI tools to reduce the risk of non-adherence in patients on anticoagulation therapy (Labovitz et al. 2017).

4.9.1 Premature Discontinuation of Treatment

4.9.1.1 Background

Treatment discontinuation in clinical trials is a widespread phenomenon, spanning various therapeutic areas (Nantz, Liu-Seifert et al. 2009). Research indicates that patients who prematurely discontinue treatment are prone to symptom relapse and other adverse effects (Shelton 2001). Moreover, patient attrition poses challenges to the analysis of clinical studies, jeopardizing internal validity and restricting result generalizability due to reduced sample size. Consequently, heightened awareness and monitoring of patient responses to treatment and adverse reactions throughout all treatment phases is crucial in ensuring trail compliance, data integrity, and patient benefit. To this end, the number of patients who discontinued treatment before the end of the treatment period as defined in the protocol is often used as a CtQ measure, and closely monitored throughout the trial conduct. Unexpected increasing trends of the CtQ can be indicative of compliance and quality issues, particularly when clinical causes can be prescribed. In cases of increased premature discontinuation, factors contributing to the adverse increase should be identified. Apart from heightened vigilance, including more frequent data review, the sponsor also needs to provide guidance for clinical investigators, including educational resources, adjustment of treatment dose, alternate interventions, to help patients to become more encouraged and engaged in their treatment with the goal of maximized patient outcomes.

4.9.1.2 Selection of CtQ

Clinical study usually involves many sites. Some are high-enrolling sites, and others may often have a few participants. In addition, since participants are normally enrolled in a study at different times, at a given time point, the drug exposure varies from participant to participant. As a result, the percentage of patients whose treatment was prematurely discontinued is not a robust measure for the CtQ. This is particularly true for small molecule drug that are being tested for frequent dosing such as once daily for an extended time. Under such circumstances, it is more sensible to use the number of participants per 1,000 doses who experience premature treatment discontinuation as a quality measure.

4.9.1.3 Zero-Inflated Model

For candidate drugs that are relatively less toxic, premature treatment discontinuation can be a rare event. Therefore, the CtQ data often exhibit an excess of zeros. The high occurrence of zero observations often invalidates the assumption that the data follow an underlying distribution such as Poisson distribution. Although a negative binomial model can account for variability due to site or time, it is insufficient to model data with frequent zero values. An alternative method based on a zero-inflated Poisson (ZIP) model is well-suited for modeling such data with excess zeros (Yang et al. 2007). This model perceives data as being generated from two processes with probability p and $1-p$, respectively. The observations from the first process are all zeros while the results from the second process assume non-zero integer values according to a probability distribution, such as Poisson.

Let X_i, $i = 1, \ldots, n$ denote the numbers of AEs per dose in the previous n report periods. Yang et al. (2007) modeled X_i through the following ZIP model:

$$X_i = WY + (1-W)Z,$$

where W is Bernoulli random variable with $P[W=1] = p$, Y is a degenerated random variable with $P(Y = 0) = 1$, $Z \sim \text{Poisson}(\lambda)$, and W, Y, and Z are independent. It can be readily verified that

$$P[X_i = x] = \begin{cases} p + (1-p)e^{-\lambda} & \text{if } x = 0 \\ (1-p)\dfrac{\lambda^x e^{-\lambda}}{x!} & \text{else} \end{cases}$$

Supposed that $(x_1, \ldots, x_n)$ are observed values of $(X_1, \ldots, X_n)$. Without loss of generality, we assume that $x_i = 0$, $i = 1, \ldots, n_1$, and $n_2 = n - n_1$. Then the likelihood function of x_i is given by

$$L = \left[p + (1-p)e^{-\lambda}\right]^{n_1} \prod_{i=n_1+1}^{n} \left[(1-p)\frac{\lambda^{x_i} e^{-\lambda}}{x_i!}\right]$$

$$= \left[p + (1-p)e^{-\lambda} \right]^{n_1} (1-p)^{n_2} \frac{\lambda^{\sum_{i=n_1+1}^{n} x_i}}{\prod_{i=n_1+1}^{n} x_i!} e^{-\lambda n_2}. \tag{4.5}$$

We assume that the parameters p and λ have a prior distribution of Beta(a, b) and gamma(α, β) respectively. Hence,

$$f(p) \propto p^{\alpha-1}(1-p)^{\beta-1}$$

$$f(\lambda) \propto \frac{\lambda^{\alpha-1} e^{-\lambda/\beta}}{\beta^{\alpha}}. \tag{4.6}$$

Combining (4.5) and (4.6), it can be shown that the posterior distribution of $\theta = (p, \lambda)$ is given by

$$f(\theta x_1, \ldots, x_n) = C_2^{-1} \sum_{j=1}^{n_1} p^{j+a-1}(1-p)^{n-j+b-1} \lambda^{\sum_{i=n_1+1}^{n} x_i + \alpha - 1} e^{-\lambda(n-j+1/\beta)}, \tag{4.7}$$

where C_2 is a normalization factor having an expression:

$$C_2 = \sum_{j=1}^{n_1} \binom{n_1}{j} \text{Beta}(j+a, n-j+b) \frac{\Gamma\left(\lambda^{\sum_{i=n_1+1}^{n} x_i + \alpha}\right)}{\left(n - j + \frac{1}{\beta}\right)^{\sum_{i=n_1+1}^{n} x_i + \alpha}}.$$

From (4.7), given the threshold value of the QTL, x_0, the predictive posterior probability for the number of participants who treatment was prematurely discontinued in the next reporting period can be calculated as:

$$p(x_0) = P\left[X \geq x_0 \right] = \int_0^1 \int_0^{\infty} \sum_{x=x_0}^{\infty} (1-p) \frac{\lambda^x e^{-\lambda}}{x!} f(\theta \mid x_1, \ldots, x_n) d\lambda dp. \tag{4.8}$$

Carrying out the above integration, we obtain

$$p(x_0) = C_2^{-1} \sum_{k=x_0}^{\infty} \sum_{j=1}^{n_1} \frac{\binom{n_1}{j} \text{Beta}(j+a, n-j+b) \Gamma(s)}{\left(n - j + \frac{1}{\beta}\right)^s}$$

with $s = \sum_{i=n_1+1}^{n} x_i + \alpha + x_0$.

Although equation (4.8) provides a closed-form for the predictive posterior probability, the calculation is not straightforward. Yang and Novick (2019) used the MCMC method to determine the cut point, x_0.

4.9.1.4 An Example

To illustrate the concept, consider an example dataset in which zero number of premature treatment discontinuation cases was reported in 14 of the last 20 report periods with the remaining six report periods yielding the set {2, 3, 3, 2, 3, 1} out of 1,000 doses administered. Informative priors were applied to this problem, with $p \sim \text{Beta}(10,3)$ and $\lambda \sim \text{Ga}(5,1)$. Most of the mass for the prior distribution for p is above 0.5 and the prior for λ was set so that the mean and variance are both equal to 5. The ensuing posterior predictive probability $p(X \geq 5) = 3.5\%$ means that the risk of a large number of participants who experience early discontinuation of treatment in a future period is non-zero, but low.

4.9.1.5 Discussion

When applied to model the early treatment discontinuation data, the ZIP may not be adequate. This is primarily because the actual variability in the data is often larger than what is expected under the Poisson assumption. This phenomenon is often referred to as overdispersion. For data collected at different times and sites, although at each fixed time or site, the metric may exhibit behavior that can be described through a Poisson distribution, the collective data may not follow any Poisson distribution. The negative binomial distribution previously discussed is a variant of Poisson distribution that has increased flexibility to address the issue of overdispersion. Mathematically it can be formulated as a mixture of Poisson distributions (Hoffman 2004). The zero-inflated negative binomial (ZINB) is suitable for modeling the early treatment termination data when both excessive number of zero observation and overdispersion issues are expected. The probability function of the negative binomial variable is given by:

$$f(x|\pi, \tau) = \frac{\Gamma(x+\tau)}{x!\Gamma(\tau)} \pi^{\tau} (1-\pi)^{x}.$$

By a reparameterization $\lambda = \frac{\tau(1-\pi)}{\pi}$, the ZINB model can also be derived from a mixed representation (Ghosh, Mukhopadhyay, and Lu 2006):

$$X \mid W \sim \mathrm{NB}(x \mid \pi, \tau) \text{with } \pi = \frac{\tau}{\left[\tau + \lambda(1-D)\right]}, D \sim \mathrm{Bernoulli}(p).$$

With this representation, the sampling of the ZINB random variable can be easily implemented in JAGS or Stan, and the above analysis can be repeated.

4.9.2 Machine Learning-Based Prediction of Medication Adherence

4.9.2.1 Background

Non-adherence presents a significant challenge across various therapeutic areas, often cited as the primary culprit behind suboptimal clinical benefits, leading to disease complications, diminished quality of life, and inefficient utilization of healthcare resources. Extensive research has delved into its underlying causes, revealing a multitude of factors spanning patient characteristics, disease complexity, treatment regimen, healthcare provider dynamics, and systemic healthcare structures.

Despite this extensive investigation, adherence rates have seen little improvement, with an estimated one-third to one-half of patients consistently failing to adhere to prescribed medications, and adherence to lifestyle modifications proving even lower. In clinical trials, non-adherence exacerbates variability, diminishes study power, and attenuates treatment effects, thereby compromising the likelihood of trial success.

In recent years, various digital technologies have been explored for monitoring and enhancing adherence in clinical trials. These tools include smartphone alerts and reminders, electronic medication tracking systems like smart pillboxes, and mechanisms for visual confirmation. Additionally, these novel methods can facilitate the electronic tracking of missed clinical visits, prompting alerts for non-adherence. In clinical research, AI/ML applications aimed at improving medication adherence leverage digital biomarkers, such as facial and vocal expressivity, enabling remote monitoring of adherence behaviors. Additionally, data collected through continuous measurement of each dosing event allows for dynamic and patient-level risk prediction rather than population-level risk ascertainment.

In the following, we discuss a ML model by Koesmahargyo et al. (2020) that was trained on the video records of patients taking medicant via a smartphone application to predict medication adherence. The algorithm utilizes both static features such as demographic, diagnosis, treatment length and dynamic features including dosing, clinical intervention, etc. to predict adherence for: (1) Remaining trial based on the first week dosing behavior; (2) Remaining trial based on the first wp weeks of dosing behavior; (3) Next week adherence; and (4) Next day adherence. The objective of the algorithm is two-fold. One is to aid decision-making for trials where researchers have the option to remove non-compliant participants early on. The second is to aid the prediction of non-adherence risk so that the investigators can proactively devise intervention strategies and potentially direct their efforts and resources to participants at high risk of non-adherence (Koesmahargyo et al. 2020).

4.9.2.2 Data Source

Data were collected from clinical trials that used the AiCure software platform for monitoring dosing adherence. The AI platform employs AI to visually verify medication ingestion. Utilizing software that can be easily downloaded as an app onto any mobile device, AiCure uses facial recognition and computer vision algorithms to accurately identify the patient, the medication, and confirm ingestion (Bain et al. 2017). The tool offers medication reminders and dosing instructions, ensuring timely administration. Notifications are promptly sent within the hour of a missed dose or nearing the end of the dosing window. Real-time data is encrypted and transmitted to web-based dashboards for continuous monitoring. Additionally, clinic staff receive automated text messages or emails alerting them to missed doses, tardiness, or deviations from correct usage (Labovitz et al. 2017). Data from a total of 4,182 participants in clinical trials in various therapeutic areas were used to develop the classification algorithm. Patient demographics and conditions are summarized in Table 4.4.

TABLE 4.4

Demographics and Clinical Information of Participants

Characteristics	Descriptive Summary (*n* = 4,182)
Age, mean ± SD	39.0 ± 11.7
Sex, *n* (%)	
Male (*n*%)	1,006 (24.1)
Female	555 (13.3)
Unknown	2,621 (62.6)
Race	(*n*%)
White	461 (11.0)
Black or African American	195 (4.67)
Multiracial	25 (0.60)
Asian	24 (0.57)
Other	22 (0.53)
Latino	15 (0.36)
American Indian or Alaskan Native	4 (0.10)
Native Hawaiian or other Pacific Islander	4 (0.10)
Unknown	3,432 (82.1)
Condition	(*n*%)
Schizophrenia	970 (23.3)
Attenuated psychosis syndrome	36 (0.86)

(*continued*)

TABLE 4.4 (Continued)

Demographics and Clinical Information of Participants

Characteristics	Descriptive Summary (n = 4,182)
Cognitive impairment in schizophrenia	538 (12.9)
Schizophrenia, schizoaffective disorder, negative symptoms	34 (0.81)
Unspecified	362 (8.7)
Major depressive disorder	777 (18.6)
Major depressive disorder with anxious distress	20 (0.48)
Depression and acute suicidal ideation behavior	21 (0.50)
Unspecified	736 (17.6)
Addiction	632 (15.1)
Opioid dependence	25 (0.60)
Opioid use disorder	5 (0.12)
Alcohol use disorder	25 (0.60)
Unspecified	577 (13.8)
HIV	86 (2.06)
Bipolar depressive disorder	27 (0.65)
Attention-deficit hyperactivity disorder	1,115 (26.7)
Healthy volunteers	340 (8.13)
Post-traumatic stress disorder	86 (2.06)
Hepatitis C	68 (1.63)
Congestive Heart Failure	39 (0.93)
Psoriasis	32 (0.77)
Epilepsy	7 (0.17)
Parkinson's Disease	2 (0.05)
Chronic Obstructive Pulmonary Disease	1 (0.02)

Source: Koesmahargyo et al. (2020).

4.9.2.3 Endpoints

The primary endpoint is daily adherence value (DAV), which is defined as

$$\text{DAV} = \frac{n_{\text{AiCure}}}{N}$$

where n_{AiCure} is the number of medications taken and confirmed by AiCure, and N is the total number of expected medications.

In addition to the software's measure of adherence, human review of dosing videos enabled the identification of non-adherent or deceptive behavior, which informed adjustments to the daily adherence calculation. These observations were documented through the issuance of "red" or "orange" alerts. A "red" alert denoted compelling visual evidence of deliberate non-adherence, while an "orange" alert signified the presence of suspicious behavior lacking clear visual confirmation of intentional non-adherence. Subsequently, the DAV was modified based on these alerts and the daily adjusted adherence value (DAAV) is estimated as follows:

$$\text{DAAV} = \frac{n_{\text{AiCure}} - n_{\text{flagged}}}{N}$$

where n_{flagged} is the total number of medications with "red" or "orange" alerts.

4.9.2.4 Cutoff for Medication Adherence

To train the AI algorithms, patients need to be labeled as either adherent or non-adherent pending on if a participant average DAAV is above the cutoff. The common clinical standards in clinical development suggest an 80% or more adherence in order to derive the intended clinical treatment benefit. As such, Koesmahargyo et al. (2020) followed the recommendation and used 80% as the cut off value.

4.9.2.4.1 Study-Specific Features

Several study-specific variables were also collected and described in Table 4.5.

4.9.2.5 Model Development

Four models were trained and tested for predicting medication adherence based on early dosing behavior and real-time prediction during study as outlined in the following:

- Predicting remaining trial adherence based on the first week

TABLE 4.5

Study-Specific Variables

Features	Description
Condition	Participant's clinical condition or disease.
Intervention	Efforts made by study coordinators to contact patients to address any non-adherence
Micro-reimbursements	A binary indicator on whether the particular study offered micro-reimbursements to patients for reporting adherence through the software platform
Trial Length	The duration in weeks in which the treatment is expected to be completed.
Dose Delay	The length of time in minutes for a patient to begin a dosing event on the smartphone application after an alarm has prompted them to do so.
Dose Length	The length of time in seconds it took for the patient to complete a dosing event

- Predicting remaining trial adherence based on the first two weeks
- Predicting next week's adherence based on the previous week
- Predicting next day's adherence based on the previous week

DAVs, adjusted adherence, intervention frequency, dose delay, and dose duration served as dynamic features, while condition, trial duration, and micro-reimbursements remained static predictors across all models. All models employed the XGBoost (Extreme Gradient Boosting) classifier, chosen for its fast computation time and superior performance on structured data compared to other decision tree-based methods. The patient-level data splitting ensured independence between training and validation sets, employing 5-fold cross-validation to evaluate each model's performance.

Performance metrics included Accuracy, Precision, Recall, True Positive Rate, False Positive Rate, and the Area Under the ROC Curve (AUC). Grid search with cross-validation optimized model parameters, minimizing search time while enhancing classification accuracy.

4.9.2.6 Results

All 4 models demonstrated strong and comparable performance as measured by the metrics of overall accuracy, AUC, precision, recall, TPR, and FPR (Table 4.6). Performance metrics presented in the table were averaged across five folds, as shown in Table 4.6. The sample sizes of the holdout set sizes for testing models were 863, 683, 6,623, and 45,312, respectively, for the models trained on the behavior of the first week and first two weeks, and next week and next prediction. The noticeable improvement in model performance was likely attributed to the increased sample sizes utilized in these evaluations.

The results in the above table are also presented in the ROC plots (Figure 4.8). For the plots, there is slight improvement in prediction accuracy of the model

TABLE 4.6

Results from the Four Classification Types

Adherence Classification	Accuracy	AUC	Precision	Recall	TPR	FPR
Remainder of study from the first 7 days	72.2%	0.8	0.76	0.74	0.74	0.25
Remainder of the study from the first 14 days	76.6%	0.83	0.78	0.78	0.78	0.18
The following week from the previous week	81.3%	0.87	0.82	0.82	0.82	0.16
The following day from the previous week	81.0%	0.87	0.84	0.82	0.82	0.17

Source: Adapted from Koesmahargyo et al. (2020).

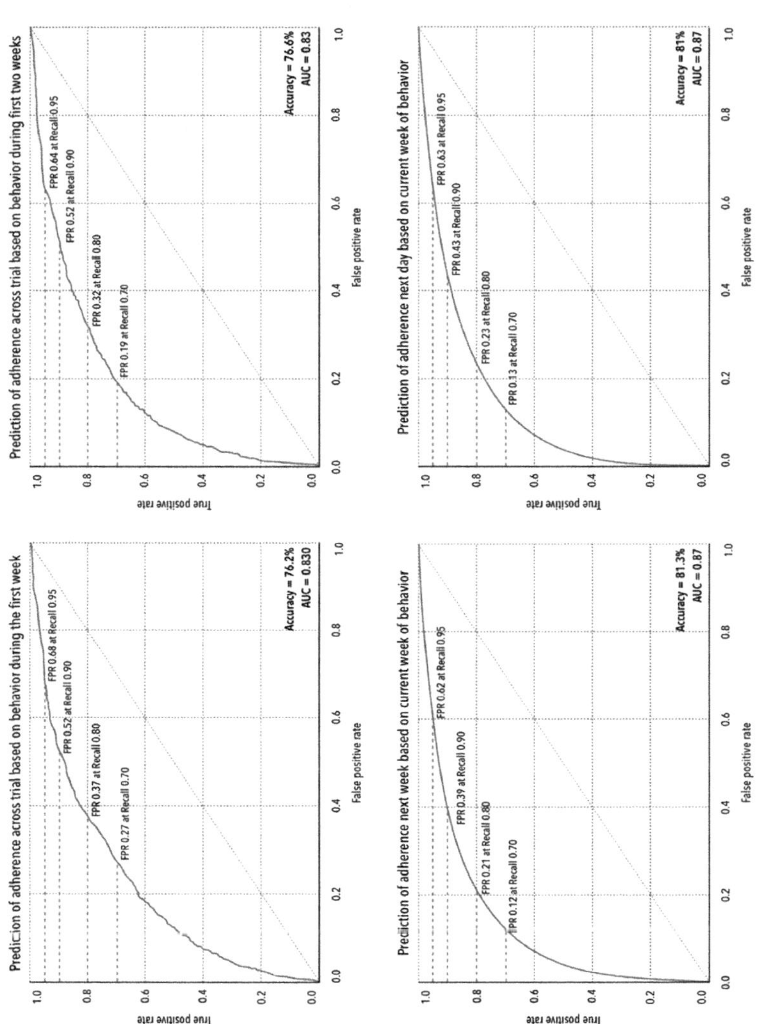

FIGURE 4.8
ROC curves with mean AUCs from cross-validation with specific recall thresholds of the four different model types: Prediction of adherence across the trial based on the first week; Prediction of adherence across the trial based on the first two weeks; Prediction of next week adherence based on current week; Prediction of next day adherence based on current week. (Adapted from Koesmahargyo et al. 2020.)

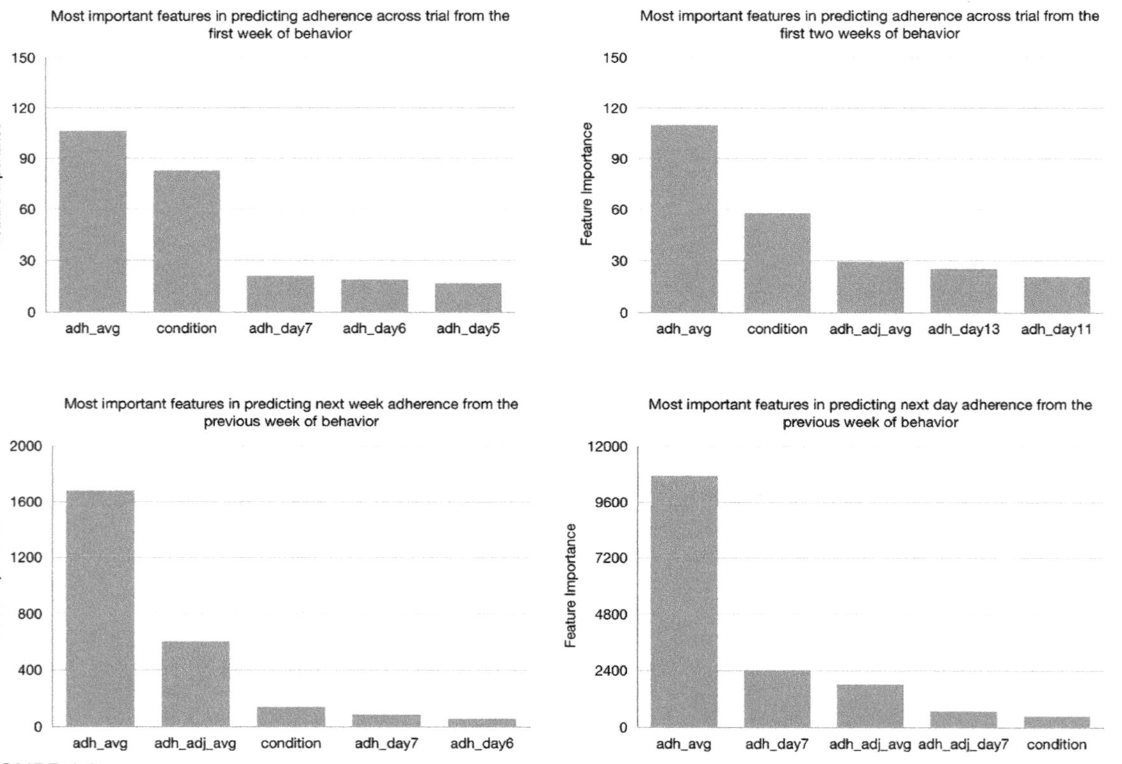

FIGURE 4.9

Highest ranking features in importance of the four adherence model types, where adh_avg = average daily adherence; adh_adj_avg = average daily adjusted adherence; condition = participant's clinical condition or disease; adh_day7 = average adherence on the 7th nominal day; adh_adj_day7 = average adjusted adherence on the 7th nominal day. (Adapted from Koesmahargyo et al. 2020.)

using the first two weeks over that using only the first week, and the model for the next week prediction over that for the next day prediction.

Koesmahargyo et al. (2020) also investigated contributions of predictive features of the four models. Feature importance from each cross-validation fold was assessed based on the average gain in accuracy across all splits associated with each feature. The results are presented in Figure 4.9. Remarkably, regardless of the adherence model type, average adherence (avg_adh) emerged as the most crucial predictor, consistently appearing among the top-ranking features in each model type. Additionally, average adherence and condition consistently featured prominently among the highest importance predictors. Noteworthy features also encompassed the average adherence values in the days leading up to the prediction (avg_day7, avg_day6, avg_day13), as well as the adjusted average adherence (adh_adj_avg).

4.9.3 Predicting Adverse Event Reporting in Clinical Trials

4.9.3.1 Background

Per ICH E6(R1), AEs, and/or laboratory abnormalities identified in the protocol as critical to safety evaluations should be reported to the sponsor according to the reporting requirements and within the time periods specified by the sponsor in the protocol. However, despite the regulatory requirements, AE under-reporting has consistently surfaced as a significant concern during health authorities' inspections and audits of Good Clinical Practices (GCP) (Ménard et al. 2019). This is of significant concern as insufficient reporting of safety incidents has the risk of jeopardizing both patient safety and data integrity. The current approach to address this issue predominantly hinges on audits or on-site source data verification to identify sites or studies exhibiting under-reporting of AEs. However, this is labor intensive and costly. In addition, the increasing number of clinical trials and sites, coupled with the increasing complexity of study designs, presents a formidable obstacle to conducting regulatory on-site visits. It is of significant interest to use predictive models to identify sites for more focused monitoring. Ménardet et al. (2019) developed a ML model based on study and patient attributes to predict expected number of AEs, at either the site, patient, or visit level. It has the potential to enhance oversight for patient's safety while reducing the need for on-site visits.

4.9.3.2 Method

To facilitate the introduction of the method by Ménard et al. (2019), we define Y_{site}, $Y_{patient}$, and Y_{visit} as the random variables describing the number of AEs reported by a given site, the number of AEs of a patient at the site, and the number of AEs collected at before a visit, respectively. Assume that Y_{visit} follows a Poisson distribution with parameter θ_{visit}. Since the number of

TABLE 4.7

Relationships among Numbers of AEs Reported by Site, Patient, and Visit

Prediction -Level	Distribution	Relationship
Visit	$Y_{\text{visit}} \sim \text{Poison}(\theta_{\text{visit}})$	
Patient	$Y_{\text{patient}} \sim \text{Poison}(\theta_{\text{patient}})$	$\theta_{\text{patient}} \sim \sum_{\text{visit} \in \text{patient}} \theta_{\text{visit}})$
Site	$Y_{\text{site}} \sim \text{Poison}(\theta_{\text{site}})$	$\theta_{\text{site}} \sim \sum_{\text{patient} \in \text{site}} \theta_{\text{patient}})$

Source: Ménard et al. (2019).

AEs of a patient is the sum of the numbers of AEs reported at each visit, Y_{patient} is also distributed according to a Poisson distribution. Due to the same additivity of Poisson distributions, Y_{site} is described through a Poisson distribution. The relations among the above three random variables and their Poisson distributional parameters are given in Table 4.7.
After the parameter θ_{site} is visited, for the number of AEs, y_{visit}, reported by the site, the probability of observing this value or below is calculated as follows:

$$S(x_{\text{site}}, y_{\text{site}}) = P(Y_{\text{site}} \leq y_{\text{site}} | x_{\text{site}}) = \sum_{k=0}^{y_{\text{site}}} \frac{\theta_{\text{site}}^k e^{-\theta_{\text{site}}}}{k!}.$$

This probability is used as a scoring function to assess if statistical significance is achieved that the site has under-reported AEs. Here, x_{site} is the observed feature vector that summarizes site and patient information at the site. It is utilized in the estimation of the model parameter θ_{site}, which is described in detail in the following sections.

4.9.3.3 Data

As described by Ménard et al. (2019), the data used for the modeling development were sourced from clinical trials sponsored by Roche/Genentech, encompassing 104 completed studies spanning various molecule types and disease areas. In total, this dataset comprised 3,231 individual investigator sites, involving 18,682 study participants who underwent a total of 288,254 study visits. Notably, all study subject data was anonymized to safeguard privacy.

To ensure data were fit for estimating the model parameters discussed in the previous section, Ménard et al. (2019) exclusively included studies which were either completed or terminated, and where AE data were fully clean through AE reconciliation and SDV as part of the study closure procedures.

TABLE 4.8

Attributes Available in Curated Dataset

Level	Source	Extracted Data
Patient	SDTM demographics	Age, sex, ethnicity
Visit	SDTM medical history	Number of co-occurring conditions
Visit	SDTM concomitant medications	Number of concomitant medications
Visit	SDTM vitals	Height, weight, blood pressure
Visit	SDTM visits	Number of previous visits
Visit	SDTM adverse events	Number of reported AEs
Study	Clinical Trial Management System	Intervention type, route of administration, use of concomitant agents, phase, randomization, blinding, molecule class, disease type

Source: Ménard et al. (2019).

From these studies, six common patient data attributes, demographics, medical history, concomitant medications, vitals, visits, and AEs, where selected, adhering to the Study Data Tabulation Model (SDTM) standard (Table 4.8).

Additionally, study-specific attributes, including study type, route of administration, concomitant agents, disease area, blinding, randomization, and study phase, were also extracted and included in the dataset. To ensure clinical relevance with respect to AE reporting, a different classification for molecule classes and disease areas were adopted while the Anatomical Therapeutic Chemical classification system was used for molecules and a simplified classification reflecting the populations enrolled in those clinical trials for disease areas.

4.9.3.4 Features Selection

When selecting features for the model development, considerations were given to whether the feature is correlated with the number of reported AEs. For example, from the plots in Figure 4.10, it is evident that ethnicity, the number of concomitant medications prior to the visit, the number of concomitant diseases, pulse, respiratory rate, and temperature were associated with the number of reported AEs.

A total of 54 features were identified for model building (Table 4.9). Some pre-processing was conducted to normalize or standardize the input features.

4.9.3.5 Model Training, Validation, and Testing

The data discussed in the previous section were split into training, validation, and testing sets, at roughly 60% / 20% / 20% ratios. Recognizing that the molecule class may have a significant influence on the number AEs, it was used as a stratification factor to ensure a representation of ever class in each of the training, validation, and testing sets. It is also worth noting that key to

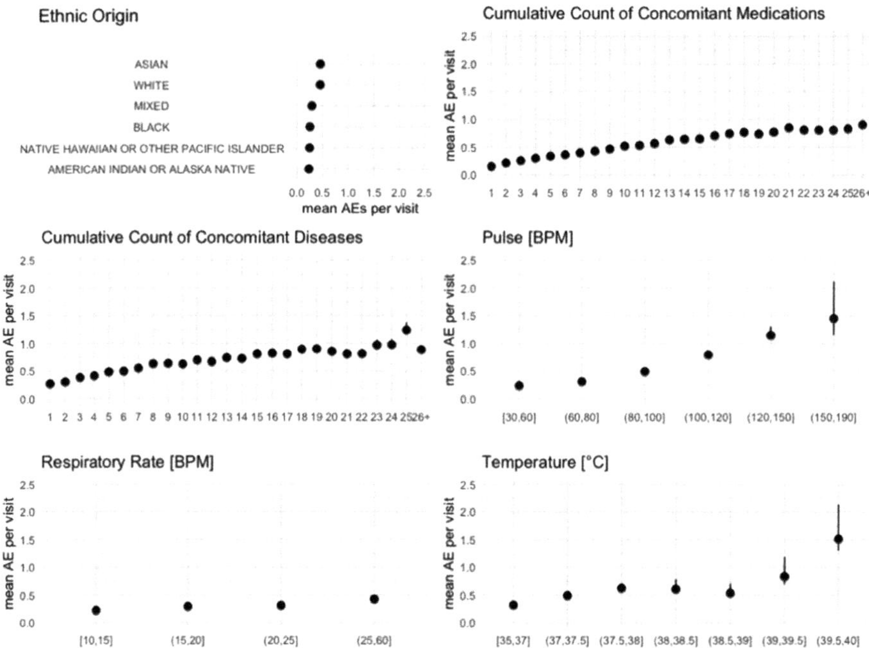

FIGURE 4.10
Results of exploratory analysis assessing association between features and the number of reported AEs. (Adapted from Ménard et al. 2019.)

TABLE 4.9
Features Used for the Final Model

Feature Identification	Description
VISIT_COUNT_CUMULATIVE	Number of study visits to date
MD_COUNT_CUMULATIVE	Cumulative count of concomitant medications prior a study visit
DISEASE_TYPE_malignancy	Disease type = malignancy
DISEASE_TYPE_lung disease other	Pulmonary disease other than malignancy
ADMIN_ORAL	Study drug administered orally
HX_COUNT_CUMULATIVE	Number of concomitant disease
AGEIC_POW	Age
WT_3V_SLOPE	Slope of the weight since last 3 visits
BP_LEVEL_VISIT_MOST_RECENT_UNKNOWN	Blood Pressure at the most recent visit was not reported
WT_VISIT_MOST_RECENT_LOG	Log of the weight reported at the most recent visit
BLINDING_Open Label	Study is open label (not blinded)
BPD_VISIT_MOST_RECENT	Blood Pressure Diastolic value (most recent visit)
PHASE_II	Study phase is II
TEMP_VISIT_MOST_RECENT	Body temperature (most recent visit)
BPS_VISIT_MOST_RECENT	Blood Pressure Systolic value (most recent visit)
ADMIN_SC	Drug administered subcutaneously

TABLE 4.9 (Continued)

Features Used for the Final Model

Feature Identification	Description
PHASE_I	Study phase is I
CONC_AGENTS	Study drug administered with concomitant agents/drugs
RANDOMIZED	Study is randomized
MOLECULE_CLASS_monoclonal antibody	Study drug monoclonal antibody
PHASE_III	Study phase is III
SEX_Female	Female subject
ADMIN_IV	Drug administered intravenously
MOLECULE_CLASS_kinase inhibitor	Study drug kinase inhibitor
MOLECULE_CLASS_sodium channel blocker	Study drug sodium – channel blocker
ADMIN_IM	Drug administered intramuscularly
ETHNIC_ORIGIN_WHITE	Subject ethnic origin = white
BP_LEVEL_VISIT_MOST_ RECENT_Normal	Blood pressure reported at most recent visit was within normal range
BLINDING_Single Blind	Study blinding of the study = single blinded
ETHNIC_ORIGIN_OTHER	Subject ethnic origin = other
MOLECULE_CLASS_interleukin	Study drug interleukin
DISEASE_TYPE_respiratory disease	Disease type = respiratory disease
BLINDING_Double Blind	Study blinding of the study = double blinded
DISEASE_TYPE_healthy participants	Study conducted on healthy subject (e.g., Phase I)
ETHNIC_ORIGIN_UNKNOWN	Subject ethnic origin = unknown (not reported)
MOLECULE_CLASS_nucleosid analog	Study drug nucleoside analog
BP_LEVEL_VISIT_MOST_ RECENT_Stage 1 hypertension	Blood pressure reported at most recent visit was within the range of stage 1
ETHNIC_ORIGIN_BLACK OR AFRICAN AMERICAN	Subject ethnic origin = black of African American
DISEASE_TYPE_infectious disease	Disease type = infectious disease
ETHNIC_ORIGIN_ASIAN	Subject ethnic origin = Asian
DISEASE_TYPE_other	Disease type other than healthy participants, malignancies, autoimmune neurodegenerative diseases, respiratory diseases, skin disorders, lung disease, infectious diseases
BP_LEVEL_VISIT_MOST_ RECENT_Stage 2 hypertension	Blood pressure reported at most recent visit was within the range of stage 2
PHASE_IV	Study phase is IV
BLINDING_Observational	Observational study
ETHNIC_ORIGIN_MIXED ORIGIN	Subject ethnic origin = mixed
MOLECULE_CLASS_PPAR modulator	Study drug – Peroxisome Proliferator Activated Receptor (PPAR) modulator
ADMIN_INTRAVITREAL	Study drug administered intravitreally
BP_LEVEL_VISIT_MOST_ RECENT_Prehypertension n	Blood pressure reported at most recent visit was within the range of pre-hypertension

(continued)

TABLE 4.9 (Continued)

Features Used for the Final Model

Feature Identification	Description
DISEASE_TYPE_neurodegenerative disease	Disease type = neurodegenerative disease
MOLECULE_CLASS_glycine reuptake inhibitor	Study drug glycine reuptake inhibitor
MOLECULE_CLASS_thiol protease inhibitor	Study drug thiol protease inhibitor
ETHNIC_ORIGIN_AMERICAN INDIAN OR ALASKA NAITIVE	Subject ethnic origin = American Indian or Alaska native
MOLECULE_CLASS_SMO antagonist	Study drug SMO antagonist
MOLECULE_CLASS_ antiinflammatory	Study drug antiinflammatory

Source: Ménard et al. (2019).

the development of the predictive model for assessing AE under-reporting is to estimate the mean number of AE, θ_{visit}. This can be accomplished by minimizing a loss function:

$$L(f) = \sum_{\text{Visit}} l\left(y_{\text{visit}}, f\left(x_{\text{visit}}\right)\right),$$

where the sum includes all visits in the training data, $l(\bullet)$ is a loss function, and $f\left(x_{\text{visit}}\right)$ is an estimate of y_{visit} conditional on the observed feature x_{visit}.

For the problem at hand, because Y_{visit} was assumed to be generated from a Poisson process, naturally the Poisson deviance is used as the loss function. Specifically,

$$l\left(y_{\text{visit}}, f\left(x_{\text{visis}}\right)\right) = 2\left(y_{\text{visist}} \log \frac{y_{\text{visit}}}{f\left(x_{\text{visist}}\right)} - y_{\text{visist}} + f\left(x_{\text{visist}}\right)\right).$$

Ménard et al. (2019) considered various potential modeling approaches to optimize the above loss function, including generalized linear models, gradient boosting models, and neural networks, and eventually selected the gradient boosting method. One of the challenges encountered during the model building was the lack of real-life studies where AE under-reporting was identified and all the above-mentioned data attributes were captured. To counter this issue, Ménard et al. (2019) resorted to simulate the positive cases by taking a sample of the test dataset and $\left(x_{\text{site}}, y_{\text{site}}\right)$ to create a sample with

TABLE 4.10

Methods for Creating Under-Reporting Dataset

Method	Definition	Description
Quantile	$\tilde{y}_{\text{site}} == \text{Poisson}_{0.01}\left(y_{\text{site}}\right)$	Use the observed AE count y_{style} at the site to estimate the mean AE count θ_{site}. Use the first percentile of the estimate Poisson distribution as the under-reported AE count.
Ratio	$25\%\,\text{under reporting}: \tilde{y}_{\text{site}} = 0.75 \times Y_{\text{site}}$ $50\%\,\text{under reporting}: \tilde{y}_{\text{site}} = 0.5 \times Y_{\text{site}}$ $67\%\,\text{under reporting}: \tilde{y}_{\text{site}} = 0.33 \times Y_{\text{site}}$ $90\%\,\text{under reporting}: \tilde{y}_{\text{site}} = 0.10 \times Y_{\text{site}}$	The lowered AE count is calculated as a percentage of the observed AE count at the site. However, only include sites with $y_{\text{site}} \geq 8$ in the under-reporting sample.
Zero	$\tilde{y}_{\text{site}} = 0$	The lowered AE count is set at zero and only include sites with no more than 10 participants but at least 6 reported AEs in the under re-porting sample.

Source: Ménard et al. (2019).

under-reporting AEs, $\left(x_{\text{site}}, \tilde{y}_{\text{site}}\right)$, where $\tilde{y}_{\text{site}}$ is the lowered AE count, using one of the three methods descripted in Table 4.10.

The final simulated sample of under-reporting AEs, Ménard et al. (2019) opted for filtering out sites where the disparity between y_{site} and $\tilde{y}_{\text{site}}$ is negligible, thus not raising quality concerns but potentially introducing irrelevant noise in model evaluation.

4.9.3.6 Results

The simulated samples coupled with those in the testing set which were not used for generating the positive cases were combined for testing the ML model. The overall discriminating power of the model was measured by the area under the ROC curve. For the case where the positive cases were generated using the quantile method, the model scored an area under the ROC curve of 0.67 (Figure 4.11). The model achieved an area under the ROC curves of 0.62, 0.79, 0.89, and 0.92, respectively, for the fixed ratio method, with the test dataset including 25%, 50%, 67%, and 75% under-reporting at the site level, respectively (Figure 4.12). Lastly, an area under the ROC curve of 0.97 was attained for the test dataset utilizing the "zero" method (Figure 4.13).

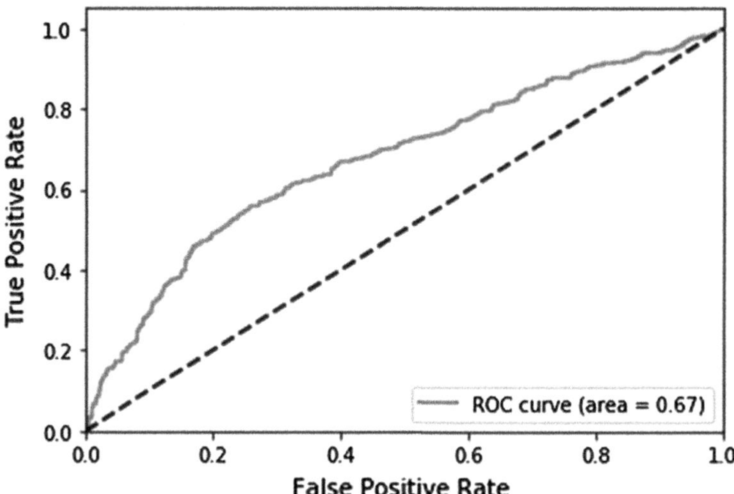

FIGURE 4.11
Receiver operating characteristic (ROC) curve for the statistical scenario. (Adapted from Ménard et al. 2019.)

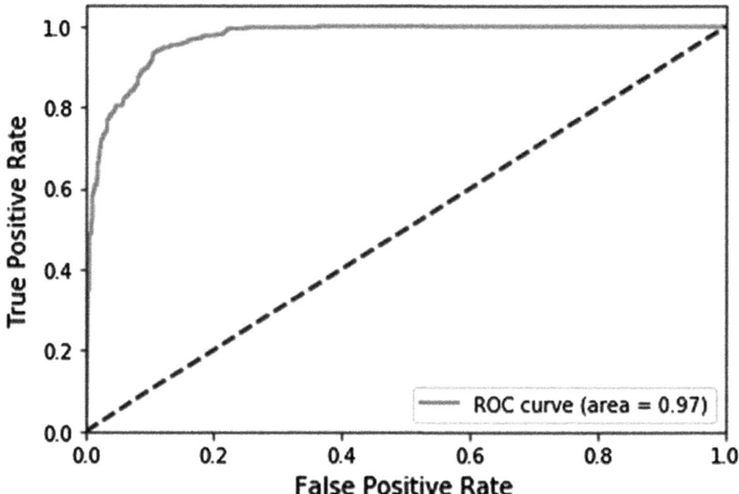

FIGURE 4.12
Receiver operating characteristic (ROC) curve for the zero scenario for small investigator sites. (Adapted from Ménard et al. 2019.).

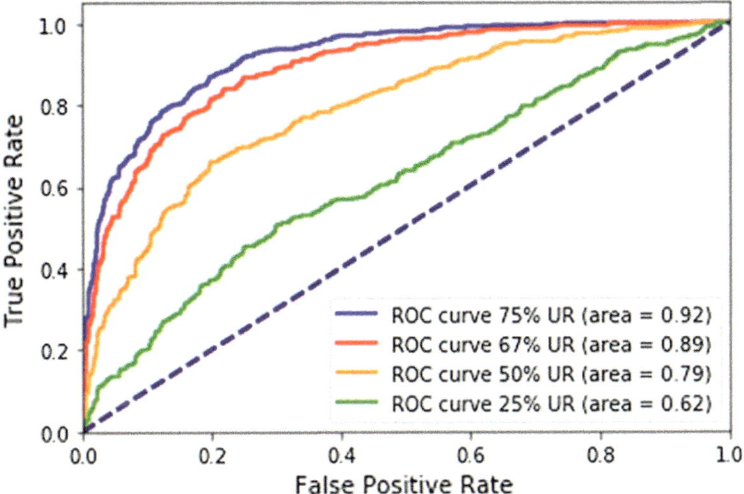

FIGURE 4.13
Receiver operating characteristic (ROC) curve for the percentage scenarios. UR = under-reporting. The curves from the top to the bottom correspond to 75%, 67%, 50%, and 25% UR, respectively. (Adapted from Ménard et al. 2019.)

TABLE 4.11
Performance Metrics for Sites Grouped by Different Alert Levels

	Alert Level 3	Alert Level 2–3	Alert Level 1–3	Alert Level 0
FPR	0.14	0.22	0.25	0.75
Zero scenario TPR	0.95	0.99	0.99	0.01
75% under-reporting TPR	0.8	0.9	0.91	0.09
67% under-reporting TPR	0.72	0.84	0.86	0.14
50% under-reporting TPR	0.5	0.64	0.66	0.36
25% under-reporting TPR	0.31	0.37	0.39	0.61

Source: Ménard et al. (2019).

4.9.3.7 Implementation

To facilitate the implementation of the predictive model, Ménard et al. (2019) developed an interactive dashboard using Tableau® to visualize and monitor data from ongoing clinical studies was developed. The dashboard also incorporated an alert level flagging system, based on the site's significance level score described in Section 4.9.3.2. This system allows for ranking sites by risk of under-reporting. under-reporting. However, determining a reasonable cutoff for flagging high-risk sites was necessary. To strike the best balance between maximizing the true positive rate (TPR) and minimizing the false positive rate (FPR), the Youden's index was used across each simulation scenario. This method maximizes the statistics, sensitivity + specificity – 1, which corresponds to the vertical distance a point on the ROC curve and the 45° diagonal line. Subsequently, three consecutive threshold values were determined, categorizing sites into four groups (AL3, AL2, AL1, AL0), where AL3 indicates the highest risk and AL0 the lowest.

Table 4.11 lists the TPR and corresponding FPR for all simulation scenarios. The FPR of each alert level indicates the minimum percentage of flagged sites within a dataset from ongoing studies. To maintain a consistent TPR metric while flagging a specific percentage of sites, we adjust the FPR accordingly. For example, focusing on AL3, reviewing the top ~14% of sites with the highest under-reporting risk as predicted by our model would identify various percentages of sites with different degrees of under-reporting.

These alert levels are prominently displayed on the dashboard alongside other key site information, providing the sponsor with comprehensive and near-real-time oversight of safety reporting quality. Suspected under-reporting studies and sites trigger additional quality activities, such as audits. Moreover, auditors utilize the tool to select sites and/or patients for review during study or investigator site audits.

4.10 Concluding Remarks

In an era of increasing complexity in clinical trials, traditional quality management approaches are proving insufficient to meet the demands of ensuring data integrity, regulatory compliance, and patient safety. The transition to a QbD model represents a paradigm shift from reactive to proactive trial oversight, enabling a more robust, risk-based approach to trial design and execution. This chapter has outlined the benefits of adopting QbD principles, emphasizing the need for integrated, forward-thinking strategies that anticipate risks, optimize processes, and leverage innovative technologies.

By integrating advanced tools like AI, ML, and advanced analytics into trial workflows, the clinical trial landscape can be transformed. These technologies not only reduce inefficiencies and operational risks but also improve data quality and participant protection, as demonstrated through various use cases. As regulatory environments continue to evolve, the adoption of QbD principles and the incorporation of cutting-edge technologies will be essential to ensuring that clinical trials remain both cost-effective and capable of generating reliable, high-quality evidence.

In conclusion, embracing the proactive, risk-based approach of QbD has the potential to significantly improve the quality and integrity of clinical trials. As the industry continues to innovate and adapt, QbD principles will become increasingly critical to the future of clinical development, supporting a more efficient and reliable path from trial design to regulatory approval.

5

Clinical Trial Optimization

5.1 Introduction

Clinical trial is a multifaceted process involving strategic, scientific, and operational considerations. The primary goal of clinical trial optimization is to select the right patient, right treatment, right dosing regimen, and right strategy including study design and operational feasibility to effectively and efficiently develop and deliver novel medicines to unmet medical needs. Often, key decisions are made quickly in the absence of critical and relevant data. In addition to the rising expectations about timelines, clinical development is also facing increasing demands from patients and payers to generate targeted and individualized data that guide the optimal treatment. In the meanwhile, clinical trials have increasingly become complex as sponsors, under the pressure accelerate clinical development, often aim to address multiple questions such as selection of dose and schedule, the optimal patient population and treatment strategy, in one trial. Against this backdrop, over the past decade, significant progress has been made in harnessing innovative study designs that enable incorporation of learning from the historical and ongoing trials into clinical development to improve and accelerate timelines. However, despite this noted advance, the current clinical trial design approaches continue encountering challenges in efficiently identifying the most promising treatments and optimizing patient outcomes. The advent of digital technology and data analytics, particularly artificial intelligence (AI) and real-world data (RWD), has emerged as powerful tools to optimize clinical trials with noticeable impact on operational excellence and acceleration. This chapter is focused on leveraging AI analytics and novel sources of data to design more precise and efficient clinical trials with greater probability of success. These new opportunities are highlighted through several case examples.

DOI: 10.1201/9781003226086-5

5.2 Clinical Development Planning

5.2.1 Background

As reported in published literature, about one in every five clinical trials failed to be fully enrolled, roughly 40% of cancer-related clinical trials terminate prematurely (Stensland et al. 2014), and about 60%–70% of drugs that make it to Phase II did not transition to Phase III (Feijoo et al. 2020), resulting in significant losses of time, money, and human effort. Suboptimal trial design, inefficient enrollment processes, and poor accrual rates are primary causes of trial failures. Many trials did not meet projected enrollment timelines. As a remedy, many investigational sites were required to fulfill timeline goals, causing excessive contracting and monitoring resources. Nonoptimized designs also caused frequent protocol amendments. All of these speak to the pressing need to identify more efficient approaches to facilitating and enabling clinical development planning at both program and trial levels.

Clinical trial planning is a complex process, involving many stakeholders and key decision-making. While clinicians and statisticians concern the selection of patient population, endpoints, and a study design that has high probability to generate sufficient evidence in support of regulatory approval, clinical operations specialists desire simplicity in trail design, and commercial teams prefer easy administration of the drug and shorter duration of the trial. Clinical trial planning is aimed at balancing, aligning the interests of all stakeholders, and making sound choices of key parameters that have the potential to impact trial timeline and outcomes. Clinical development planning is a critical step toward optimized trial designs. The outcome of this process is a robust clinical program/trial is characterized by a reasonable chance of successfully leading to the submission of the drug, if effective, or early stopping if this is not the case (Benda et al. 2010).

5.2.2 Statistical Perspective

Statisticians play a critical role in the optimization of clinical trials, ensuring that studies are designed and conducted efficiently to yield reliable and valid results. They are responsible for developing the statistical methodologies that underpin trial design, such as endpoints, sample sizes, randomization processes, and statistical analysis methods. Often utilizing theoretical derivations, modeling and simulations (M&Ss), and advanced computational techniques, they conduct quantitative evaluations of various design options and recommend the ones with high chance of success within the confine of recourses and timelines.

For illustration, many diseases can manifest through various clinical events, symptoms, and functional changes. For example, in oncology drug development, the clinical benefit of a drug can be assessed through clinical events such as disease progression and death, symptoms like pain, assessments of function (such as the ability to walk or exercise), or surrogate endpoints (reduction in tumor size) anticipated to predict a clinical benefit. In situations where efficacy cannot be adequately determined based on a singular disease aspect, studies often either employ an endpoint consolidating multiple disease facets into one or demonstrate effects across multiple endpoints. Alternatively, an effect observed on any of several endpoints might suffice to support the approval of a marketing application.

When designing such trials, various factors need to be carefully considered to find the optimal strategy among many competing options. Figure 5.1 presents an example of different testing strategies and associated impact of a oncology trial where five primary endpoints (PEs), objective response rate (ORR), duration of response (DoR), progression-free survival (PFS), overall survival (OS), and a patient reported outcome (PRO) were measured.

A key consideration is the adjustment of multiplicity testing. Failure to address multiplicity concerns can elevate the risk of arriving at erroneous conclusions regarding the drug's effects. Regulatory apprehensions regarding multiplicity chiefly arise during the assessment of clinical trials intended to validate effectiveness, supporting drug approval and claims within FDA-approved labeling. Nonetheless, this issue holds significance across trials throughout the drug development continuum. Given the sheer number of combinations of strategies, a more systematic approach is beneficial in determining the optimal approach for a particular trial.

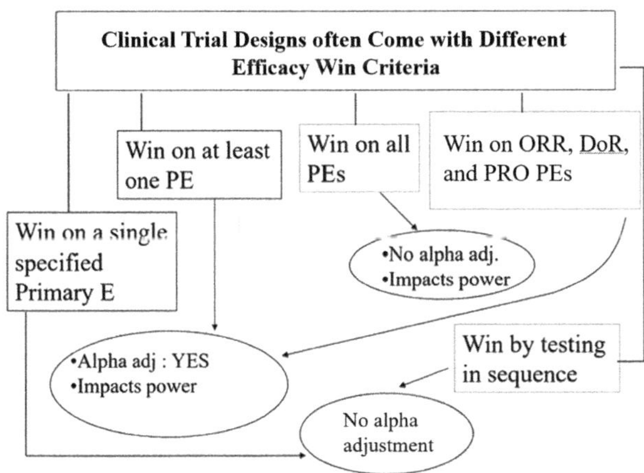

FIGURE 5.1
An example of testing strategies of an oncology clinical trial. (Adapted from Huque 2015.)

5.2.3 Clinical Scenario Evaluation

Traditionally, finding the optimal design strategy has relied on empirical methods, including reviews of competitive intelligence, regulatory trend, market needs, and expert opinions. However, advances in statistical and computational methods offer opportunities to employ more quantitative and rigorous approaches such as novel statistical design and analysis methods, M&S to inform decision-making. To bring all these aspects together, Benda et al. (2010) proposed a general Clinical Scenario Evaluation (CSE) framework that tailors the planning of clinical development through a systematic and hierarchical process. Specifically, CSE considers key clinical development objectives, alternate and competing strategies including design and analysis to address the objectives, underlying assumptions, and quantitative metrics to facilitate decision-making (refer to Figure 5.2). The optimal option is selected based on M&S.

Dmitrienko (2016) refined the key components of the CSE method, renaming Assumption as Data Model, Option as Analysis Model, and Metrics as Evaluation Model, respectively. The CSE was further expounded by Dmitrienko and Pulkstenis (2017) at the trial level, through practical use cases centered on the applications of the CSE for trial planning, optimization, and go/no go decision-making and provided various case examples to systematically guide these decisions in a reproducible manner, aiming to offer an evidence-based approach to trial and program design.

CSE aims to break down the complex process of clinical trials into manageable components: (1) data models, (2) analysis models, and (3) evaluation models. Data models pertain to the assumptions and processes involved in

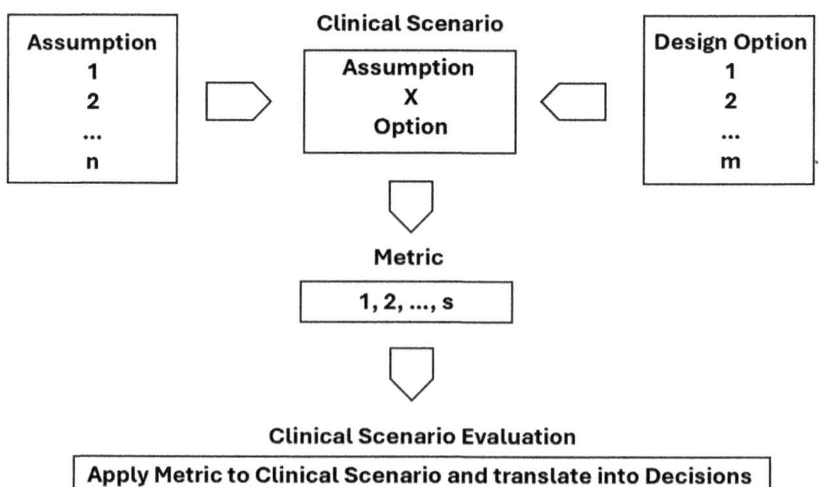

FIGURE 5.2
Key elements of clinical scenario evaluation. (Adapted from Benda et al. 2010.)

generating trial data. Analysis models specify the strategies used during data analysis, while evaluation models assess the performance of these strategies under predefined criteria. By combining data and analysis models, clinical scenarios are constructed for simulations to systematically evaluate the operating characteristics of proposed designs and methods, emphasizing realistic assumptions for the trial.

5.2.4 Optimization Strategy

In literature, various clinical trial optimization methods have been proposed. In broad strokes they fall into two categories: Frequentist method and Bayesian approach.

5.2.4.1 Frequentist Method

Leveraging the CSE framework, the former determines the optimal design by maximizing the pre-specified success criteria over combinations of design scenarios and analysis strategies (Dmitrienko and Pulkstenis 2017). Oftentimes, simulations are conducted to optimize the evaluation criteria.

Consider the following example (Figure 5.3), where a pivotal phase 3 trial is evaluated for an oncology drug with the intent for marketing approval. It is a parallel randomized trial using group sequential designs with PEs of ORR and PFS.

The parameters related to the data model and analysis model are listed in Table 5.1.

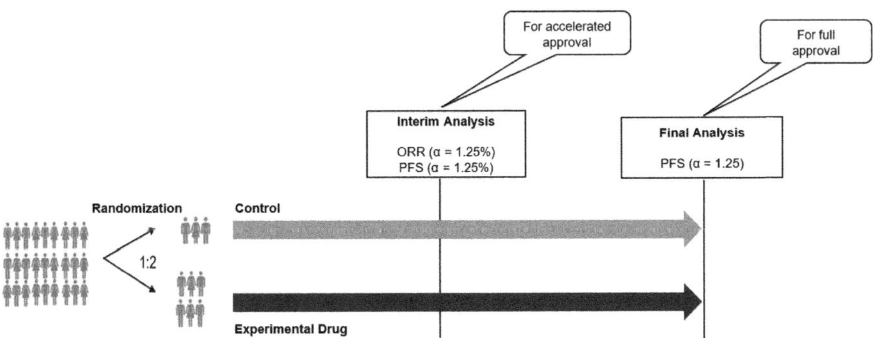

FIGURE 5.3
A randomized controlled design with two primary endpoints, ORR and PFS. One interim analysis is planned with a positive outcome of ORR for accelerated approval and positive results of both ORR and PFS for full approval. The total Type I error ($\alpha = 2.5$) is equally split for testing ORR at the interim analysis and PFS at the conclusion of the study. A fixed-sequence test is used for the interim analysis.

TABLE 5.1

List of Parameters Evaluated for Selecting Optimal Testing Procedure

Model	Parameter
Data	ORR rate
	median PFS
	Randomization ratio between treatment and control arms
	Number of interim analyses
	Decision boundary
Analysis	Analysis methods for ORR and PFS
	Multiplicity adjustment: alpha-cycling vs. alpha splitting

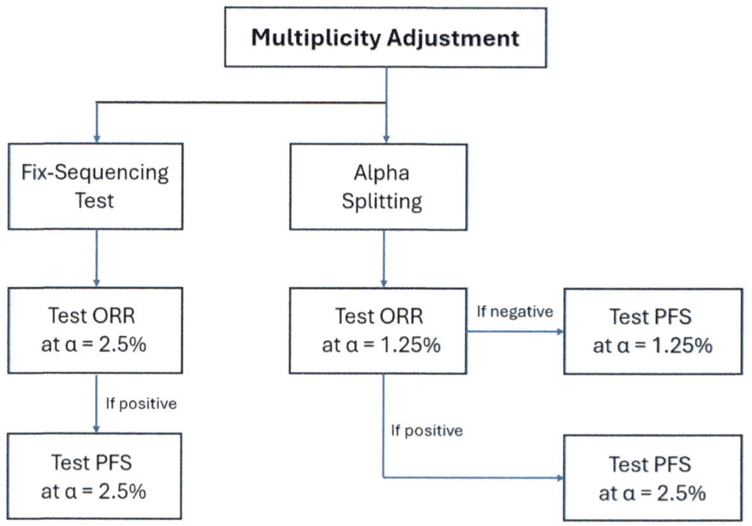

FIGURE 5.4
Two test procedures.

The objective is to find the parameter combination of the data and analysis model such that it renders approximately 90% power for achieving statistical significance for both ORR and PFS with a total sample size of 300. The alpha cycling or alpha splitting analysis method depicted below is one of the analysis parameters (Figure 5.4).

The above objective can be achieved through large-scale simulations. For illustration, Table 5.2 presents the result of simulation of the fixed-sequence testing scenario for the analysis method.

Dmitrienko and Pulkstenis (2017) discussed two commonly used optimization strategies aimed for selecting analysis models that meet the predefined success criteria.

TABLE 5.2

Fixed-Sequence Testing

		Treatment Effect									
	Endpoint	Exp. Drug	Control	Alpha	Accrual Time (Month)	Follow-up Time (Month)	Total Duration	No. of Events / Maturity	Sample size	Critical Value	Power
IA1	ORR	0.8	0.46	0.025	24	6	30	NA	153		>90%
	PFS (mo)	18	11.2	0.00041	24	6	30	86/28%	190	0.465	10%
IA2	PFS	18	11.2	0.007	42	0	42	149/49%	267	0.655	62%
FA	PFS (mo)		HR = 0.622	0.023	48	9	55	213/70%	306	0.748	89%

5.2.4.1.1 Direct Optimization

The first, called direct optimization, is well-suited for situations where a design is chosen to meet a single or multiple success criteria. For example, for oncology trials, it is always of interest to select a design that yields high probabilities of success for ORR and OS.

Now let λ and θ be the indexes of the analysis model and data model, respectively. They represent different analysis and data model options of interest. Further, let ψ_i be denote the evaluation criterion of the ith objective, $i = 1,\ldots,m$. Define

$$\psi_M(\lambda|\theta) = \min_{i=1,\ldots,m} \psi_i(\lambda|\theta)$$

and

$$\psi_A(\lambda|\theta) = \sum_{i=1}^{m} w_i \psi_i(\lambda|\theta),$$

where w_i's are positive values such that $\sum_{i=1}^{m} w_i = 1$.

The optimization can be carried out either based on maximizing either $\psi_M(\lambda|\theta)$ or $\psi_A(\lambda|\theta)$. The method can be readily extended to optimize the above criteria under the constraint:

$$\psi_c(\lambda|\theta) \in C,$$

where C is a set of acceptable values for the objective function $\psi_c(\lambda|\theta)$.

A simple example is the case where $m = 1$, and $\psi_M(\lambda|\theta) = \psi_1(\lambda|\theta)$ is power function while $\psi_c(\lambda|\theta)$ is Type I error, and $C = \{x : x \leq 0.05\}$.

5.2.4.1.2 Tradeoff-Based Optimization

The second strategy is referred to as tradeoff-based optimization (Dmitrienko and Pulkstenis 2017). It is often utilized to select the optimal designs for clinical trials with competing objectives. For example, while a high dose may increase the chance of demonstrating efficacy, it may also generate more safety concerns. For two sets of competing goals, let $\psi_1(\lambda|\theta)$ and $\psi_2(\lambda|\theta)$ denote the two corresponding success criteria. The optimal design is chosen such that it maximizes the following objective function

$$\psi_T(\lambda|\theta) = g\left(\psi_1(\lambda|\theta), \psi_2(\lambda|\theta)\right)$$

where $g(x, y)$ is a monotonically increasing function with respect to both x and y, with a special case being given by $g(x, y) = w_1 x + w_2 y$, $w_1 > 0$, $w_2 > 0$, $w_1 + w_2 = 1$.

5.2.4.1.3 Optimization Method

The success criteria discussed above are typically complex functions of the parameters of analysis model. Hence, oftentimes, closed-form solutions are unattainable. Instead, in most of the clinical applications, Monte Carlo simulations are carried out, and the optimal analysis parameters are determined through a search over a grid of all possible values of the parameters. Knowing that the parameters that optimize the objective function may not be easy to implement in practice, it is desirable to identify a "confidence region" for the optimal target parameters, comprising of values of the parameters that render the objective function sufficiently close to its optimal value. A design with parameters within this confidence region that is practically to implement will be chosen as the optimal solution. Assume that λ^* is an optimal value of the target parameter for the evaluation criterion $\psi(\lambda|\theta)$. That is,

$$\lambda^* = \arg\max_{\lambda \in \Delta} \psi(\lambda|\theta),$$

where Λ is a set of possible values of λ.

Dmitrienko and Pulkstenis (2017) suggested a method that defines a $100\eta\%$ optimal region $I_\eta(\theta)$ for λ^*such that

$$I_\eta(\theta) = \{\lambda \in \Lambda : \psi(\lambda|\theta) \geq \eta\psi(\lambda|\theta),$$

where η is a pre-selected level satisfying $0 < \eta < 1$.

5.2.4.1.4 Sensitivity Assessment

Clinical applications of the afore-mentioned analysis model optimization strategies need to consider uncertainty about the data model. For this purpose, sensitivity analysis is performed to evaluate the impact of deviation from the target data model on the optimal analysis model. Dmitrienko and Pulkstenis (2017) described two approaches, namely qualitative and quantitative analyses. Qualitative method assesses how well the optimal region $I_\eta(\theta)$ overlap with each other for a range of data model, indexed by $\theta_1, \ldots, \theta_m$ with θ_1 being the main target data model and the others the alternative data models. The closer the joint optimal regions $\bigcap_{i=1}^{m} I_n(\theta_i)$ to the target data model parameter space, the less sensitive the optimal analysis model is to the uncertainty around the data model.

In contrast, the quantitative approach evaluates the performance of an optimal analysis model over a range of data models randomly generated from the main data model. This method relies on an underlying assumption that the individual parameters of the main data model follow certain distributions. The bootstrapping sampling technique is used to generate a random sample $\theta'_1, \dots, \theta'_k$ from the joint distribution. Conditional on this sample, a set of optimal criterion values can be obtained as follows:

$$\psi_i = \psi(\lambda^* \mid \theta'_i), \quad i = 1, \dots, k$$

from which an empirical distribution can be constructed. By repeating the bootstrap sampling for a number of time, characteristics of the empirical distribution such as mean, mode, standard deviation can be estimated to describe the impact of uncertainty of data model parameters on the optimal analysis model. Furthermore, other performance metrics can also be assessed. For example, the probability of performance loss owing to data model uncertainty can be estimated by (Dmitrienko and Pulkstenis 2017):

$$\frac{\text{No. of } \left\{ i \in \{1, \dots, k\} : \psi_i = \psi(\lambda^* \mid \theta'_i) \le (1-r)\, \psi(\lambda^{\wedge *} \mid \theta) \right\}}{k},$$

where r is a positive number between 0 and 1.

5.2.4.2 Decision-Theoretic Approach

In clinical trials, many decisions impact the attainment of optimal data and analysis models. For example, the selection of sample size, trial adaptation, and study power significantly influence trial outcomes. Decision theory provides a statistical framework to formalize decision-making under uncertainty. This methodology facilitates the selection of an optimal decision among various possible actions by evaluating the consequences of each action across all potential scenarios and selecting the design that optimizes the expected utility function. To implement the method, it is critical to define a utility function that assigns specific values to the consequences of all possible actions and to take actions that maximize this utility. The methods can generally be formulated as follows:

Let $Y = (y_1, \dots, y_n)$ be the outcomes of a trial that follow a distribution of $F_\theta(y)$ with unknow parameter θ. The parameter θ represents the true value of Y. Let $A(Y)$ be the action based on the outcomes Y and a specific analysis model, and $U(A(Y), \theta)$ be the utility function. Since both Y and θ are unknown at the planning stage of the trial, one may estimate the expected utility by integrating over the prior distribution of θ, $\pi(\theta)$, and the distribution of Y, $F_\theta(y)$,

$$U_M = \iint U(A(y), \theta)\, \pi(\theta)\, dF_\theta(y)\, d\theta.$$

The optimal analysis model is the one that maximizes the above expected utility. Various methods have been proposed in literature. Hee et al. (2016) provided a comprehensive review of these methods. The decision-theoretic approaches are applicable to optimizing both single-stage, multi-stage, multi-arm, and enrichment designs.

For single-stage designs, the methods primarily concern the optimal selection of sample size given the cost associated with patient recruitment, treatment, and monitoring. For example, Krisam and Kieser (2014) explored a decision-theoretic approach for single-stage designs, minimizing a quadratic loss function. They derived optimal decision functions for selecting either a pre-specified subpopulation or the overall population and investigated the impact of errors in determining a patient's biomarker status.

In the multi-stage scenario, the decision-theoretic framework can be employed to optimally select adaptation rules at interim analyses (Ondra et al. 2016; Hee et al. 2016). Beckman et al. (2011) developed a Bayesian decision-analytic approach to determine whether a subsequent Phase III trial should be enriched, stratified within the entire population, adaptive, or not conducted at all based on available Phase II data. Similarly, Götte et al. (2014) derived optimal rules for selecting a patient population at interim analyses or futility assessments. They aimed to maximize a utility function representing the expected probability of a correct selection, proposing adaptation rules based on estimated effect sizes in the subpopulation and its complement. Graf et al. (2015) utilized a decision-theoretic approach to assess fixed-sample and adaptive population enrichment designs controlling the family-wise error rate (FWER). They defined utility functions representing various objectives and optimized adaptation strategies and sample sizes across stages, identifying scenarios where single-stage enrichment designs or trials in the entire population are preferable to adaptive enrichment designs.

These methodologies extend to support decision-making when a series of studies are under consideration. For a comprehensive review of the applications of decision-theoretic approaches to various clinical trial settings, see Hee et al. (2016).

5.3 AI in Clinical Trial Optimization

In recent years, the advances of AI and digital technology have spurred a boom of healthcare date ecosystem and made electronic health records (EHRs) and claims data broadly available in certain geographic region including the United States. AI-powered tools, such as large language models, can systematically extract and curate unstructured data from these systems into high-quality machine-readable structured data. The regulatory policies toward the use of AI and RWD in clinical development have evolved significantly. It is possible to use evidence generated from sources outside the traditional RCTs for regulatory filings (refer to Chapter 8). The confluence of AI and novel data sources present multiple opportunities for clinical trial optimizations. As applications of AI and RWD in enhancing the efficiency of clinical operations are thoroughly discussed in Chapter 4, in the following sections, we focus on using AI and RWD to inform clinical trial design.

5.3.1 Cohort Optimization

5.3.1.1 Challenges in Defining Eligibility Criteria

Cohort, also known as study population, of a clinical trial is defined through a set of eligibility criteria. This is a challenging task as overly restrictive eligibility criteria could lead to slow enrollment and compromise the generalizability of study findings (He et al. 2020). On the other hand, too broad eligibility criteria could increase the heterogeneity of the study population, diluting treatment effect and making it harder to achieve study objectives. It may even increase safety risk for certain subgroups of the population. At the same time, regulatory authorities have promoted enrollment practices that would lead to clinical trials that better reflect the population most likely to use the drug upon marketing approval (FDA 2020d).

Traditionally, the selection of these criteria primarily relied on expert opinions. In the absence of deep understanding of the target population, this practice often led to many protocol amendments to alleviate the situations of slow enrollment. In addition, some of the eligibility criteria such as life expectancy of no less than 16 weeks are very subjective, resulting in unexpected number of early dropouts before the data for key decision-making such as dose-escalation were collected (Yu et al. 2020).

5.3.1.2 AI-Based Methods

RWD sources, including EHRs, patient registries, and wearable devices, provide valuable insights into real-world patient characteristics, treatment patterns, and outcomes. Leveraging RWD, trial sponsors can optimize inclusion and exclusion criteria to enhance patient recruitment, minimize protocol deviations, and improve the generalizability of trial results. In recent years, various AI solutions have been suggested by researchers to optimize eligibility criteria, reducing guesswork and manual labor. For example, at Stanford University, a team led by James Zou, a biomedical data scientist, pioneered Trial Pathfinder, a system designed to evaluate the impact of adjusting trial eligibility criteria on study efficiency and effectiveness (Liu et al. 2021). In a notable study, they applied Trial Pathfinder to drug trials targeting a specific type of lung cancer. The findings revealed that aligning criteria adjustments with Trial Pathfinder's recommendations could potentially double the pool of eligible patients without elevating the hazard ratio of OS. Moreover, this system demonstrated efficacy across various cancer types, mitigating adverse outcomes by extending treatment eligibility to individuals with more severe conditions, who stand to benefit the most from the drugs. Notably, several pharmaceutical firms, including Roche, Genentech, and AstraZeneca, have embraced Trial Pathfinder.

Recent advancements, such as AutoTrial developed by Sun's lab in Illinois, introduced a method for training large language models. This enables researchers to input trial descriptions, prompting the model to generate appropriate criterion ranges, such as body mass index thresholds, facilitating further automation and efficiency in trial design (Hutson 2024). Similarly, Kim et al. (2020) explored how eligibility criteria influence recruitment and clinical outcomes in COVID-19 clinical trials by analyzing EHR data. Their findings indicated that modifying the thresholds of typical eligibility criteria in COVID-19 trials could increase the occurrence of outcome events while requiring fewer participants.

There are other efforts centered around exploring eligibility criteria that require alternate and less invasive procedures to reduce patient and clinical staff burden, thus enhancing site and patient participation in the trial (Yang and Deepak 2023; Weissler et al. 2021).

5.3.1.3 An Example

5.3.1.3.1 Background

COPD is a highly heterogenous disease with many different phenotypes. An anti-IL5 receptor monoclonal antibody was developed from treating eosinophilic asthmatics (Gossag et al. 2015). However, at the outset of the clinical program, only accurate and reliable methods to identify eosinophilic asthmatics had been limited to procurement of induced sputum samples from patients (Molfino et al. 2012). Two percent of eosinophils in sputum as the cut-off for classifying patients as sputum eosinophilic asthmatics (Szefler et al. 2012). However, the sputum induction procedure is a tedious and complex process that requires skilled technicians and equipment that are not readily available in clinical practice. Using eosinophil count in sputum as part of the trial eligibility criteria would significantly increase the burden of both the site staff and patients.

5.3.1.3.2 Predictive Algorithm

Khatry et al. (2015) developed a prediction algorithm for diagnosing sputum-eosinophilic asthma patients. The algorithm uses absolute counts of peripheral blood eosinophils, neutrophils, and lymphocytes in routinely conducted complete blood count with differentials to classify patients with asthma as "non-eosinophilic" (<2% predicted sputum eosinophil counts) or "eosinophilic" (≥2% predicted sputum eosinophil counts) to tailor treatment regimens. A linear discriminant analysis (LDA) multivariate model was used to develop the ELEN Index. The method used two mathematically weighted ratios of three blood cell populations, Eosinophils/Lymphocytes and Eosinophils/Neutrophils (ELEN), as predictor variables in two equations shown in the following:

Score for Sputum EOS % < 2.0:

$$= a + \left[b \times \frac{\text{Blood EOS}}{\text{Blood Lymphocyte}} \right] - \left[c \times \ln\left(\text{Blood EOS / Blood Neutrophil}\right) \right]$$

Score for Sputum EOS % ≥ 2.0:

$$= d + \left[e \times \frac{\text{Blood EOS}}{\text{Blood Lymphocyte}} \right] - \left[f \times \ln\left(\text{Blood EOS / Blood Neutrophil}\right) \right]$$

Coefficients of the model parameters were estimated with training data from a clinical trial (Table 5.3), and 95% confidence intervals (CIs) of model coefficients were estimated by bootstrap re-sampling ($n = 10{,}000$).

The ELEN index calculates probability-based discrimination scores of binary group association and uses a decision rule to assign each individual case to either the sputum-eosinophilic group or to the sputum non-eosinophilic group.

5.3.1.3.3 Validation

Both internal and external statistical validation of the prediction model was carried out to test generalizability of the algorithm with two independent asthma datasets, cohort 1 and cohort 2 and pooled data of cohorts 1 and 2. The results are summarized in Table 5.4.

TABLE 5.3

ELEN Index Model Coefficients and 95% CIs

Coefficient	Current Model	Mean	Median	95% CI (Lower)	95% CI (Upper)
a	−9.5243	−23.5236	−11.8804	−74.4666	−6.6279
b	70.0975	135.0464	103.2067	45.2753	412.8505
c	3.779	−11.3741	−4.3005	−38.5399	−2.2609
d	−14.5853	−30.2162	−19.9893	−95.2441	−10.2884
e	101.2198	176.1841	65.2795	65.2729	247.1979
f	3.9567	−11.6615	−39.5223	−39.5223	−2.3559

Source: Yang and Deepak (2023).

TABLE 5.4

Prediction and Validation Accuracy in Categorizing Asthma Patients into Eosinophilic and Non-Eosinophilic Phenotypes

	Prediction (N = 23)	**Validation 1 (*N* = 23)**	**Validation 2 (*N* = 99)**	**Validation 3 (*N* = 75)**	**Validation 4 (*N* = 174)**
Performance Characteristics	**Pre-AC Data from CP 138**	**Jacknife (Leave-One-Out)**	**Cohort 1**	**Cohort 2**	**Cohort 1 & 2**
Specificity (%)	93	93	79	84.8	83
Sensitivity (%)	63	63	74	64.3	70.5
Overall Accuracy (%)	83	83	74.7	73.3	74
NPV[a] (%)	82.4	82.4	42	65	54.4
PPV[b] (%)	83.3	83.3	94	84.4	90.5
Prevalence[c] (%)	35	35	81	56	70

Source: Yang and Deepak (2023).

Notes:

[a] NPV: Negative predictive value·

[b] PPV: Positive predictive value;

[c] Prevalence: Proportion of EOS% ≥ 2.0.

5.3.2 Personalized Trial Planning

5.3.2.1 AI Methods in Clinical Scenario Evaluation

M&S play a significant role in clinical trial optimization, particularly during the planning stage where limited data regarding the asset under evaluation is available. Despite its significance, most of the current M&S approaches used for trial planning rely on findings at a population level such as mean response rate and median survival from historical clinical studies as parameters design parameters; they overlook crucial factors that are beyond the disease or treatment such as patient heterogeneity but pivotal in shaping patient prognosis. As a result, the simulated population is often not representative of the target population. For patients with heterogeneous conditions, especially chronic ailments like asthma, a multifaceted approach integrating diverse data is needed to predict future outcomes, in response to therapeutic intervention. AI-driven strategies offer unique opportunities to optimize trial designs by integrating data from diverse sources including patient characteristics, molecular features, disease phenotype, and other relevant factors in M&S for CSE, thus gaining better understanding of the performance of the chosen study design and yielding greater fidelity in decision-making (Barret et al. 2023).

5.3.2.2 Digital Twins

Patients with the same disease can exhibit varying symptoms, progressions, and responses to therapies. Computational models, particularly those developed using machine learning (ML) techniques, offer promising avenues to address patient heterogeneity when conducting clinical trial scenario evaluations. Digital twins are virtual models that replicate real-world entities and processes, and they have been increasingly applied to clinical trial optimization. These sophisticated methods simulate individual patient data to mirror patient characteristics, disease progression, and treatment responses for different design options, helping researchers identify the most effective and efficient protocols. By modeling various scenarios, they can predict outcomes, optimize inclusion and exclusion criteria, and reduce the number of necessary trial participants. These methods can also be utilized to identify potential risks and adverse events before they occur in real patients and explore personalized therapies. This proactive approach can enhance patient outcomes and improve the overall success rate of clinical trials.

Many diseases such as Alzheimer's Disease (AD), Mild Cognitive Impairment (MCI), and Chronic Obstructive Pulmonary Disease (COPD) exhibit diverse patterns of progression and therapeutic responses. The heterogeneity of these diseases makes diagnosis, management, and treatment challenging, necessitating improved M&S methods for assessing clinical design options. Various disease progression models have been developed for simulating AD, MCI, COPD, and other diseases. Fisher et al. (2019) used an unsupervised ML model called a Conditional Restricted Boltzmann Machine (CRBM) to simulate detailed patient trajectories that closely assemble to the actual data. Similarly, Walsh et al. (2020) reported that the digital twins generated by a CRBM model of Multiple Sclerosis (MS) were also statistically indistinguishable from their actual subject counterparts along a number of measures. Gold et al. (2014) trained hierarchical longitudinal model for simulating COPD patient experiences to informed trial design and planning with biomarkers.

These digital twin approaches hold tremendous promise for addressing patient heterogeneity and advancing precision medicine. By developing models capable of simulating detailed patient trajectories, research can gain deeper insights into disease progression and tailor treatments more effectively to individual patients' needs as clinal trials are being planned.

5.3.3 Precision Medicine

5.3.3.1 Background

Precision medicine, also known as personalized medicine, is a medical approach that tailors treatment to individual patient characteristics, such as genetic makeup, lifestyle, and environmental factors. According to the

National Research Council, precision medicine aims to "classify individuals into subpopulations that differ in their susceptibility to a particular disease, in the biology or prognosis of those diseases they may develop, or in their response to a specific treatment" (National Research Council 2011). The need for precision medicine arises from the limitations of the traditional one-size-fits-all approach, which often leads to suboptimal outcomes due to the variability in patient responses to treatments. By considering individual differences, precision medicine can improve diagnostic accuracy, treatment efficacy, and overall patient outcomes. It is for this reason, whether to enrich a study or not, based on precision medicine principles, is a key decision to make during clinical trial planning.

5.3.3.2 AI-Driven Precision Medicine

Numerous factors play a pivotal role in shaping a patient's clinical outcome, encompassing intrinsic characteristics of the patient, the disease or medical condition itself, and the impact of any administered treatments. Within this framework, certain intrinsic characteristics manifest as prognostic biomarkers, indicating the likelihood of a clinical event, disease recurrence, or progression in patients with the specific disease or medical condition of interest. Others manifest as predictive biomarkers, signaling individuals who are predisposed to experiencing either beneficial or adverse effects from exposure to a medical product or environmental agent compared to similar individuals lacking the biomarker.

Both prognostic markers can be used to increase the probability of success of clinical trials. For example, in randomized trials, using prognostic markers as stratification factors can help reduce imbalance among treatment arms, thus enhancing the detection of drug effect. On the other hand, predictive markers can be used to identify "responders" of a medical intervention, thus increasing treatment effect size and clinical trial outcomes.

Current prognostic models such as the Acute Physiology and Chronic Health Evaluation (APACHE) are mostly restricted to only a handful of variables (Obermeyer and Ezekiel 2016). The predictive ability of these tools can be limited as they often fail to account for the nonlinearity and interactions among the input predictors (Yang 2024). In contrast, AI models trained on data from disparate sources can identify combinations of variables most likely to influence predictive scores, capture response heterogeneity, and improve the probability of success of a trial. These predictive tools can be used to guide clinical decision-making (Topol 2019a). In addition, the systematic analysis of all available data sources and the predicted responses of various subgroups can help companies fine-tune their portfolio strategies, in using the right drug for the right patient (Anagnostopoulos).

In recent years, there has been a growing reliance on RWD to identify prognostic indicators or baseline characteristics for prognostic enrichment.

Furthermore, predictive markers derived from real-world sources are being employed to facilitate enriched trial designs.

5.3.3.3 Strategy for Biomarker Discovery

In literature, the strategy for biomarker discovery has been extensively explored (Sechidis et al. 2018; Huang and Chiu 2023). In general, biomarker discovery is a data-driven process, augmented with analytical methods. Figure 5.5 presents a typical discovery process.

As the first step, univariate analysis is carried out to assess associations between markers and clinical activities. Since complex diseases often exhibit nonlinear relationships between biomarkers and outcomes, multivariate methods such as multivariate regression models and ensemble ML algorithms offer an advantage in identifying prognostic biomarkers by capturing these nonlinearities and interactions between biomarkers. ML models are well-suited for these purposes as they do not require pre-specification of the models describing the data. ML model development typically involves three critical steps: model training, tuning, and validation, using independent datasets (Yang 2024).

Another class of methods for predictive marker discovery is based on subgroup identification (Lipkovich et al. 2017). These methods recursively partition data space using interaction tests between treatment and covariates. Additionally, to ensure the reliability of the biomarkers, the results need to be validated through independent data.

In practice, multiple markers are often combined into a composite score, either through a parametric model as described previously or an ML algorithm to achieve better results. When the clinical outcome is binary, the composite score is compared to a cutoff value to determine the predicted outcome. Various methods have been suggested to choose the optimal cutoff value. They include maximizing the Youden index (sensitivity + specificity − 1) (Youden 1950), minimizing the Euclidean distance between ROC curve and the theoretical optimum outcomes with both sensitivity and specificity being equal to 1, maximizing the product of sensitivity and specificity, and minimizing a loss function (Cantor et al. 1999; McNeil et al. 1975; Metz 1978; Zweig and Cambell 1993).

5.3.4 Endpoint Optimization

In many clinical trials, particularly those for rare diseases, there is often no single, established endpoint. In other cases, it can be challenging to determine which endpoint would best measure the drug's activity. Some endpoints, such as survival, may take a long time to collect, leading to extended development timelines. Others may reliably detect progression in certain types of patients but not others or may be particularly invasive, increasing the

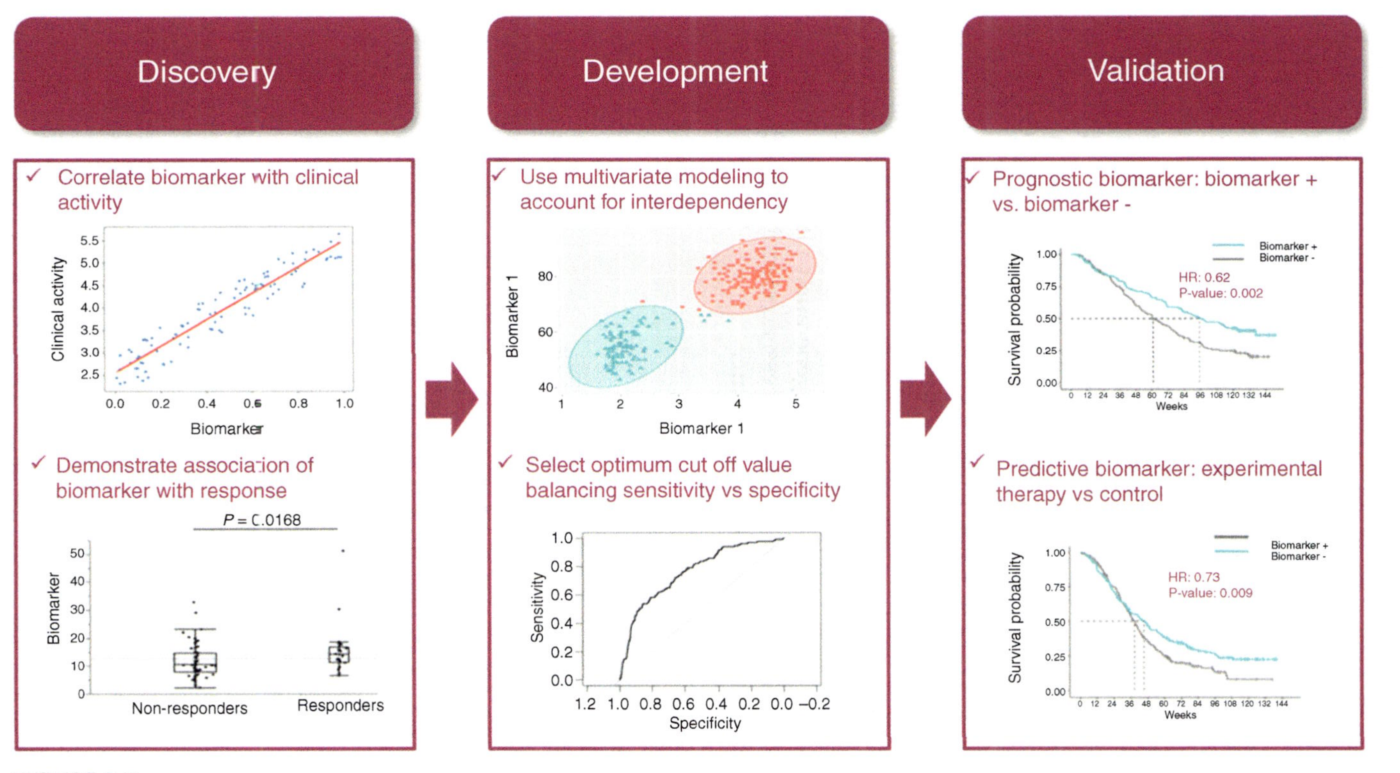

FIGURE 5.5
Process for biomarker discovery.

trial's burden on patients (Anagnostopoulos et al. 2023). AI and ML can analyze large, complex datasets to identify new, more sensitive biomarkers and surrogate endpoints that provide earlier and more accurate indications of treatment efficacy and safety. For instance, these technologies can integrate and analyze multi-omics data (genomics, proteomics, metabolomics), imaging data, and real-world evidence from EHRs to uncover subtle patterns and correlations that might be missed by conventional analysis (Obermeyer and Emanuel 2016). AI and ML also have the potential to generate novel clinical trial endpoints that can be used to improve the speed and quality of clinical evaluations (Lee and Lee 2020). For example, Kihara et al. (2019) and Lee et al. (2018) used AI models to predict central macular thickness results, conventionally measured by optical coherence tomography (OCT), which is known to be time-consuming, and less sensitive and specific. Moreover, AI-driven endpoints can enhance the granularity of patient monitoring, leading to more personalized and adaptive trial designs. This not only increases the speed of evaluation by quickly identifying promising therapies but also improves the quality and quantity of endpoints, ensuring more robust and comprehensive assessments of clinical outcomes (Topol 2019b). The implementation of AI and ML in endpoint generation ultimately fosters a more efficient and effective clinical trial process, accelerating the development of new therapies and improving patient care.

5.4 Case Examples

This section presents several case examples, showcasing applications of AI and RWD for clinical trial optimization.

5.4.1 Patient Level in Silico Simulation to Assist in COPD Trial Planning

5.4.1.1 Background

COPD is a complex and heterogeneous condition characterized by a spectrum of clinical phenotypes, ranging from chronic bronchitis to emphysema (Barnes and Celli 2009; Allen-Ramey et al. 2012). The global prevalence of COPD is on the rise, with projections indicating a significant impact on healthcare costs worldwide in the coming decade (Toy et al. 2010; Mannino and Buist 2007; Lopez et al. 2006; Richard et al. 2011). Guidelines provided by the Global Initiative for Chronic Obstructive Lung Disease (GOLD) outline diagnostic procedures and criteria for assessing disease progression in COPD (Vestbo et al. 2012). While COPD shares symptoms with asthma, it is typically diagnosed later in life and often occurs in individuals with a history of smoking.

Exacerbations, marked by episodes of reduced lung function and a decline in quality of life, are common among COPD patients and are frequently triggered by bacterial and viral infections (Vestbo et al. 2012; Gómez and Rodriguez-Roisin 2002). The frequency of exacerbations has been linked to accelerated lung function decline and the presence of comorbidities (Barnes and Celli 2009). Unfortunately, patients experiencing frequent and severe exacerbations requiring hospitalization have limited treatment options available to slow disease progression. Notably, the treatment of exacerbations accounts for a significant portion of COPD-related healthcare costs per patient per year (Toy et al. 2010). Advancements in identifying distinct subgroups within the COPD population hold promise for the development of targeted therapies (Silverman et al. 2011).

Traditionally, the inflammatory processes in airway diseases have been categorized into four groups: neutrophilic, eosinophilic, neutrophilic and eosinophilic, and paucigranular, providing insights into the underlying biology of asthma (Simpson et al. 2006). While COPD has historically been associated with a predominantly neutrophilic inflammatory response, recent studies suggest that a subset of individuals may exhibit eosinophilic airway inflammation during stable periods (Saha and Brightling 2006; Siva et al. 2007) and exacerbations (Bafadhel et al. 2012a, 2012b). Estimates suggest that between 17% and 28% of exacerbations exhibit an eosinophilic inflammatory profile (Bafadhel et al. 2011).

Assessment of the cellular inflammatory phenotype in COPD has shown promising clinical implications. Patients with elevated sputum eosinophil percentages have demonstrated a more favorable response to corticosteroid therapy, both inhaled and oral formulations (Siva et al. 2007; Brightling et al. 2000; Leigh et al. 2006). Similar findings have been observed in individuals with COPD/Asthma overlap syndrome (Kitaguchi et al. 2012). However, current methods for determining airway inflammatory phenotypes, such as invasive procedures or impractical techniques, pose challenges in routine clinical practice. Recent studies have suggested the potential utility of blood eosinophil percentage or exhaled nitric oxide levels as biomarkers to predict clinical responses in COPD (Bafadhel et al. 2012a, 2012b; Soter et al. 2013), hinting at the emergence of biomarkers for eosinophilic airway inflammation in COPD.

Integrating biomarkers into COPD drug development offers great potential but also presents challenges and risks. Identifying the target patient population most likely to benefit from anti-inflammatory targeted treatments is complex due to the diverse cellular inflammatory phenotypes in COPD, potential overlap between phenotypes, and the intricate mechanisms of targeted drugs. While some therapies targeting specific inflammatory phenotypes have shown promise, challenges remain in accurately categorizing phenotypes over time and determining biomarker responsiveness and accuracy.

Clinical trial simulation offers a valuable approach to understanding the impact of early planning decisions on late-stage opportunities and success in COPD drug development. By exploring various scenarios, these simulations can assess the potential advantages of incorporating biomarkers. Key parameters include COPD clinical phenotypes and their prevalence, biomarker performance in predicting relevant phenotypes, and differential treatment responses based on phenotype or surrogate markers.

Advanced computational tools are essential for analyzing the consequences of decision criteria throughout different stages of development, considering the uncertainties associated with these parameters (Holford et al. 2010; Girard 2005). While simulations complement scientific studies, they do not replace independent replication. Therefore, parallel efforts should be monitored closely during clinical development. Simulation plans can adapt to evolving beliefs and evidence, providing a framework for efficient exploration of adjustments.

Gold et al. (2014) introduced the Clinical Trial Object Oriented Research Application (CTOORA), a computer-aided clinical trial simulator designed to inform COPD trial planning incorporating biomarkers. CTOORA generates serial projections of trial outcomes under various hypothetical scenarios, aiding in identifying key characteristics of biomarker-based diagnostics necessary to confer a meaningful advantage in clinical trials. CTOORA serves as a robust tool for guiding clinical trial planning with biomarkers, facilitating successful development from early to late stages.

5.4.1.2 Method

The CTOORA was developed to simulate COPD patient outcomes and assess the potential benefits of incorporating biomarkers in targeted drug development. CTOORA utilizes a hierarchical longitudinal model comprised of a series of modules designed to simulate various aspects of COPD patient experiences throughout enrollment and follow-up in a clinical trial as depicted in Figure 5.6.

CTOORA integrates three main components: (1) Clinical input features, such as rates of disease progression, variability in pulmonary function, and exacerbation history; (2) Biological characteristics, including the prevalence of clinical phenotypes, sensitivity, and specificity of biomarker-based diagnostics for predicting phenotypes; and (3) Differential clinical responses to targeted treatments within each phenotype.

Simulated COPD patients with diverse clinical backgrounds and biological profiles are enrolled in a simulated trial using a mock protocol and randomized to different treatment arms (Eddy and Schlessinger 2003; Weinstein et al. 2003). During the simulated trial, these COPD patients encounter exacerbations, adverse events, and fluctuations in pulmonary function, influenced by their phenotype and assigned treatment arm (Hurst

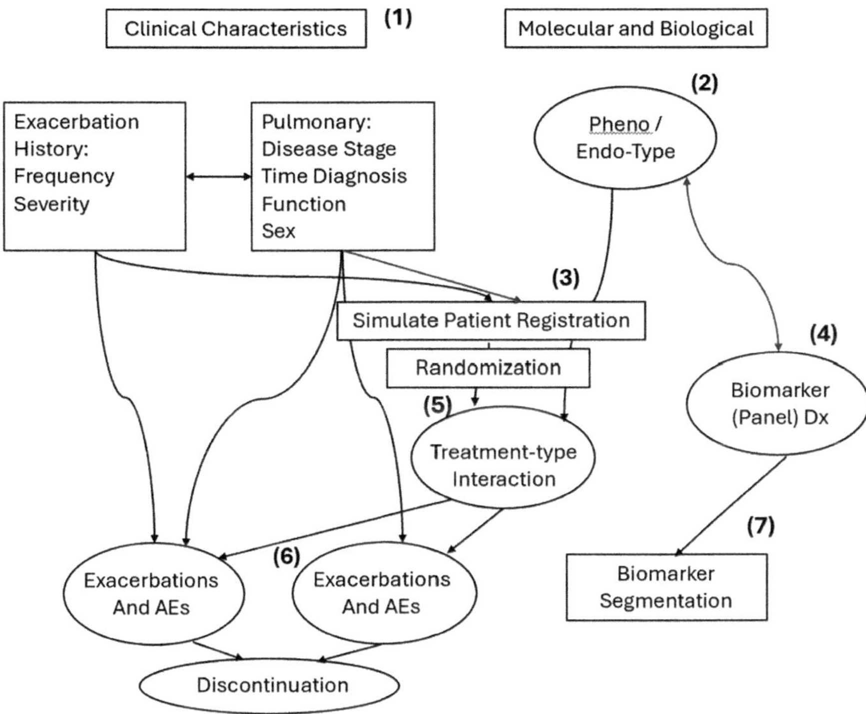

FIGURE 5.6
Data flow of CTOORA. (Adapted from Gold et al. 2014.)

et al. 2010; Magda 2010). Discontinuations from the trial are modeled as a stochastic process, taking into account disease progression and adverse events (Westfall et al. 2008).

Clinical and biomarker endpoints are assessed at interim time points and at the conclusion of the simulation trial. This process is repeated across multiple replicate trials to assess trial performance in achieving endpoints, estimate population parameters, or gain insights into the COPD population with a desired level of accuracy.

5.4.1.3 Simulation Plan

We conducted a series of early-to-late phase COPD clinical trials comparing a targeted treatment against standard of care (SoC) in a moderate to severe COPD patient population to ascertain the utility of incorporating a biomarker. The PE focused on exacerbations, aiming to assess reduction in the rate of exacerbations per year (RRE) from a baseline mean of 1.5 exacerbations per year (Vogelmeier et al. 2010; Calverley et al. 2009). Through simulation, Gold et al. (2014) explored various factors including treatment response by

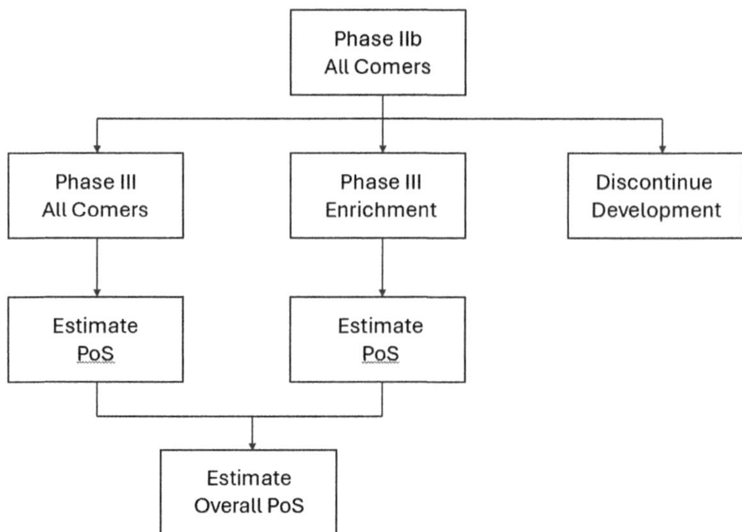

FIGURE 5.7
Simulation plan. (Adapted from Gold et al. 2014.) PoS = Probability of success.

phenotype, early phase study size, biomarker-based diagnostic (Dx) test sensitivity, and the biomarker's ability to predict phenotype. These factors influenced the likelihood of adopting a biomarker in Phase III and subsequently impacted the overall probability of late-stage success.

The simulation sequence commenced with a Phase IIb two-arm trial involving all comers, followed by decisions regarding progression to Phase III with or without a biomarker, or discontinuation as depicted in Figure 5.8. We assumed at Phase IIb initiation that dose selection was finalized, a known phenotype had been established for drug development, and a biomarker predictive of phenotype was available (though its clinical validity remained untested). After each Phase IIb trial iteration, RRE was evaluated in the full cohort and within biomarker-based Dx test positive and negative segments as exploratory endpoints. Progression to Phase III without a biomarker occurred if a significant RRE was observed across the entire COPD cohort, or enrichment for Dx test positive subjects was pursued if the significant RRE was confined to the Dx test positive patients. Decision probabilities were estimated across 5,000 simulated Phase IIb trial iterations (Figure 5.7).

5.4.1.4 Probability of Success

The probability of Phase IIb success was defined as the probability of progression to Phase III. Phase III success entailed observing significant reductions in exacerbation rates (RREs) in two independent trials, encompassing both all comers and enrichment designs. The overall probability of Phase III success was determined as the weighted average probability of success across all

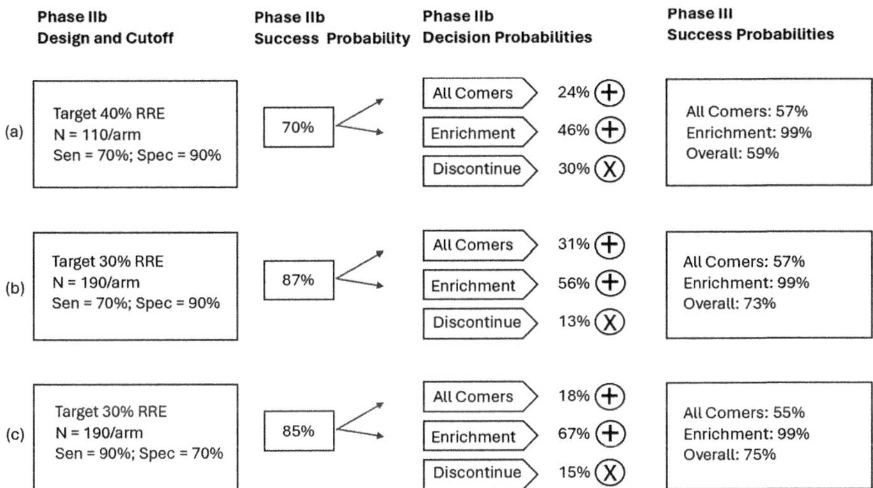

FIGURE 5.8
Selected Simulation Results, Phase IIb Decision and Phase III Success Probabilities (a) Phase IIb simulated with N = 110 subjects per arm and biomarker Dx test sensitivity of 70%. There was a 46% probability of enriching with a biomarker in Phase III, and a 59% overall probability of Phase III success. (b) Increasing Phase IIb study size to N = 190 subjects per arm increased probability of enriching with biomarker in Phase III to 56%, and overall probability of Phase III success to 73%. (c) As in (b) although with increased Dx test sensitivity of 90%. Probability of enriching with biomarker in Phase III increased to 67%, while the overall probability of Phase III success of 75% was negligible from (b) at 73%. Phase II success probability = Phase IIb decision probability to proceed to Phase III. (Adapted from Gold et al. 2014.)

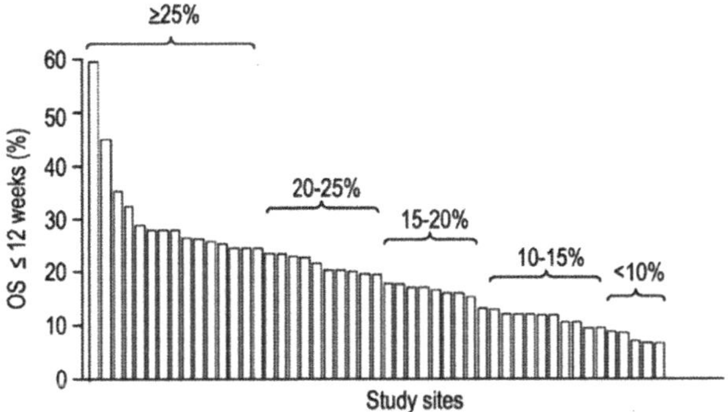

FIGURE 5.9
The plot shows 45% of sites from a large IO trial have more than 20% patients who dropped out of study within 12 weeks. (Adapted from Yu et al. 2020.)

comers and enrichment trials, factoring in the decision probabilities from the end of Phase IIb, whether to proceed with either design in Phase III (Figure 5.9). Throughout the simulations, Gold et al. (2014) deemed the biomarker advantageous if its adoption and enrichment in Phase III resulted in an overall probability of late-stage success equal to or exceeding 70%.

5.4.1.5 Results

5.4.1.5.1 Simulating Treatment Response

Initially, in the base case scenarios, the RRE was fixed at 25% for the targeted drug in the simulated Phase IIb all-comers population. It was assumed that clinical efficacy was never worse than SoC despite phenotype differences, so a simulated patient either responded more favorably to the targeted treatment than SoC or not, but symptoms were not aggravated by the targeted treatment. Fixing the RRE at 25% meant that the RRE among the Dx test positive subjects varied, ranging from 30% to 55%, depending on phenotype prevalences and Dx test sensitivity, as listed in Table 5.8.

The option to enrich with a biomarker-based Dx in Phase III increased the overall probability of late-stage success, depending on several simulated factors, particularly the Phase IIb study size. For instance, increasing the Phase IIb study size from $N = 110$ to $N = 190$ subjects per arm increased the chances of enrichment with the biomarker-based Dx in Phase III and markedly increased the overall probability of late-stage success from 55%–60% to 72%–75%, as shown in Figure 5.8a–c and Table 5.5.

The simulations were repeated, fixing the RRE for the targeted drug in the Phase IIb all-comers population at 20%, as shown in Table 5.6. The overall probabilities for late-stage success were below 50% for most scenarios, ranging from 32% to 57% with $N \leq 190$ subjects per arm. A study size of $N = 190$ subjects per arm was insufficient to raise the chances of enrichment in Phase III above 57%.

CTS showed that the biomarker was helpful when the RRE in the simulated COPD population was moderate, a gray zone for decision-making. In the simulations, the RRE in the simulated COPD Phase IIb all-comers population was fixed to the targeted treatment. Had the RRE been fixed in the Dx test positive and negative segments, the phenotype prevalence would have appeared to play a larger role in overall success, although the factors leading to the probability of overall success would have been less clear.

5.4.1.5.2 Early Phase Study Size

Increasing the Phase IIb study size proved beneficial for both biomarker adoption and late-stage success. When the study size was set at $N = 110$ subjects per arm, Phase IIb stopping rates were relatively high: ≥28% for a RRE of 25% and ≥43% for a RRE of 20%, as detailed in Tables 5.5 and 5.6. Expanding the study size to $N = 140$ subjects per arm moderately boosted the likelihood of adopting a biomarker and enriching in Phase III from 44%–51% to 49%–58%, as shown in Table 5.5. Further increasing the study size to $N = 190$ subjects per arm significantly enhanced the probability of biomarker enrichment in Phase III to between 55% and 68%, and subsequently, the overall probability of Phase III success rose to 72%–75%. However, for a lower RRE, as seen in Table 5.6, the study size had to be more than doubled

TABLE 5.5

Simulated Phase IIb Decision and Phase III Success Probabilities with a Biomarker, RRE = 25% to Targeted Treatment in the All-Comer Population

Sample Size (N/Arm)	P	Biomarker Dx Test					Ph IIb Decision Probability			Ph III Success Probability		
		SE	SP	Test Pos.%	PPV	NPV	All Comers	Enrich	Stop	All Comers	Enrich	Overall
110	30%	70%	90%	28%	75%	88%	24%	48%	28%	54%	99%	60%
110	30%	80%	80%	38%	63%	90%	21%	49%	30%	54%	99%	60%
110	30%	90%	70%	48%	56%	94%	18%	50%	32%	55%	98%	59%
110	40%	70%	90%	34%	82%	82%	24%	46%	30%	57%	99%	59%
110	40%	80%	80%	44%	73%	86%	20%	48%	32%	56%	99%	59%
110	40%	90%	70%	54%	67%	91%	16%	51%	33%	55%	97%	58%
110	50%	70%	90%	40%	88%	75%	23%	44%	33%	57%	99%	57%
110	50%	80%	80%	50%	80%	80%	19%	48%	33%	56%	96%	57%
110	50%	90%	70%	60%	75%	88%	14%	50%	36%	58%	94%	55%
140	30%	70%	90%	28%	75%	88%	27%	53%	20%	54%	99%	67%
140	30%	80%	80%	38%	63%	90%	24%	53%	23%	54%	99%	65%
140	30%	90%	70%	48%	56%	94%	18%	58%	24%	55%	98%	67%
140	40%	70%	90%	34%	82%	82%	26%	51%	23%	57%	99%	65%
140	40%	80%	80%	44%	73%	86%	23%	54%	23%	56%	99%	66%
140	40%	90%	70%	54%	67%	91%	17%	59%	24%	55%	97%	67%
140	50%	70%	90%	40%	88%	75%	26%	49%	25%	57%	99%	63%
140	50%	80%	80%	50%	80%	80%	20%	54%	26%	56%	96%	63%
140	50%	90%	70%	60%	75%	88%	16%	58%	26%	58%	94%	64%
190	30%	70%	90%	28%	75%	88%	31%	57%	12%	54%	99%	73%
190	30%	80%	80%	38%	63%	90%	27%	59%	14%	54%	99%	73%
190	30%	90%	70%	48%	56%	94%	21%	64%	15%	55%	98%	74%
190	40%	70%	90%	34%	82%	82%	31%	56%	13%	57%	99%	73%
190	40%	80%	80%	44%	73%	86%	24%	61%	15%	56%	99%	74%
190	40%	90%	70%	54%	67%	91%	18%	67%	15%	55%	97%	75%
190	50%	70%	90%	40%	88%	75%	30%	55%	15%	57%	99%	72%
190	50%	80%	80%	50%	80%	80%	24%	61%	15%	56%	96%	72%
190	50%	90%	70%	60%	75%	88%	17%	68%	15%	58%	94%	74%
285	30%	70%	90%	28%	75%	88%	40%	56%	4%	54%	99%	77%
285	30%	80%	80%	38%	63%	90%	31%	64%	5%	54%	99%	80%
285	30%	90%	70%	48%	56%	94%	23%	71%	6%	55%	98%	82%
285	40%	70%	90%	34%	82%	82%	37%	59%	5%	57%	99%	79%
285	40%	80%	80%	44%	73%	86%	29%	66%	6%	56%	99%	81%
285	40%	90%	70%	54%	67%	91%	20%	73%	7%	55%	97%	82%
285	50%	70%	90%	40%	88%	75%	37%	57%	6%	57%	99%	78%
285	50%	80%	80%	50%	80%	80%	27%	67%	6%	56%	96%	79%
285	50%	90%	70%	60%	75%	88%	18%	76%	6%	58%	94%	82%

Source: Gold et al. (2014).

TABLE 5.6

Simulated Phase IIb Decision and Phase III Success Probabilities with a Biomarker, RRE = 20% to Targeted Treatment in the All-Comers Population

Sample Size (N/Arm)	P	Biomarker Dx Test					Ph IIb Decision Probability			Ph III Success Probability		
		SE	SP	Test Pos.%	PPV	NPV	All Comers	Enrich	Stop	All Comers	Enrich	Overall
110	30%	70%	90%	28%	75%	88%	20%	37%	43%	30%	98%	42%
110	30%	80%	80%	38%	63%	90%	17%	37%	46%	29%	94%	40%
110	30%	90%	70%	48%	56%	94%	15%	39%	46%	31%	86%	38%
110	40%	70%	90%	34%	82%	82%	19%	36%	45%	32%	94%	40%
110	40%	80%	80%	44%	73%	86%	17%	38%	45%	31%	88%	39%
110	40%	90%	70%	54%	67%	91%	14%	39%	47%	32%	79%	35%
110	50%	70%	90%	40%	88%	75%	19%	33%	48%	31%	88%	35%
110	50%	80%	80%	50%	80%	80%	17%	37%	46%	32%	79%	35%
110	50%	90%	70%	60%	75%	88%	13%	39%	48%	32%	71%	32%
140	30%	70%	90%	28%	75%	88%	23%	41%	36%	30%	98%	47%
140	30%	80%	80%	38%	63%	90%	20%	44%	36%	29%	94%	47%
140	30%	90%	70%	48%	56%	94%	16%	45%	39%	31%	86%	44%
140	40%	70%	90%	34%	82%	82%	20%	40%	40%	32%	94%	44%
140	40%	80%	80%	44%	73%	86%	18%	44%	38%	31%	88%	44%
140	40%	90%	70%	54%	67%	91%	14%	45%	41%	32%	79%	40%
140	50%	70%	90%	40%	88%	75%	21%	37%	42%	31%	88%	39%
140	50%	80%	80%	50%	80%	80%	17%	42%	41%	32%	79%	39%
140	50%	90%	70%	60%	75%	88%	13%	44%	43%	32%	71%	35%
190	30%	70%	90%	28%	75%	88%	25%	50%	25%	30%	98%	57%
190	30%	80%	80%	38%	63%	90%	21%	51%	28%	29%	94%	54%
190	30%	90%	70%	48%	56%	94%	18%	53%	29%	31%	86%	51%
190	40%	70%	90%	34%	82%	82%	25%	48%	27%	32%	94%	53%
190	40%	80%	80%	44%	73%	86%	21%	51%	28%	31%	88%	51%
190	40%	90%	70%	54%	67%	91%	16%	54%	30%	32%	79%	48%
190	50%	70%	90%	40%	88%	75%	26%	44%	30%	31%	88%	47%
190	50%	80%	80%	50%	80%	80%	21%	49%	30%	32%	79%	45%
190	50%	90%	70%	60%	75%	88%	15%	53%	32%	32%	71%	42%
285	30%	70%	90%	28%	75%	88%	31%	56%	13%	30%	98%	64%
285	30%	80%	80%	38%	63%	90%	24%	59%	17%	29%	94%	62%
285	30%	90%	70%	48%	56%	94%	20%	63%	17%	31%	86%	61%
285	40%	70%	90%	34%	82%	82%	30%	54%	16%	32%	94%	60%
285	40%	80%	80%	44%	73%	86%	24%	60%	17%	31%	88%	60%
285	40%	90%	70%	54%	67%	91%	19%	65%	17%	32%	79%	57%
285	50%	70%	90%	40%	88%	75%	29%	53%	18%	31%	88%	56%
285	50%	80%	80%	50%	80%	80%	23%	60%	17%	32%	79%	55%
285	50%	90%	70%	60%	75%	88%	16%	65%	19%	32%	71%	51%

Source: Gold et al. (2014).

to N = 285 subjects per arm to raise the probability of biomarker adoption well above 50%.

5.4.1.5.3 *Impact of Biomarker-Based Dx Sensitivity on Study Success*

The choice of biomarker-based Dx test sensitivity significantly influences Phase IIb and overall success. In practice, Dx test sensitivity is known only within a margin of error. It is advisable to simulate a range of Dx test performances to understand their impact on late-stage success. For more information on Dx test sensitivity, specificity, positive predictive value (PPV), and negative predictive value (NPV), and the relationships between these metrics, refer to the supplement.

Simulation scenarios indicated that increasing Dx test sensitivity enhanced the likelihood of obtaining a positive Dx test result and, consequently, the power to detect a significant RRE among Dx test positive subjects in Phase IIb. This improvement increased the chances of adopting the biomarker and enriching Phase III. However, higher Dx test sensitivity reduced the PPV of the Dx test, thereby decreasing the probability of Phase III enrichment success from 99% to 94% for an RRE of 25% in the simulated all-comers population, as shown in Table 5.6. The reduction was more pronounced for a lower RRE, as seen in Table 5.6.

For example, with N = 190 subjects per arm, 40% phenotype A prevalence, and a Dx test sensitivity of 70%, the overall probability of Phase III success was 73%, as detailed in Table 5.8 and Figures 5.7, 5.8, and 5.9. Increasing Dx test sensitivity to 90% raised the probability of biomarker enrichment in Phase III from 56% to 67%, while the overall probability of success remained at 75%. Additionally, the NPV increased from 82% to 91%, meaning fewer simulated patients with the target phenotype were denied access to the targeted drug. A sensitivity of 95% (with a specificity of 55%) resulted in similar overall pivotal study success but tended to increase the chance of a positive Dx test result to greater than 65%.

5.4.1.5.4 *Simulating a Less Predictive Biomarker*

A simulation was conducted with a Dx test that is less predictive of phenotype, characterized by a poorer trade-off in sensitivity and specificity, as listed in Table 5.7. Increasing the Phase IIb study size to N = 190 subjects per arm was insufficient to raise the probability of biomarker enrichment in Phase III above 55% or the overall probability of Phase III success above 64%. To elevate the overall probability of success with the biomarker option to between 67% and 71%, a Phase IIb study size of N = 285 subjects per arm was necessary.

TABLE 5.7

Simulated Phase IIb Decision and Phase III Success Probabilities with a Biomarker, RRE = 25% to Targeted Treatment in the All-Comer Population with Reduced Sensitivity and Specificity

Sample Size (N/Arm)	P	Biomarker Dx Test					Ph IIb Decision Probability			Ph III Success Probability		
		SE	SP	Test Pos.%	PPV	NPV	All Comers	Enrich	Stop	All Comers	Enrich	Overall
110	30%	60%	80%	32%	56%	82%	29%	35%	36%	54%	98%	50%
110	30%	70%	70%	42%	50%	84%	25%	38%	37%	54%	95%	50%
110	30%	80%	60%	52%	46%	88%	22%	42%	36%	55%	91%	50%
110	40%	60%	80%	36%	67%	75%	29%	35%	36%	57%	97%	50%
110	40%	70%	70%	46%	61%	78%	24%	39%	37%	56%	93%	50%
110	40%	80%	60%	56%	57%	82%	21%	41%	38%	55%	89%	48%
110	50%	60%	80%	40%	75%	67%	28%	34%	38%	57%	94%	48%
110	50%	70%	70%	50%	70%	70%	24%	38%	38%	56%	90%	48%
110	50%	80%	60%	60%	67%	75%	20%	42%	38%	58%	86%	48%
140	30%	60%	80%	32%	56%	82%	32%	39%	29%	54%	98%	46%
140	30%	70%	70%	42%	50%	84%	28%	43%	29%	54%	95%	43%
140	30%	80%	60%	52%	46%	88%	25%	46%	29%	55%	91%	40%
140	40%	60%	80%	36%	67%	75%	33%	39%	28%	57%	97%	46%
140	40%	70%	70%	46%	61%	78%	27%	42%	31%	56%	93%	44%
140	40%	80%	60%	56%	57%	82%	23%	48%	29%	55%	89%	39%
140	50%	60%	80%	40%	75%	67%	33%	38%	29%	57%	94%	46%
140	50%	70%	70%	50%	70%	70%	27%	42%	31%	56%	90%	43%
140	50%	80%	60%	60%	67%	75%	21%	48%	31%	58%	86%	39%
190	30%	60%	80%	32%	56%	82%	39%	42%	19%	54%	98%	62%
190	30%	70%	70%	42%	50%	84%	34%	47%	19%	54%	95%	63%
190	30%	80%	60%	52%	46%	88%	26%	53%	21%	55%	91%	62%
190	40%	60%	80%	36%	67%	75%	39%	42%	19%	57%	97%	63%
190	40%	70%	70%	46%	61%	78%	31%	50%	19%	56%	93%	64%
190	40%	80%	60%	56%	57%	82%	25%	55%	20%	55%	89%	63%
190	50%	60%	80%	40%	75%	67%	39%	42%	19%	57%	94%	62%
190	50%	70%	70%	50%	70%	70%	32%	50%	18%	56%	90%	63%
190	50%	80%	60%	60%	67%	75%	25%	55%	20%	58%	86%	62%
285	30%	60%	80%	32%	56%	82%	49%	43%	8%	54%	98%	69%
285	30%	70%	70%	42%	50%	84%	41%	50%	9%	54%	95%	70%
285	30%	80%	60%	52%	46%	88%	31%	59%	9%	55%	91%	71%
285	40%	60%	80%	36%	67%	75%	48%	44%	9%	57%	97%	69%
285	40%	70%	70%	46%	61%	78%	40%	52%	9%	56%	93%	70%
285	40%	80%	60%	56%	57%	82%	30%	61%	9%	55%	89%	71%
285	50%	60%	80%	40%	75%	67%	48%	43%	9%	57%	94%	67%
285	50%	70%	70%	50%	70%	70%	39%	51%	10%	56%	90%	68%
285	50%	80%	60%	60%	67%	75%	30%	61%	9%	58%	86%	70%

Source: Gold et al. (2014).

5.4.2 Patient Selection in Oncology Trials

5.4.2.1 Challenges in Patient Selection

5.4.2.1.1 Phase I Trials

As discussed in Section 1.2.2.1, the primary aim of Phase I trials is to identify the maximum tolerated dose (MTD) of a new drug or combination product for further evaluation in Phase II studies. In oncology Phase I trials, the assessment is often carried out in patients with advanced malignancy for whom treatment options are limited. Since at this stage of clinical development the benefit of novel therapy is unknown, from an ethical standpoint, it is important to decide who should or should not be entered into these studies. Additionally, from a logistic perspective, it is of interest to enroll patients who can stay in the trial until the key study endpoints, such as dose limiting toxicity, are assessed. Traditionally these issues were addressed through the use of eligibility criteria. For example, these criteria conventionally include reasonable levels of performance status such as ECG of 0 or 1 and expected life expectancy ≥3 months or 12 weeks (Arkenau et al. 2008). Despite these measures, approximately one-third of patients fail to meet all necessary eligibility criteria at screening and approximately15% to 20% die within the first 90 days of phase I trial (Arkenau et al. 2008). As noted by Yu et al. (2020), 45% sites from a large IO trial had more than 20% patient drop out of study within 12 weeks of beginning of the study, even though ideally most trials of this type last much longer than 2–3 months (Figure 5.9).

In addition, premature patient dropout may delay dose escalation, thus diminishing the efficiency of Phase I trials. An unexpected rapid disease progression or exacerbation of another underlying pathological condition in a significantly large number of patients participating in a phase I clinical trial can diminish the ability to detect early efficacy signals of the therapeutic being tested.

5.4.2.1.2 Late Phase Trials

The early mortality (EM) issue in the early immune checkpoint inhibitor (ICI) studies has also been consistently observed in late stage randomized trials where ICI was assessed as monotherapy (Champiat et al. 2018). This is evidenced by a higher mortality rate in the early follow-up period in the ICI group than in the control. As a result, the Kaplan–Meier (KM) curves of the ICI and control groups cross each other. An example is provided in Figure 5.10. Therefore, it is important to predict EM risks of patients. This would prevent such patients from receiving ICI treatment, which they do not derive benefit from, and allow them to consider other alternate treatment options.

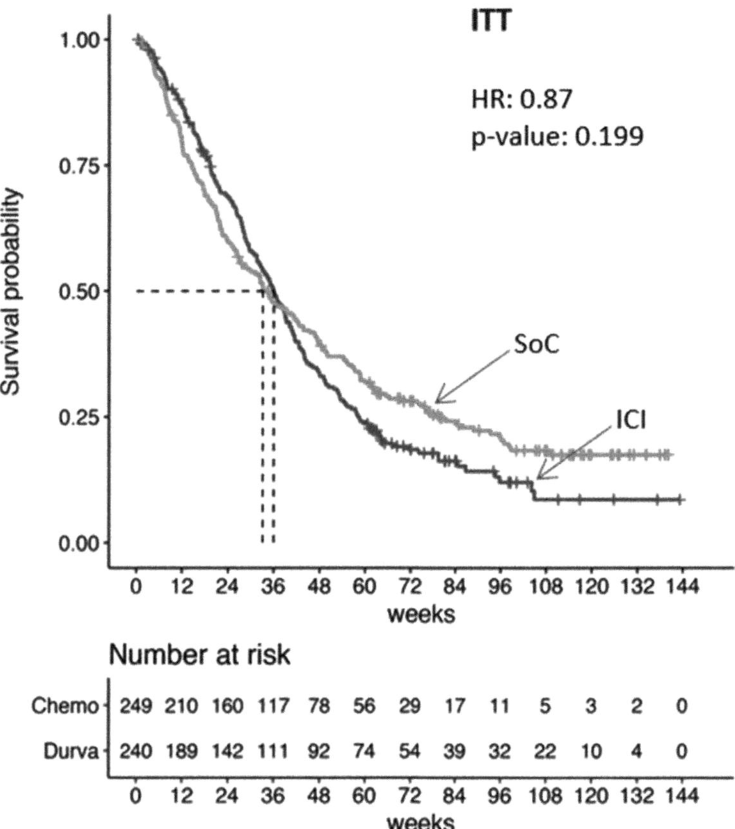

FIGURE 5.10
The plot shows an intention-to-treat (ITT) survival probability analysis over a period of 144 weeks (top) and a table of patients at risk over the same period (bottom) for two treatment schemes, ICI vs. standard of care (SoC). The 2 KM curves cross each other. (Adapted from Yu et al. 2020.)

5.4.2.2 Prognostic Scores

To address the above issues, several prognostic indices have been developed and validated based on retrospective analyses of Phase I cancer trials. Of note are the RMH (Arkenau et al. 2009), GRIm (Bigot et al. 2017), and LIPI scores (Benitez et al. 2020). These scores predict patient's survival using laboratory values collected at baseline and are summarized in Figure 5.11.

The RMH was based on the prognostic variables, albumin, LDH, and number of metastatic sites. It was originally developed using data from Phase I oncology studies of targeted therapies (Arkenau et al. 2009), and retrospectively validated by Wheler et al. (2012) based on data from separate trials across tumor types. Patients with RMH score ≥2 had poorer OS. The GRIM

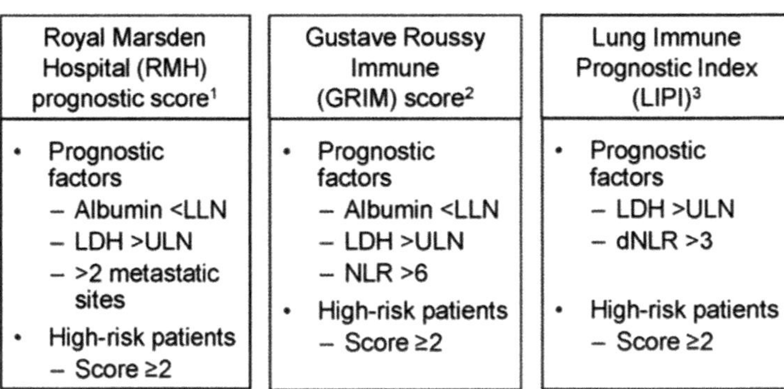

FIGURE 5.11
Prognostic scores predicting patient overall survival using baseline lab values. (Adapted from Yu et al. 2020.) dNLR, derived neutrophil and neutrophile minus leukocytes ratio; LDH, lactate dehydrogenase; NLR, neutrophil and lymphocyte ratio; ULN, upper limit of normal range.

score utilizes baseline albumin, LDH, and NLR to predict EM. The score was developed and prospectively validated in two studies by Minami (2019) and Bigot et al. (2017). Patients with a GRIM score exceeding 1 were shown to have poorer OS. The LIPI index predicts patients' prognostic prospects based on albumin and dNR. Retrospective cohorts of NSCLC patients treated with ICI and chemotherapy were used in training and validating the index. Poorer OS was seen with patients who had LIPI score ≥ 2. As discussed by Yu et al. (2020) in the original publication, the LIPI score was applied retrospectively in two independent cohorts of NSCLC patients treated with ICI (N = 305) or chemotherapy (N = 162) (Mezquita et al. 2018). The authors found that LIPI score >1 was associated with poor OS in ICI treated patients but not in chemotherapy treated patients. However, a subsequent analysis of LIPI performed by the FDA on pooled clinical trial data from studies evaluating 1,368 second line metastatic NSCLC patients who received ICI and 1,072 patients who received chemotherapy did not confirm these results (Kazandjian et al. 2018). This study indicated that LIPI may exert a prognostic impact irrespective of therapeutic modalities (ICI or chemotherapy) for 2nd line metastatic NSCLC. Finally, another study investigated the LIPI score in the context of metastatic patients with various solid tumors enrolled in phase 1 trials. While this analysis was performed on a very heterogeneous cohort, a LIPI score >1 was also associated with poor OS, suggesting its prognostic role is not limited to NSCLC patients (Varga et al. 2019).

5.4.2.3 3i Score

The three prognostic scores, particularly RMH and GRIM, were developed using data from early oncology trials, which were not necessarily dealing with the EM issue in ICI trials. Taking advantage of a large dataset from various ICI trials, Yu et al. (2020) developed a prognostic index using ML algorithms to predict the EM in the ICI context. It uses pre-treatment measurements of routinely collected blood-based factors to predict a patient's risk of EM, and to optimize benefit and risk profile for treatment of patients with ICI . This method is referred to as Immune Immediacy Index or 3i Score. The detailed development of the method is described in the following.

5.4.2.3.1 *Model Requirements*

The key development objective for the 3i score was to predict EM in ICI trials. EM is defined as deaths within 12 weeks of randomization or start of treatment. It is also desirable to use routinely collected baseline laboratory test results as model inputs to enable use of the method in real-world clinical practice. ML methods have been increasingly used in healthcare for predicting safety risk. When appropriately trained, the models are less sensitive to variability in data and can yield accurate predictions. However, deploying ML models in a real-world setting is complex due to the need for robust performance and interface software. Consequently, acceptable performance metrics throughout training, tuning and testing, as well as flexible deployment options are key considerations in the development of the 3i Score. In addition, careful treatment should also be given to the choices of datasets for training, tuning and testing.

5.4.2.3.2 *Model Training, Tuning, and Testing*

Development of supervised ML models consists of three critical steps, namely training, tuning and testing as shown in Figure 5.12.

During training, the model's parameters, such as tree depth, are optimized to encode relationships between input and output. In an iterative manner, the model accepts an input from the training dataset and produces an output which is compared against ground truth, and the resulting error guides an

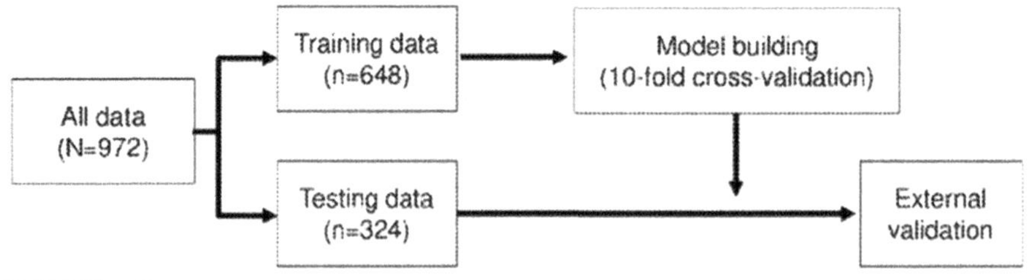

FIGURE 5.12
Diagram of 3i Score development process consisting of identification of datasets for training and testing, internal modeling validation including tuning, and external testing.

update of the model parameters. Over time, the model learns representations of input that leads to desired target output using the training dataset. The same ML technique can yield a different model in terms of architecture, weights and performance depending on what training constraints have been chosen.

The optimal training constraints, called hyperparameters, are different in each ML situation and are set before training, as opposed to other model parameters that are developed during training. Hyperparameter optimization, also called tuning, is a step in ML development that is often found to have a profound effect on a model's ability to generalize successfully. A choice of hyperparameters can lead to a model that will overfit or underfit on the same training set. One common approach to exploring a space of hyperparameter ranges is a grid search where each time a set of hyperparameters is sampled, the model is trained on a training dataset and then evaluated on an independent dataset (tuning dataset). Performance evaluation on a tuning set guides the selection of optimal training constraints that produce a model robust enough to function beyond the training set.

Model testing is performed to evaluate the final model from the training dataset using hyperparameters of choice. Testing is performed on an independent dataset, which was neither used in training nor in tuning, to evaluate how well the model generalizes and derives patterns beyond the data it has encountered before.

5.4.2.3.3 Final 3i Score Model

The following six key predictors: NLR, NEUT, ALB, LDH, GGT, AST, and the tumor type were retained in the final model. The model produces a score (i.e., value between 0 and 1) representing the probability of death occurring in <12 weeks for each patient. The 3i score is then converted to a status that assigns patients to prognostic or risk categories of high or low.

The cut-off was determined to allow 10% false positive rate (FPR) at predicting EM in the training dataset and calculated as 0.649. Patients with a score above the cut-off of 0.649 are identified as high risk of EM and patients at or below the cut-off (0.649) are identified as low risk of EM. For patients missing any of the six lab test values, the 3i score will not be calculated. This decision was supported by the expectation that globally, patients will have data on these variables readily available as they are routinely collected standard laboratory measures.

When applied in the training set (*N* = 2213 patients), the true positive rate (TPR) at 12 weeks was 67% and the FPR was 10%. Performance as defined by pAUC (0.7, 0.9) was 0.83. The median OS of patients identified with a high risk 3i score status (486/2,213 patients) was 9.29 weeks (95% Cl: 6.29 weeks, 9.86 weeks) and the median OS of patients with a low risk 3i score status was 61.43 weeks (95% Cl: 57.71, 66.14). Figure 5.13 displays the KM curves of the

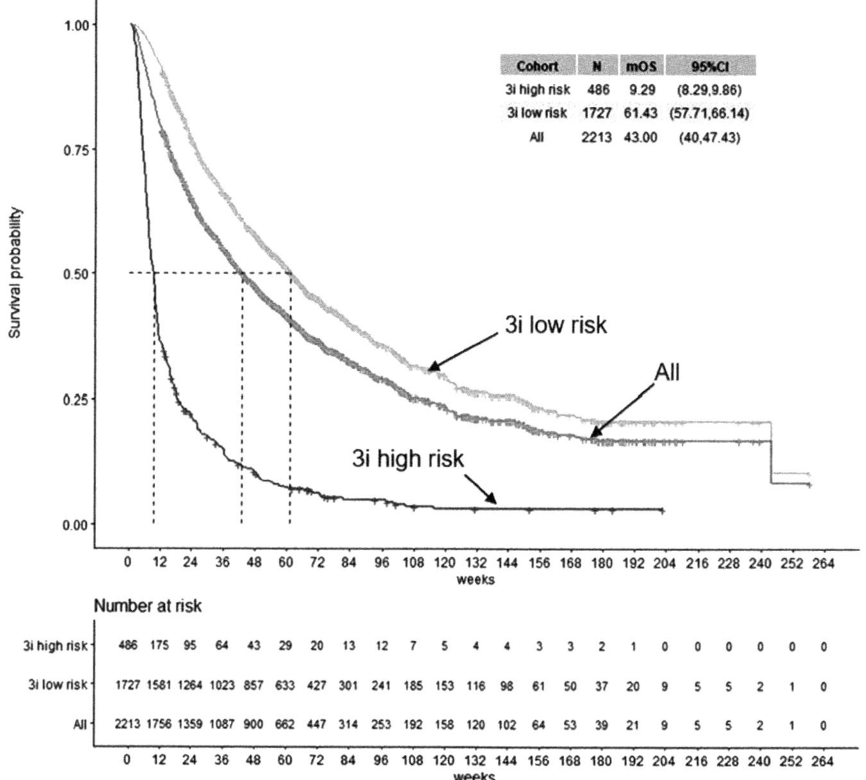

FIGURE 5.13

Kaplan–Meier curves of low-risk and high-risk patients identified by 3i Score, and all pooled patients of the training dataset. (Adapted from Yu et al. 2020.)

low- and high-risk patients identified by 3i Score and pooled all patients, along with the summary statistics. ULN = 240 was used for LDH for calculating GRIm and LIPI scores.

Comparison of 3i, GRIm, and LIPI in Tuning Set

MYSTIC is a Phase III randomized, open-label, multi-center, global study of MEDI4736 in combination with tremelimumab therapy or MED14736 monotherapy versus SOC platinum-based chemotherapy in first line treatment of patients with advanced or metastatic non-small-cell lung cancer (N = 1,118). The primary analysis population is the ITT population of patients with PD-L1 ≥ 25%. Data from MYSTIC were used as the tuning set. The KM curves of patients in MYSTIC study who received durvalumab, and who were identified as high being at risk by the 3i Score, GRIm, and LIPI scores and associated summary statistics are shown in Figure 5.14.

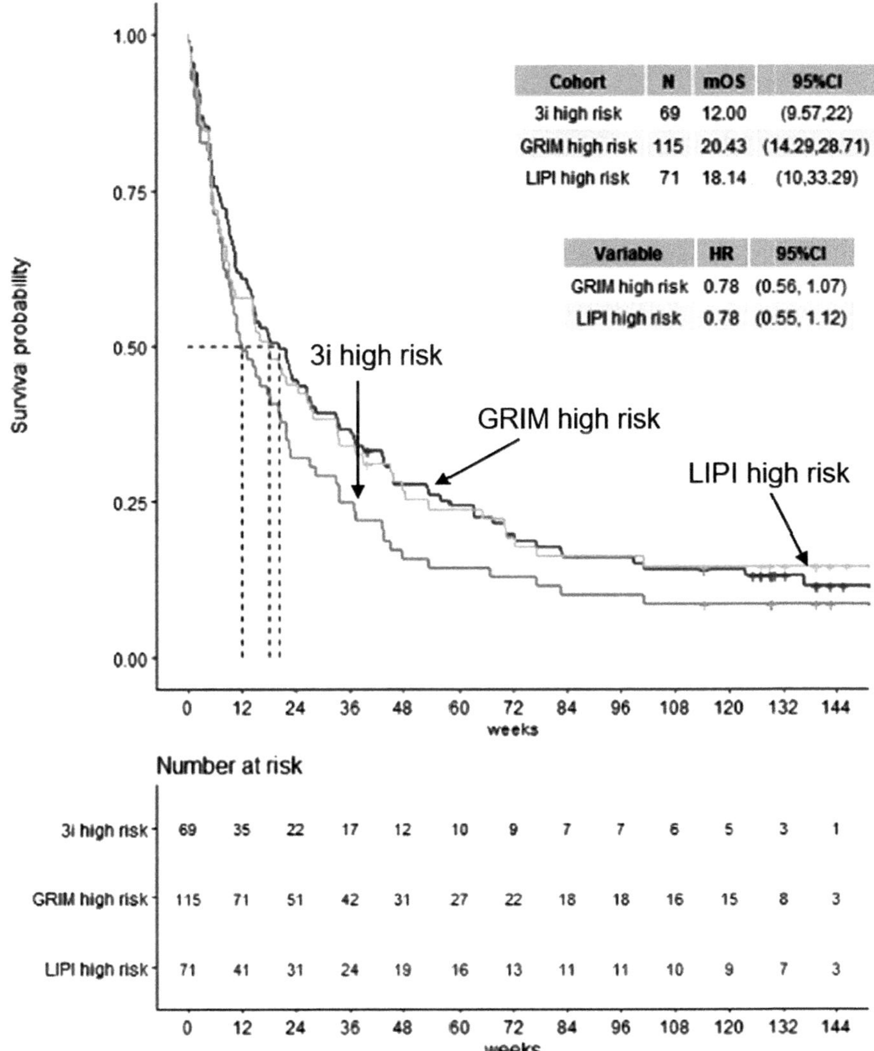

FIGURE 5.14
Kaplan–Meier curves of high-risk patients by 3i score, GRIM, and LIPI models in MYSTIC ITT patients randomized to durvalumab. $N = 374$, patients with non-small cell lung cancer. (Adapted from Yu et al. 2020.)

From the above plot, the patients identified as high risk by the 3i score had shorter median OS (69 patients, median OS 12 weeks [9.57, 22]) compared with patients identified as high risk using other prognostic models developed to predict for early death in patients with advanced or metastatic cancers: GRIm (115 patients, median OS 20.43 weeks [95% Cl: 14.29, 28.71]); and LIPI score (71 patients, median OS 18.14 weeks [95% Cl: 11.43, 33.29]).

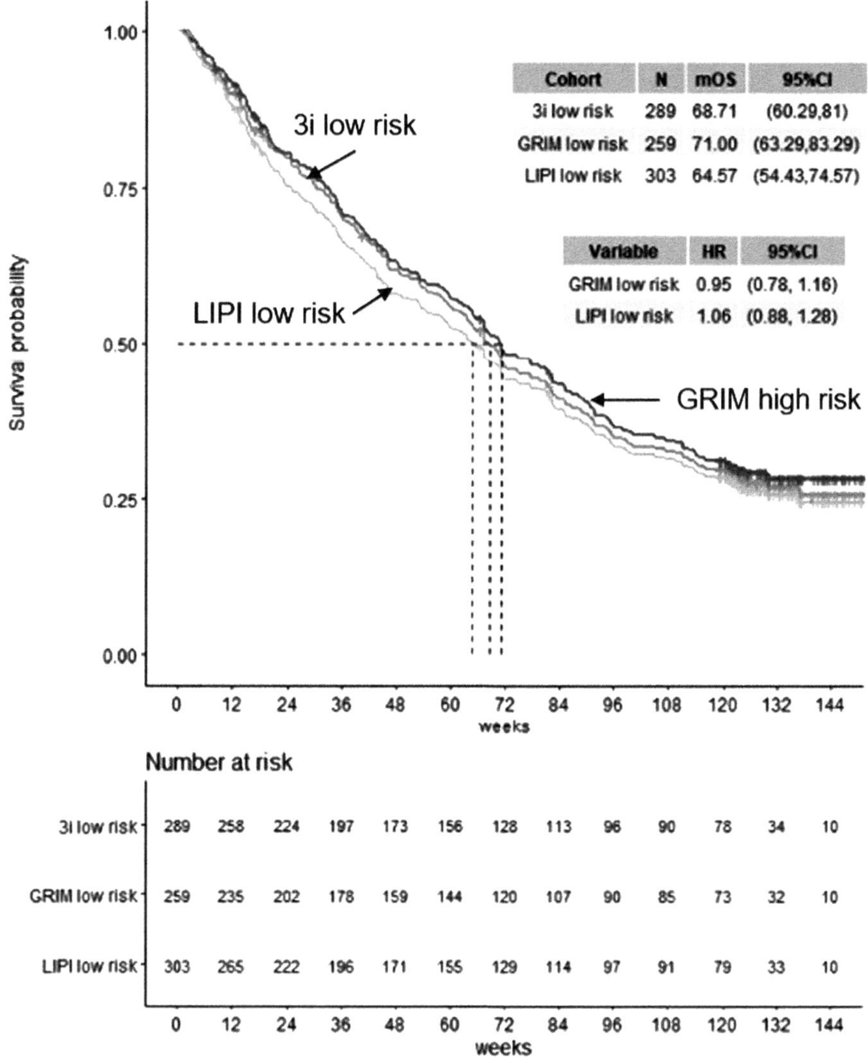

FIGURE 5.15
Kaplan–Meier curves of low-risk patients by 3i score, GRIM, and LIPI models in MYSTIC ITT patients randomized to durvalumab, $N = 374$ patients with non-small cell lung cancer. (Adapted from Yu et al. 2020.)

Figure 5.15 shows the KM curves of low-risk patients by 3i score, GRIm, and LIPI models in MYSTIC ITT patients randomized to receive durvalumab. The median OS (90% CI) are 68.71 weeks (60.29, 81), 71 weeks (63.29, 83.29), and 64.57 weeks (54.43, 74.57), respectively, which are comparable.

In the MSYTIC PD-L1 ≥ 25% subgroup, a total of 56 patients were identified as high 3i score risk, 24 in the SoC arm, and 32 in the durvalumab arm. As

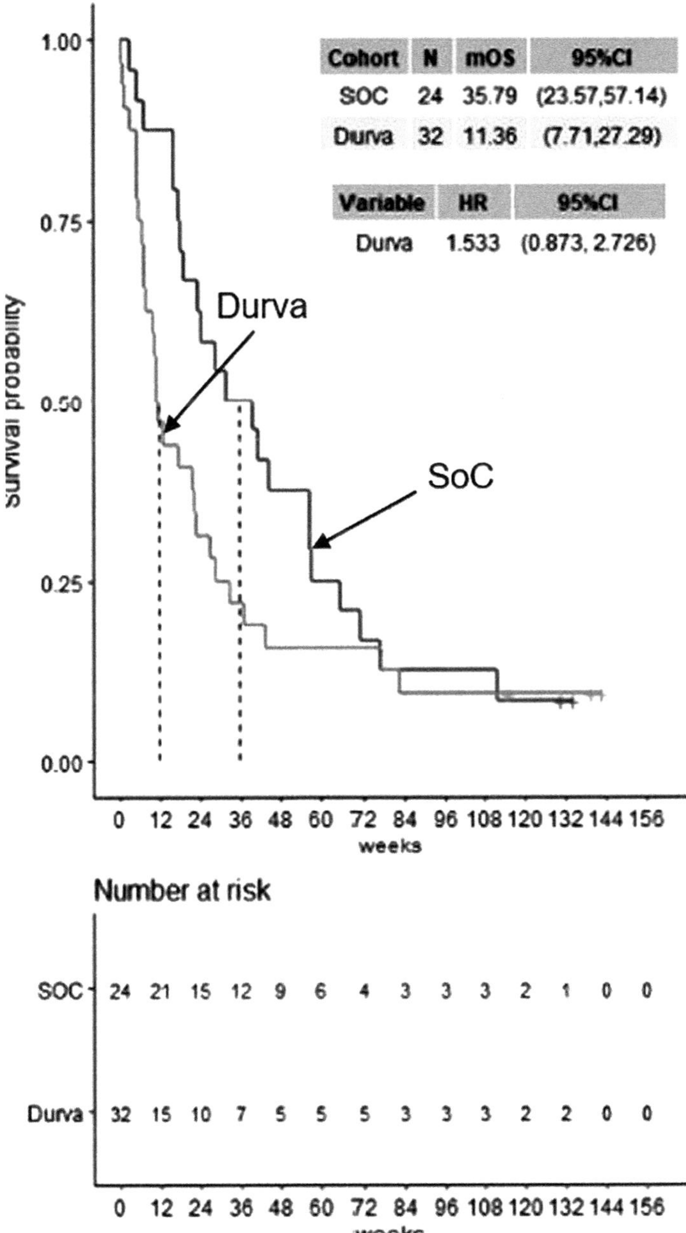

FIGURE 5.16
Kaplan–Meier estimates of overall survival in MYSTIC PD-L1 ≥ 25% 3i score high-risk subgroup. (Adapted from Yu et al. 2020.)

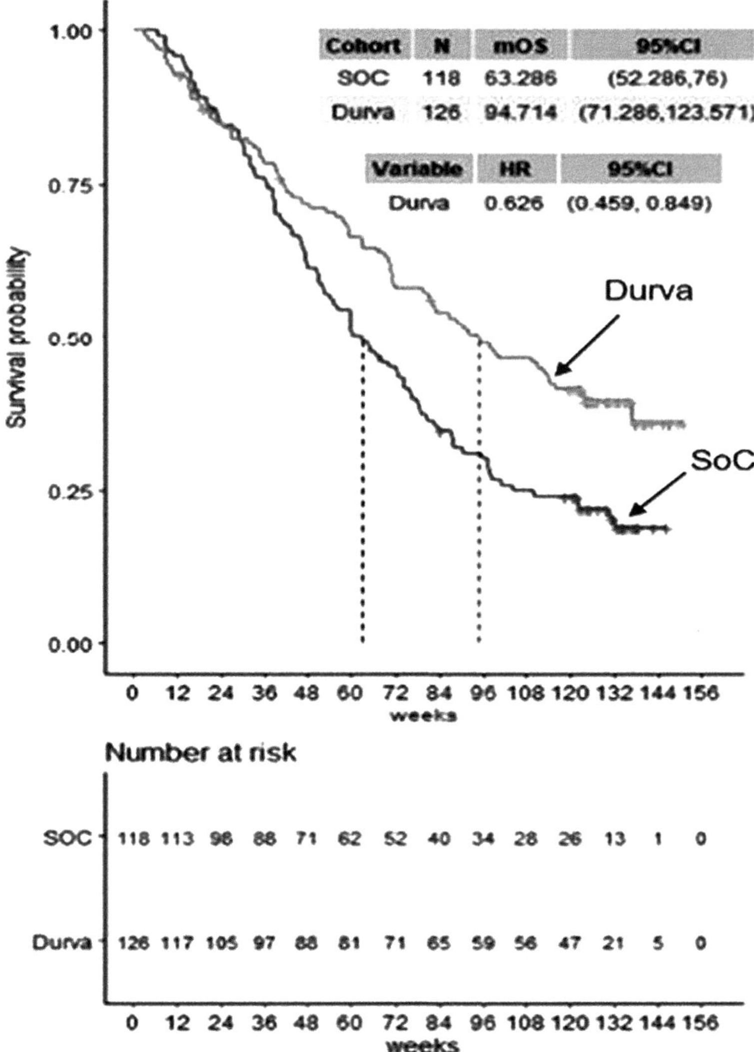

FIGURE 5.17
Kaplan–Meier estimates of overall survival in MYSTIC PD-L1 ≥ 25% 3i score low risk subgroup. (Adapted from Yu et al. 2020.)

seen from Figure 5.16, for high risk 3i score patients PD-L1 ≥ 25% subgroup, median OS was shorter in the durvalumab arm than the chemotherapy arm (11.36 weeks vs. 35.79 weeks).

Excluding patients with a high-risk status by the 3i score from the MYSTIC PD-L1 ≥ 25% subgroup reduced the previously observed crossing of the OS curves (Figure 5.17). The treatment effect of durvalumab vs. SoC was larger

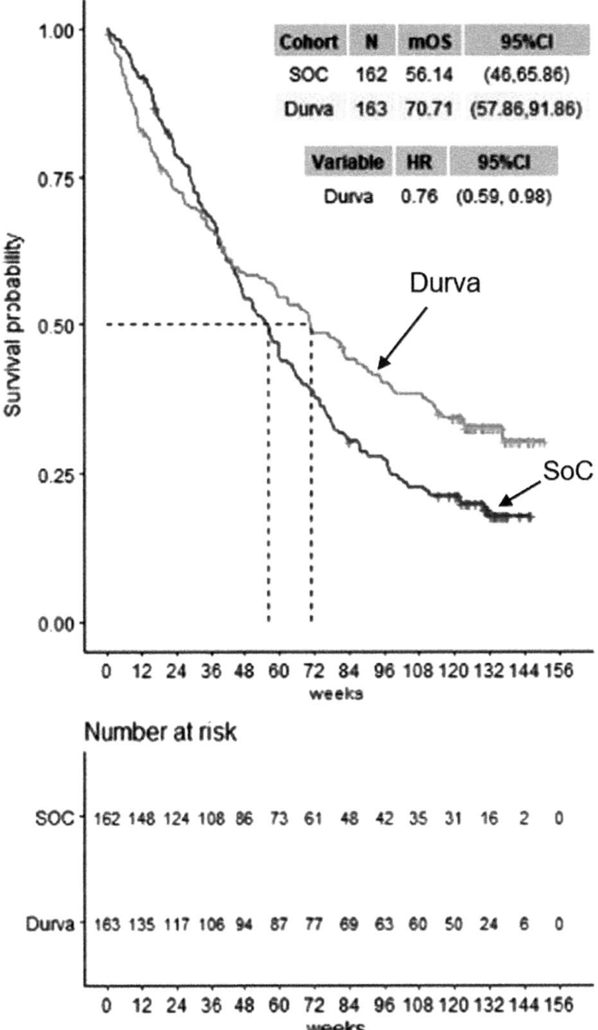

FIGURE 5.18
Kaplan–Meier estimates of overall survival in MYSTIC PD-L1 ≥ 25% 3i score subgroup. (Adapted from Yu et al. 2020.)

(HR: 0.626 [95% CI: 0.459, 0.849]) compared to the original primary analysis of the PD-L1 ≥ 25% population (HR: 0.76 [95% CI: 0.59, 0.98]) (Figure 5.18).

5.4.2.4 Summary

The challenge of EM in early ICI trials and disproportionately higher EM rate in ICI arm in late-stage trials when compared to chemotherapies has been broadly recognized. The 3i Score developed by Yu et al. (2020) was shown to have clinical utility in identifying patients with high risk of EM. Further validation using data from prospective studies may lend additional confidence in using the ML-based method for determining ICI treatment options for patients.

5.5 External Control for Phase III Trial

This section presents a case example by Kidwell et al. (2022) to illustrate the potential of borrowing information in Phase III trial for progressive supranuclear palsy (PSP).

5.5.1 Background

PSP is a degenerative neurological disorder characterized by progressive issues with balance, walking, eye movement, muscle tone, speech, and swallowing. The PSP Rating Scale (PSPRS) is a disease-specific measure of disability, evaluating 28 items across six domains: daily activities, behavior, bulbar function, ocular motor function, limb motor function, and gait/midline stability. Clinicians and regulatory agencies accept the mean change from baseline PSPRS score at week 52 as a primary outcome measure when assessing new therapies for PSP. A four-point improvement over placebo in mean change from baseline at 52 weeks is considered clinically meaningful.

Assuming a one-sided α error of 2.5% and standard deviation (SD) of 8 on PSPRS, a traditional frequentist design with 1:1 randomization would require 85 patients per arm be required for a well-controlled trial to warrant 90% power to detect a 4-point mean change from placebo. Kidwell et al. (2022) explored an alternative Bayesian design with a 2:1 randomization (85 in the treatment arm and 43 in the placebo arm). The objective was to reduce the number of subjects in the placebo arm by incorporating data from three completed randomized studies, thereby increasing efficiency. This design employs a meta-analysis approach with considerations for practical feasibility and regulatory requirements. It is ethically and practically appealing, as it reduces the number of subjects in the placebo arm.

5.5.2 Method

An informative prior for the mean change in PSPRS at week 52 for the placebo arm was derived using placebo data from three previous randomized Phase II studies in PSP and the meta-analytic-predictive (MAP) approach. The MAP framework mathematically links parameters from external data (e.g., mean change in PSPRS at week 52 in Phase II trials) to predict outcomes for the current trial, using this prediction as a prior. This approach assumes exchangeability, meaning the external control data is similar to the current trial data, allowing the trial to "borrow strength" from external sources via a hierarchical model. The degree of borrowing is influenced by the between-trial data source heterogeneity parameter, which requires careful specification to ensure accurate results, especially with a small number of external trials. A weakly informative prior for between-trial heterogeneity was recommended to cover a broad range of plausible endpoint outcomes and associated variability.

The three Phase II studies used as external information in this example are similar in terms of population, inclusion/exclusion criteria, and other important study characteristics. Table 5.8 summarizes the mean change in PSPRS score for the placebo group in the three trials. Using the MAP approach, the predictive distribution for the mean change in PSPRS score at week 52 in a new study is derived, leading to an estimated mean of 10.8 (and a 95% credible interval of 8.4 to 13.4).

For ease of use and interpretation, Kidwell (2020) approximated this predictive distribution assuming a normal density with matching mean and standard deviation for this example. The result of this process is the MAP prior with a normal (mean = 10.8, SD = 1.19) distribution. Alternatively, a mixture of normal distributions can be used to accurately approximate the predictive distribution.

Although the external and new populations are assumed to be similar or exchangeable to allow one to inform the other, they are not the same, and important differences, not known a priori, might exist. Therefore, along with informative priors, it is necessary to incorporate a certain degree of skepticism. This can be achieved using robust mixture priors, which is a combination of informative and non-informative priors. It allows for dynamic borrowing of prior information; the analysis learns how much of the external data to borrow in the prior based on the consistency between the external information and the trial data. Mixture priors have been proposed in different contexts (e.g., bridging studies, historical controls, pediatric extrapolation)

TABLE 5.8

Results of Three Phase II Trials in PSP Patients

Study	*N*	**Mean Change from Baseline to Week 52 in PSPRS**	**Standard Error**
Boxer et al. (2014)	153	10.9	0.99
Tolosa et al. (2014)	31	11.4	1.13
Höglinger et al. (2021)	59	10.5	1.00

Source: Kidwell et al. (2022).

and are relevant for rare disease studies. Further robustification of the MAP prior in our PSP example to handle conflict between external and trial placebo data is provided by including a non-informative Normal (mean = 15, SD = 10) prior with 50% weight to the MAP prior to reflect considerable heterogeneity between the Phase II and Phase III populations and to control inflation of the FPR. This prior is used as the historical data prior for the placebo group in the primary analysis. As no relevant external information is available for the experimental treatment, a non-informative prior (normal distribution with parameters mean = 0 and sd = 10) is used for the mean change from baseline to week 52 in PSPRS for the experimental treatment group.

5.5.3 Performance Characteristics of Bayesian Design

Assessing operating characteristics is crucial in clinical trial design, traditionally involving frequentist concepts like type I error and power. Figure 5.19 illustrates type I error and power of the proposed design with and without a robust MAP prior. The parameter δ is mean difference between the treatment and control arms in change from baseline in PSPRS at week 52. When $\delta = 0$ (left panel) there is no difference between placebo and treatment whereas when $\delta = 4$ (right panel) there is a treatment difference. The plot shows the

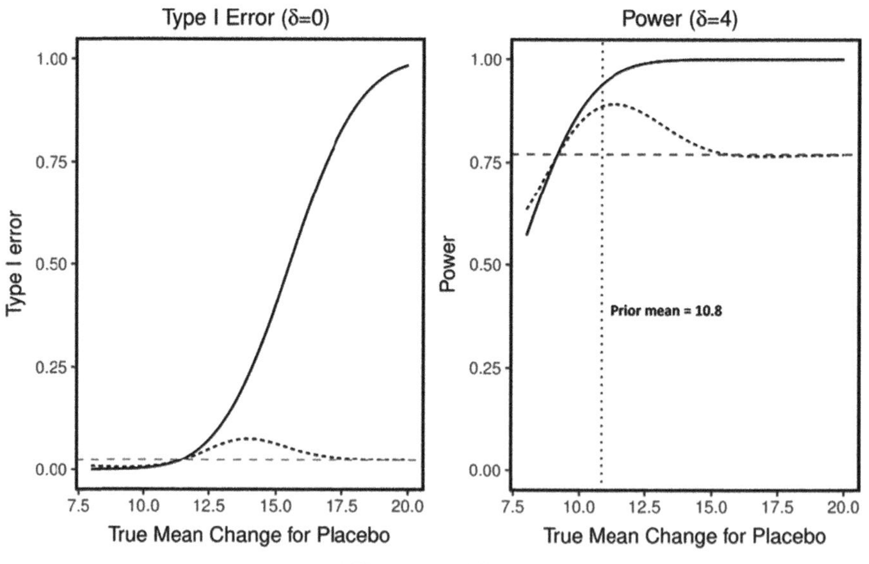

FIGURE 5.19
Frequentist Operating Characteristics (type I error (left panel) and power (right panel) of proposed design with meta-analytic predictive (MAP; solid line) and robust MAP (dotted line). (Adapted from Kidwell et al. 2020.)

type I error (0.025, left panel) and power (0.8, right panel) for the traditional frequentist design (gray dashed line). Success is defined by a significant positive treatment effect $P(\delta > 0 \mid \text{data}) > 0.975$.

The left panel of Figure 5.19 shows that the robust MAP prior (dotted line) prevents an excessive increase in type I error when there are conflicts between trial and external placebo data, with a maximum type I error of 6.3%. For power (right panel of Figure 5.1), both the MAP prior and the robust MAP prior show considerable gains over the traditional frequentist design (gray horizontal dashed line) with the same sample size in the treatment arm, particularly if the true mean change in PSPRS falls within the prior support range (9 to 13). Power gain is significant, up to 14%, compared to the traditional design when external and trial placebo data are consistent. However, in cases of conflict (true mean change PSPRS at 52 weeks > 12), the MAP prior results in a stronger type I error inflation compared to the robust version. For substantial prior-data conflict (true mean change PSPRS at 52 weeks > 14), the power of the robust MAP prior approximates the traditional design.

5.5.4 Conclusion

In summary, this example demonstrates that a reduced sample size design utilizing robust MAP priors with a weakly informative component offers good robustness and improved power in certain scenarios compared to a traditional frequentist design. The definition of acceptable frequentist metrics must be tailored to each unique trial setting, with careful consideration of potential conflict scenarios. Other methodologies, such as the power prior and commensurate prior, also discount historical data to account for heterogeneity between external and internal trial information, sharing the same general feature as the MAP prior.

5.6 Concluding Remarks

The landscape of clinical trial optimization is rapidly evolving, driven by the convergence of strategic, scientific, and operational imperatives. The increasing complexity of clinical trials, combined with the urgency to deliver novel treatments to address unmet medical needs, underscores the necessity for innovative approaches. The traditional methodologies, while foundational, are often inadequate in the face of rising expectations for timelines and the demand for individualized data from patients and payers.

The integration of AI and RWD represents a paradigm shift in clinical trial design and execution. These advancements have enabled the incorporation of insights from historical and ongoing trials, thereby improving the efficiency and effectiveness of clinical development. By leveraging AI analytics and novel data sources, we can design more precise and efficient clinical trials that not only enhance operational excellence but also significantly accelerate the development process.

The case examples discussed in this chapter illustrate the tangible benefits of these innovative approaches. They highlight the potential of AI and RWD to address critical challenges in clinical trial optimization, such as the selection of the right patient populations, treatments, and dosing regimens. These examples also demonstrate the ability of these technologies to streamline operational processes, reduce costs, and increase the probability of trial success.

Despite these advancements, challenges remain. The implementation of AI and RWD in clinical trials requires careful consideration of ethical, regulatory, and practical aspects. It necessitates a collaborative effort among stakeholders, including researchers, clinicians, patients, and regulatory bodies, to ensure that these technologies are applied responsibly and effectively.

In conclusion, the future of clinical trial optimization lies in the continued exploration and integration of AI and RWD. As we embrace these tools, we move closer to achieving a more efficient, effective, and patient-centered clinical development process. The promise of these technologies is vast, and their potential to transform clinical trials is undeniable. Through sustained innovation and collaboration, we can overcome current challenges and unlock new opportunities to improve patient outcomes and advance medical science.

6

Model-Informed Decision-Making in Clinical Trials

6.1 Introduction

Bringing a new medicine to the market is both time-consuming and expensive. Despite the advance in science and technology in the past two decades that render better understanding of disease biology and more effective means in patient selection, the attrition of late-stage clinical trials remains high. To counter the trend, newer approaches and technologies have been incorporated into drug development to improve both productivity and efficiency of clinical trials, to streamline drug development. In recent years, there has been broader adoption of model-informed drug development (MIDD) in guiding key decision-making at each stage of clinical development. These approaches provide quantitative frameworks for integrating information from preclinical and clinical data sources to inform drug decision-making. The scope of these applications includes informing dose selection and trial design, assisting evaluation of drug safety and effectiveness, and facilitating key decision-making in drug development. However, despite these efforts, most of these approaches remain reliant on limited data from internal studies and traditional structured model types. Recently, there is emerging evidence suggesting that artificial intelligence (AI) and machine learning (ML) can be used to improve the current methods, by leveraging data from diverse sources and the predictive power of models and algorithms beyond the traditional approaches. This chapter focuses on discussion of various statistical decision frameworks for different phases of clinical trials introduction of the emerging applications of ML methods to improve the predictive power of the current methods.

DOI: 10.1201/9781003226086-6

6.2 Decision-Making in Clinical Development

Clinical development entails numerous pivotal decisions, spanning overall clinical strategy, trial design, patient selection and recruitment, and comprehensive evaluations of efficacy and safety. The overarching goal of these decisions is to efficiently identify and advance promising compounds while discontinuing the development of inferior ones. Each decision holds significant implications for patients, sponsors, and the healthcare community at large. Adding to the burden is the need to make correct decisions as expeditiously and efficiently as possible (Dmitrienko and Pulkstenis 2017). These decisions may entail advancing to the subsequent developmental stage, terminating development, or acquiring additional data. Given the potential opportunity costs associated with each choice, the ramifications of incorrect decisions are significant for patients, sponsors, and the healthcare community at large.

However, decision-making in clinical trials represents a complex and multifaceted process, characterized by inherent uncertainty and risk. The high failure rate of Phase III trials and the inconsistency between Phase II and Phase III data underscore the complexity and challenges in this process (Arrowsmith and Miller 2013; Paul et al. 2010; Kirby et al. 2016; Kola and Landis 2004; Chuang-Stein and Kirby 2017). Therefore, it is imperative to establish a robust decision-making framework to maximize the likelihood of making correct decisions at each stage of development. Consequently, the decision-making process must exhibit robustness to maximize the probability of accurate decisions at every juncture.

6.3 Structured Framework

At the outset of a new drug development, establishing a structured framework, such as a Target Product Profile (TPP), is essential for guiding the development process, providing a clear line of sight to the overarching goal of bringing a safe and effective drug to market (FDA 2007b). The TPP outlines the desired attributes of a drug, including its intended use, target populations, and key performance criteria, ensuring alignment among stakeholders. This alignment mitigates risks associated with clinical trials by setting well-defined objectives and expectations, facilitating more efficient decision-making. Furthermore, a TPP enables a rigorous and quantitative assessment of a drug's safety and efficacy as it provides standardized benchmarks against which clinical data can be evaluated. This structured approach is critical for identifying potential issues early in the development process, thereby reducing the likelihood of costly late-stage failures (CMR Institute n.d.). By incorporating insights from regulatory guidelines and previous clinical experiences, a TPP serves as a dynamic tool that adapts to emerging data, ultimately supporting the successful navigation of the complex drug development landscape. The development of a TPP is a critical first step of clinical development.

6.3.1 Target Product Profile

A TPP defines the desired characteristics of a target product intended to treat a particular disease(s). The TPP outlines the intended use of the product, target populations, and safety and efficacy-related characteristics among others. The TPP can serve multiple purposes, including providing a structure framework and planning tool to guide the design, conduct, and analysis of clinical trials to maximize efficiency and facilitating dialogue with regulatory authorities. The TPP embodies a notion of "begin with the end in mind" and is an important step in the entire clinical development process. For a TPP to be clinically relevant and commercially viable, it needs to be deeply rooted in a good understanding of the target population against the backdrop of current and future competition. It also needs to reflect healthcare stakeholder's perspectives on the value of the new intervention. A TPP is a dynamic document that requires regular review and revision. Changes to the TPP may necessitate re-evaluation of functional strategies. The TPP serves as a vital tool for guiding and aligning product development efforts across various domains. Its collaborative development process ensures thorough analysis and buy-in from all stakeholders, ultimately enhancing the product's chances of success in the market.

A TPP is a dynamic document that requires regular review and revision. Changes to the TPP may necessitate re-evaluation of functional strategies. The TPP serves as a vital tool for guiding and aligning product development efforts across various domains. Its collaborative development process ensures thorough analysis and buy-in from all stakeholders, ultimately enhancing the product's chances of success in the market.

Table 6.1 lists the key components of a TPP. A detailed discussion of select aspects of TPP are provided in the following sections.

6.3.1.1 Key Indications

The first and most time-intensive step involves deep research to identify potential indications and claims for the drug. This process includes due diligence, market data analysis, and leveraging consumer research to address unmet patient needs. Evaluation of existing literature and competitors aids in understanding performance and treatment trends

TABLE 6.1

Key Characteristics of Target Product Profile

Component	Description
Clinical Characteristic	Indications or diseases the drug will treat, the patient population it will serve, intended efficacy, safety profile, dosing regimen, and potential side effects.
Nonclinical Aspect	Preclinical data, pharmacokinetics, pharmacodynamics, and any unique mechanisms of action that differentiate the drug from existing treatments.
Commercial and Market Considerations	Pricing, market size, competition analysis, and potential market positioning to meet commercial success.
Regulatory Expectations	Anticipated regulatory pathways, including FDA or other regulatory agency approval requirements, and any special considerations needed for approval.
Manufacturing Requirements	Specifications for drug formulation, packaging, stability, and quality control standards

6.3.1.2 Safety and Efficacy Claims

This aspect of a TPP prioritizes indications and claims with high potential rewards and minimal risks. It outlines existing data required for trial design and identifies any additional preclinical data needed for regulatory compliance.

Defining the desired efficacy in a TPP lacks a universal approach. Different methods exist, such as setting a simple threshold, establishing a base case with a minimally acceptable efficacy, or adopting multilevel approaches with a base case, a minimal acceptable efficacy (Lower Reference Value, LRV), and a Target Value (TV) (Lalonde et al. 2007). Here, the LRV is the smallest treatment effect that is clinically meaningful for advancing the compound to the next stage of development, whereas the TV is the desired treatment effect associated with solid competitiveness, sometimes leading to a change in the standard of care.

As noted by Pulkstenis et al. (2017), the advantage of employing a multilevel TPP lies in addressing the challenge of a drug's value not being adequately represented by a single threshold. The value curve encounters interpretability issues near the threshold due to its steepness as a binary event. Thus, a multilevel TPP offers additional granularity, facilitating decision-making, although it may not suit all contexts (e.g., noninferiority). Choosing between a "binary" and "multilevel" TPP depends on evaluating the drug's value across various efficacy values. A binary TPP is suitable when a single well-established threshold exists, whereas a multilevel TPP is preferable when discriminating across a range of efficacy levels, from marginal to solid efficacy, is desired.

Establishing a TPP typically involves synthesizing information from various sources, including available data, the competitive landscape, and regulatory precedents. Once defined, accruing evidence can be evaluated against the TPP to inform decision-making. This framework results in three TPP "case profiles": the Minimal Case Profile (M), where $\theta \leq \theta$ LRV; the Lower Case Profile (L), where θ LRV $< \theta < \theta$ TV; and the Target Case Profile (T), where $\theta \geq \theta$ TV, defining the desired efficacy range as $L \leq T$.

6.3.1.3 *Potential Economic Value*

Understanding treatment patterns and anticipating future healthcare system changes are crucial. Assessment of the current standard of care, global differences, market size, competitor pricing, and reimbursement environment aids in predicting the drug's relevance over time and expected return on investment.

6.3.1.4 *Best Differentiating Features or Outcomes*

Identifying unique features supported by existing data and assessing their ability to justify premium pricing is essential. Clearly outlining the product's benefits is critical for effective marketing strategies.

6.3.1.5 *Exclusivity Strategy*

This section aims to position the device for marketing exclusivity. It encompasses product availability, reimbursement data, regulatory classification, trademarking, branding strategy, and intellectual property considerations.

6.3.2 Clinical Development Plan

A Clinical Development Plan (CDP) is crucial for drug development as it provides a comprehensive roadmap outlining the strategy and stages required to bring a new drug from discovery through to market approval. The CDP includes detailed plans for clinical trial phases and studies, regulatory interactions, and post-marketing commitments. Its relationship with the TPP is synergistic – while the TPP defines the desired characteristics and goals for the drug, the CDP translates these targets into actionable steps and timelines. The CDP embodies a holistic approach by integrating scientific, regulatory, and commercial considerations, ensuring that each phase of development is aligned with the overall objectives. This structured framework helps to anticipate and mitigate risks, allocate resources efficiently, and adapt to new information or regulatory changes, thereby increasing the probability of clinical trial success. By aligning the development process with clearly defined targets and a strategic execution plan, the CDP ensures that all aspects of drug development are cohesively managed, facilitating a smoother path to market.

6.3.2.1 Intended Use

The CDP outlines the strategic approach for conducting clinical trials to evaluate the safety and efficacy of a new medical product, in alignment with the defined TPP. By integrating the objectives and parameters set forth in the TPP, the CDP ensures that the development process remains focused and efficient, ultimately maximizing the likelihood of regulatory approval and successful market entry. Key elements of a CDP are listed in Table 6.2.

6.3.2.2 Trial Designs

Select appropriate trial designs (e.g., randomized controlled trials, adaptive designs) that address specific questions outlined in the TPP. Determine sample size calculations and statistical methodologies to ensure the trials are adequately powered to detect meaningful differences.

6.3.2.3 Study Objectives and Endpoints

Define clear objectives for each clinical trial, aligning with the desired outcomes outlined in the TPP. Specify primary and secondary endpoints that accurately measure the safety, efficacy, and other relevant parameters of the product.

TABLE 6.2

Components of Clinical Development Plan

Component	Description
Clinical Trial Strategy	It defines the clinical trials needed to assess the safety, efficacy, and other critical aspects of the investigational product. This includes outlining the phases of clinical trials, the sequence in which they'll be conducted, and the rationale behind each trial's design.
Trial Objectives and Endpoints	It details the specific objectives, primary and secondary endpoints, and measurable outcomes for each clinical trial within the program.
Patient Population and Recruitment Strategy	It identifies the target patient population, enrollment criteria, and strategies for patient recruitment and retention.
Regulatory and Compliance Considerations	It outlines the regulatory requirements and strategies for interactions with regulatory agencies. It includes plans for meeting regulatory standards and obtaining approvals.

6.3.2.4 Patient Population

Identify and characterize the target patient population based on the indications specified in the TPP. Consider inclusion and exclusion criteria to ensure the enrollment of patients who are representative of the intended market.

6.3.2.5 Safety and Monitoring

Develop comprehensive safety monitoring plans to detect and manage adverse events during the clinical trials. Implement data safety monitoring boards and interim analyses to safeguard participant well-being and maintain trial integrity.

6.3.2.6 Regulatory Strategy

Develop a regulatory strategy that aligns with the requirements of regulatory agencies and the expectations outlined in the TPP. Plan for interactions with regulatory authorities to seek guidance, provide updates, and address any concerns throughout the development process.

6.3.2.7 Timeline and Milestones

Establish realistic timelines and milestones for each phase of the clinical development process, considering factors such as patient recruitment, data collection, and regulatory submissions. Regularly monitor progress against milestones and adjust timelines as needed to ensure timely completion of the development program.

6.4 Statistical Frameworks for Clinical Decision-Making

The broad adoption of MIDD paradigm in the past decade has brought innovative statistical modeling and simulation to the forefront of clinical trials. In recent years, the utilization of statistical methodologies to guide drug development decision-making processes is experiencing a notable surge across the lifecycle of drug development. This surge is particularly evident in clinical trials, where decisions often hinge on complex data analysis and predictive modeling. Within this context, significant attention has been directed toward refining statistical techniques aimed at predicting trial outcome, ensuring clinical assurance, and harnessing Bayesian methodologies to bolster decision-making based on individual trials, incorporating informative priors on treatment effects and placebo.

One of the primary areas of focus in statistical decision-making lies in estimating the probability of technical success (PTS). This involves employing sophisticated statistical models to assess the likelihood of success or failure of a given technical endeavor, be it the identification of maximum tolerated dose (MTD), transition from Phase II to Phase III, or a positive pivotal trial. By leveraging historical data, simulation techniques, and advanced probability distributions, statisticians can provide stakeholders with valuable insights into the potential outcomes of their initiatives, enabling more informed decision-making and risk management strategies.

Furthermore, ensuring clinical assurance in medical research and healthcare interventions is paramount. Statistical methods play a crucial role in this regard by facilitating the design and analysis of clinical trials, the evaluation of treatment efficacy, and the validation of medical devices and procedures. By employing rigorous experimental designs, robust statistical tests, and meticulous data analysis techniques, researchers can ascertain the safety and effectiveness of new medical interventions, thereby guiding regulatory approval processes and clinical practice guidelines.

More recently, Bayesian methods have gained traction as powerful tools for supporting decision-making in various contexts. Unlike traditional frequentist approaches, Bayesian statistics allow for the incorporation of prior knowledge or beliefs into the analysis, thereby enabling more nuanced and context-specific inference. This is particularly valuable in the realm of clinical trials, where informative priors on treatment effects and placebo responses can enhance the efficiency and accuracy of statistical inference, especially in cases where sample sizes are limited, or data are sparse. By leveraging Bayesian frameworks, researchers and practitioners can make more informed decisions based on a comprehensive synthesis of existing evidence and expert judgment.

However, to truly enhance evidence-based decision-making, it is imperative to establish a formal quantitative framework that aligns with the objectives of the current development stage. This entails defining clear metrics and performance indicators that reflect the specific goals, constraints, and priorities of the project or initiative at hand. Whether it be maximizing profitability, minimizing risk, optimizing resource allocation, or achieving regulatory compliance, the chosen metrics should provide actionable insights and guide decision-makers toward the most favorable outcomes.

Moreover, this quantitative framework should be dynamic and adaptive, capable of evolving in response to changing circumstances, new information, and shifting stakeholder preferences. This necessitates ongoing monitoring, evaluation, and refinement of the statistical models and decision-making processes, ensuring their relevance and effectiveness over time. By fostering a culture of continuous improvement and learning, organizations can harness the full potential of statistical methods to drive innovation, mitigate risks, and achieve sustainable success.

In the following sections, we discuss various statistical frameworks for making decisions during clinical development. We discuss the traditional approaches and latest developments.

6.4.1 Identification of Maximum Tolerated Dose

6.4.1.1 Background

Phase I dose-escalation clinical trials aim to determine the MTD of a new drug for a specific mode of administration. Ethical considerations necessitate minimizing the risk of subjecting patients to excessive toxicity. Therefore, dose-escalation studies are often conducted sequentially, beginning with a low dose and gradually escalating to higher doses. Implicit in this approach is the assumption that the likelihood of encountering toxicity rises with dosage. These trials may involve healthy volunteers if the drugs are relatively non-toxic, whereas drugs with known toxicities are typically tested in the target patient population. Numerous factors are considered when designing Phase I dose-escalation trials, encompassing the selection of a starting dose, the definition of dose-limiting toxicity (DLT), the dose escalation decision rules, and the determination of the MTD upon study completion. The quality of the decision of the MTD is usually assessed through simulations, in which the probability of correctly identifying the true MTD is assessed. Other metrics such as over- or under-exposure may also be considered, along with PK, PD, and drug activity data.

6.4.1.2 3+3 Design

According to a survey conducted by Rotgako et al. (2007), the "3 + 3 design" stands out as the most utilized approach. Its primary objective is to ascertain MTD, defined as the highest dose at which a predetermined proportion of patients encounter a DLT. Typically, the MTD falls within the range of 20% to 33%. A DLT refers to an adverse reaction to the treatment that raises safety concerns and is often specific to the study. Figure 6.1 illustrates a schematic overview of the rules for dose-escalation and MTD selection without dose de-escalation. In essence, at dose level i, a cohort of three patients is enrolled. If none of the three patients experience a DLT, dose escalation occurs, and three patients are enrolled in the subsequent higher dose level (i + 1). If one out of the three patients exhibits a DLT, three more patients are added to the current dose group. If none of these additional patients experience a DLT, dose escalation continues; otherwise, dose level (i + 1) is deemed the MTD.

Despite their conceptual simplicity and ease of implementation, algorithm-based designs suffer from several drawbacks. Chief among these are: (1) Under-testing at MTD and overexposure to suboptimal doses: Existing evidence suggests that algorithm-based designs tend to test only a few patients at the MTD level, while subjecting a disproportionate number of patients

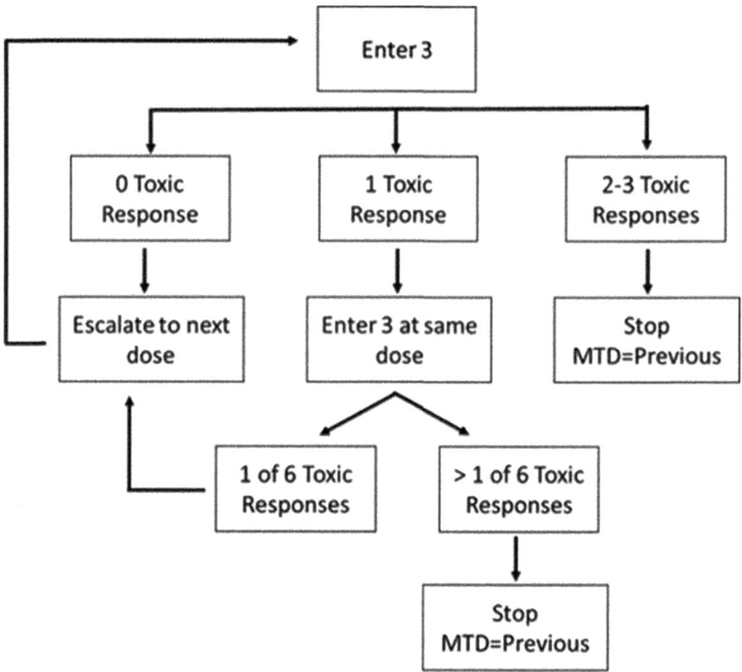

FIGURE 6.1
Decision rules from dose-escalation and MTD selection in 3+3 design.

to suboptimal doses (Heyd and Carlin 1999; O'Quigley et al. 1990); (2) Low probability of identifying the MTD: Research indicates that the likelihood of identifying the true MTD using algorithm-based designs is notably low (Thall and Lee 2003); (3) High uncertainty in MTD estimation due to small sample sizes: The limited sample sizes inherent in algorithm-based designs lead to considerable uncertainty in estimating the MTD (Goodman, Zahurak, and Piantadosi 1995); and (4) Poor operating characteristics: Algorithm-based designs often exhibit poor operating characteristics, failing to achieve the desired target interval for the DLT rate, such as the widely accepted range between 16% and 33% in the 3+3 design (Lin and Shih 2001; He et al. 2006).

The Continual Reassessment Method (CRM) was introduced by O'Quigley et al. (1990) to overcome the limitations associated with the 3+3 designs. Operating under the assumption that the likelihood of a toxic response increases with dosage, CRM employs a parametric model to characterize the dose-toxicity relationship. Utilizing a Bayesian framework, CRM continually updates this relationship based on the posterior distribution of model parameters following each patient's response.

In CRM, the initial patient is allocated to the dose nearest to the estimated MTD according to prior probabilities. Subsequent patients are then assigned doses based on the updated dose-toxicity relationship, aiming to converge toward the MTD.

Compared to traditional algorithm-based approaches, CRM offers several advantages, as highlighted by O'Quigley et al. (1990) and O'Quigley and Chevret (1991). It minimizes the risk of assigning patients to suboptimal or harmful doses, improves efficiency in MTD determination, and alleviates ethical concerns. Furthermore, CRM doesn't require precise modeling across the entire dose range; instead, it focuses on accurately estimating the MTD and gathers more information around this point. Consequently, even relatively simple one-parameter models can perform comparably to more complex models, especially when doses deviate from the target MTD (O'Quigley et al. 1990).

While the CRM has gained popularity since its original publication by O'Quigley, Pepe, and Fisher, it has also sparked several controversies. Notably, it has been criticized for potentially exposing patients to higher doses compared to traditional algorithm-based designs (Ratain et al. 1993). In response, various modifications have been proposed. Some authors advocate for selecting the lowest dose as the starting point and limiting dose escalation to no more than one level increase (Korn et al. 1994; Moller 1995). Additionally, Goodman et al. (1995) suggested including two or three patients at each dose level. For further detailed discussions and alternative suggestions, refer to O'Quigley and Chevret (1991), O'Quigley (1992), O'Quigley and Zhen (1996), Piantadosi and Liu (1996), and Heyd and Carlin (1999).

6.4.1.3 Model-Informed Designs

Implementing Bayesian CRM methods presents several challenges. Firstly, it requires a high level of statistical expertise and continual adjustment of dose-escalation rules based on real-time data from the trial, making practical implementation demanding. Secondly, investigators must define a target toxicity probability, which can be a complex task. Additionally, the sensitivity of the methods to the selection of priors further complicates implementation.

Among these challenges, defining both the target toxicity and the dose-toxicity relationship poses a particularly daunting task for investigators. An alternative approach proposed by Ji et al. (2007) addresses this issue by allowing investigators to specify a range of target toxicity probabil ities, referred to as a toxicity probability interval. This method represents a hybrid of Bayesian CRM and the 3+3 design, offering a potential solution to the complexities associated with defining target toxicity and dose-toxicity relationships simultaneously. Additionally, similar to the CRM method, another notable feature of the mTPI design is its integration of Bayesian statistical methods. By leveraging these methods, the mTPI design can incorporate prior data and expert opinions into the analysis. This integration enhances the accuracy of MTD estimates and boosts the overall efficiency of the study.

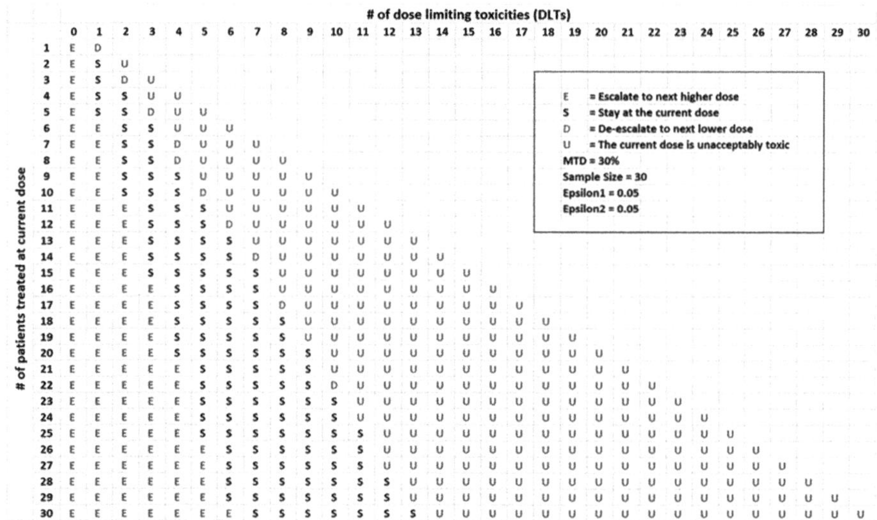

# of dose limiting toxicities (DLTs)

# of patients treated at current dose	0	1	2	3	4	5	6	7	8	9	10	11	12	13	14	15	16	17	18	19	20	21	22	23	24	25	26	27	28	29	30
1	E	D																													
2	E	S	U																												
3	E	S	D	U																											
4	E	S	S	U	U																										
5	E	S	S	D	U	U																									
6	E	E	S	S	U	U	U																								
7	E	E	S	S	D	U	U	U																							
8	E	E	S	S	D	U	U	U	U																						
9	E	E	S	S	S	U	U	U	U	U																					
10	E	E	S	S	S	D	U	U	U	U	U																				
11	E	E	E	S	S	S	U	U	U	U	U	U																			
12	E	E	E	S	S	S	D	U	U	U	U	U	U																		
13	E	E	E	S	S	S	S	U	U	U	U	U	U	U																	
14	E	E	E	S	S	S	S	D	U	U	U	U	U	U	U																
15	E	E	E	S	S	S	S	S	U	U	U	U	U	U	U	U															
16	E	E	E	E	S	S	S	S	U	U	U	U	U	U	U	U	U														
17	E	E	E	E	S	S	S	S	D	U	U	U	U	U	U	U	U	U													
18	E	E	E	E	S	S	S	S	S	U	U	U	U	U	U	U	U	U	U												
19	E	E	E	E	S	S	S	S	S	U	U	U	U	U	U	U	U	U	U	U											
20	E	E	E	E	S	S	S	S	S	S	U	U	U	U	U	U	U	U	U	U	U										
21	E	E	E	E	E	S	S	S	S	S	U	U	U	U	U	U	U	U	U	U	U	U									
22	E	E	E	E	E	S	S	S	S	S	D	U	U	U	U	U	U	U	U	U	U	U	U								
23	E	E	E	E	E	S	S	S	S	S	S	U	U	U	U	U	U	U	U	U	U	U	U	U							
24	E	E	E	E	E	S	S	S	S	S	S	U	U	U	U	U	U	U	U	U	U	U	U	U	U						
25	E	E	E	E	E	S	S	S	S	S	S	S	U	U	U	U	U	U	U	U	U	U	U	U	U	U					
26	E	E	E	E	E	E	S	S	S	S	S	S	U	U	U	U	U	U	U	U	U	U	U	U	U	U	U				
27	E	E	E	E	E	E	S	S	S	S	S	S	U	U	U	U	U	U	U	U	U	U	U	U	U	U	U	U			
28	E	E	E	E	E	E	S	S	S	S	S	S	S	U	U	U	U	U	U	U	U	U	U	U	U	U	U	U	U		
29	E	E	E	E	E	E	S	S	S	S	S	S	S	U	U	U	U	U	U	U	U	U	U	U	U	U	U	U	U	U	
30	E	E	E	E	E	E	E	S	S	S	S	S	S	S	U	U	U	U	U	U	U	U	U	U	U	U	U	U	U	U	U

E = Escalate to next higher dose
S = Stay at the current dose
D = De-escalate to next lower dose
U = The current dose is unacceptably toxic
MTD = 30%
Sample Size = 30
Epsilon1 = 0.05
Epsilon2 = 0.05

FIGURE 6.2
Dose assignment as a function of the number of patients in a cohort and the number of patient experience dose limiting toxicities.

As demonstrated by Ji et al. (2007), the determination of whether to deescalate, maintain, or escalate doses can be made based on the number of patients (n_j) in the current cohort (j) and the number of observed toxicities within the cohort. Consequently, it becomes possible to devise a table outlining dose-assignment actions prior to the commencement of the trial (Figure 6.2).

In more recent developments, another Bayesian-designed approach, the Bayesian Optimal Interval (BOIN) design, was proposed by Liu and Yuan (2015). Like the mTPI design, it employs Bayesian statistical methods to steer the dose escalation process and integrate prior data and expert opinion into the analysis. However, the BOIN design diverges from the mTPI design by utilizing a BOIN to direct the dose escalation process instead of a modified TPI (Yuan et al. 2016, 2019; Zhou et al. 2018; Zhou et al. 2018; Zhou et al. 2021; Zhou et al. 2021) The BOIN design has earned recognition as a "fit-for-purpose" method by the FDA (2021b).

6.4.1.3.1 Performance Characteristics

For any decision-making framework, it is important to evaluate its operating characteristics, which characterize a clinical trial's ability to inform decision-making. Since the primary objective of a Phase I dose-escalation trial is to determine the MTD, the performance of a decision-making framework is assessed against this metric. This can be accomplished through simulations for various possible toxicity profiles. For illustration, consider a BOIN design for a Phase I study where five dose levels are evaluated for the identification

TABLE 6.3

Dose-Toxicity Relationships for Simulations

	Dose Level				
Scenario	**1**	**2**	**3**	**4**	**5**
1	0.30	0.46	0.50	0.54	0.58
2	0.26	0.30	0.47	0.54	0.60
3	0.04	0.15	0.30	0.48	0.68
4	0.02	0.07	0.12	0.30	0.45
5	0.02	0.06	0.10	0.13	0.30

TABLE 6.4

Operating Characteristics of the BOIN Design

	Dose Level					Number of Patients	% Early Stopping
	1	**2**	**3**	**4**	**5**		
Scenario 1							
True DLT Rate	0.3	0.46	0.5	0.54	0.58		
Selection %	66.9	17.1	2.7	0.7	0.1		12.5
% Pts Treated	62.5	30.6	5.8	1	0.2	16.1	
Scenario 2							
True DLT Rate	0.16	0.3	0.47	0.54	0.6		
Selection %	25.8	54.2	16.7	2.5	0.1		0.7
% Pts Treated	35.7	41.9	18.8	3.3	0.3	20.6	
Scenario 3							
True DLT Rate	0.04	0.15	0.3	0.48	0.68		
Selection %	1.2	26.1	52.7	19.2	0.8		0
% Pts Treated	16.7	30.6	35.2	15.4	2	22.6	
Scenario 4							
True DLT Rate	0.02	0.07	0.12	0.3	0.45		
Selection %	0.1	2.3	20.9	52.2	24.5		0
% Pts Treated	13.8	17.5	25.2	29	14.5	23.6	
Scenario 5							
True DLT Rate	0.02	0.06	0.1	0.13	0.3		
Selection %	0	0.7	4.9	25.3	69.1		0
% Pts Treated	13.4	15.7	18	22.6	30.2	23.8	

Note: "% Early Stopping" refers to early stopping due to excessive DLT.

of MTD. Assuming the target toxicity level is 0.30, a simulation study was conducted using five scenarios where the MDT are at dose 1 (D1), dose 2 (D2), dose 3 (D3), dose 4 (D4), and dose 5 (D5), respectively (Table 6.3).

For each scenario, 1,000 runs of simulations were carried out and the performance of both the BOIN method and the traditional 3+3 design were estimated. The results are presented in Tables 6.4 and 6.5. It is evident that

TABLE 6.5

Operating Characteristics of 3+3 Design

	Dose Level					Number of Patients	% Early Stopping
	1	2	3	4	5		
Scenario 1							
True DLT Rate	0.3	0.46	0.5	0.54	0.58		
Selection %	34.2	7.1	0.9	0.5	0		57.3
% Pts Treated	63.6	29	6.1	1.1	0.2	8.2	
Scenario 2							
True DLT Rate	0.16	0.3	0.47	0.54	0.6		
Selection %	42.7	28.7	4.4	1	0.1		23.1
% Pts Treated	44.2	36.4	15.8	3.1	0.5	11.3	
Scenario 3							
True DLT Rate	0.04	0.15	0.3	0.48	0.68		
Selection %	23.2	43.3	27.3	4.5	0.1		1.6
% Pts Treated	26.4	33.4	27.3	11.1	1.7	14.9	
Scenario 4							
True DLT Rate	0.02	0.07	0.12	0.3	0.45		
Selection %	5.6	13.5	48.3	26.9	5.5		0.2
% Pts Treated	18.5	21.5	26.2	23.6	10.2	18.2	
Scenario 5							
True DLT Rate	0.02	0.06	0.1	0.13	0.3		
Selection %	3.2	8.9	13.5	42.4	31.6		0.4
% Pts Treated	17	19	20.4	22.7	20.8	19.2	

Note: "% Early Stopping" refers to early stopping due to excessive DLT.

while on average the BOIN used more subjects than the 3+3 method, it has a much higher probability to detect the MTD.

6.4.2 Decision-Making in Proof-of-Concept Trials

In the new drug development process, one of the most important milestones for a drug candidate is the establishment of Proof-of-Concept (PoC). PoC trials, typically conducted in early Phase II stage, aim to provide preliminary evidence of drug efficacy to drive "Go/No-Go" decisions. Successful outcomes from these trials not only guide decision-making regarding further development but also attract investment and support for subsequent larger-scale trials. Conversely, negative results may prompt re-evaluation of the therapeutic approach or discontinuation of development, thus mitigating resource allocation toward ineffective interventions. Overall, PoC trials play a pivotal role in de-risking the development process and shaping the trajectory of novel therapeutics toward eventual regulatory approval and clinical implementation. PoC trials can be either single arm or randomized studies, depending on the availability of standard of care comparators. Oftentimes, it is powered to test the hypothesis of drug activity characterized by the primary efficacy endpoints.

6.4.2.1 *Traditional Approach*

Traditionally, the decision regarding whether the primary efficacy endpoint is achieved in a PoC is made through hypothesis-testing. Consider an oncology study where the patient response is binary. The study is designed to test the hypotheses concerning a new therapy:

$$H_0: p \leq p_0 \text{ vs. } H_1: p > p_1,$$

where p is the overall response rate of the new therapy, p_0 is the response rate of the standard of care, p_1 is usually the target rate which represents a meaningful improvement of the new therapy over the standard of care. Both rates are typically defined in the TPP.

The outcomes of the hypothesis testing entail whether moving the drug for further development or not. In this context, the rejection of the null hypothesis results in a "Go" decision, leading to dose optimization in Phase IIb studies to long-term large-scale Phase III confirmatory trials while failing to reject the null hypothesis leads to a "No-Go" decision. The associated sample size and statistical method are chosen such that the risks of false "Go" and false "No-Go" will be capped by two pre-specified quantities α and β. This framework is depicted in Table 6.6.

6.4.2.2 *Evidence-Based Framework*

6.4.2.2.1 *Limitations of Current Methods*

While the p-value based approaches are conventionally used in assisting "Go/No-Go" decisions from PoC trials, they do not provide a good measure for the strength of evidence, particularly as related to the TPP. Therefore, there is a need for clear and evidence-based decision-making. In addition, the

TABLE 6.6

Go or No-Go Decision Based on Traditional Hypothesis Testing

		Decision[a]	
		Not Reject H0 (No-Go)	**Reject H0 (Go)**
TRUE	H_0	$1-\alpha$	α
	H_a	β	$1-\beta$

Note:
[a] α = Type I error; β = 1-Type II error.

traditional approach provides a dichotomous Go/No-Go decision-making framework which is not applicable to situations where further efforts are needed in gathering additional information through new studies before a Go/No-Go decision is made.

6.4.2.2.2 Framework for Multilevel Decision

Several authors have proposed multilevel decision-making frameworks (Lalonde et al. 2007; Sargent and Godberg 2001; Paul et al. 2010). These multilevel outcome designs typically include a trial outcome for results that fall between extremely promising and extremely poor, termed as a Consider decision. In cases where results are within the Consider zone, the decision on whether and how to proceed in development is permitted to rely on factors beyond the proportion of patients achieving a response.

These authors emphasize that these multilevel outcome designs better reflect real-life development decision-making, where moderate efficacy results must be weighed against various other development considerations such as safety, secondary efficacy endpoints, ease of administration, manufacturing concerns, internal priorities, and competitive landscape. Moreover, multilevel outcome approaches offer the advantage of requiring smaller sample sizes while maintaining statistical Type I and Type II error rates. This control over error rates with a smaller sample size is achieved because trial outcomes within the Consider zone do not align with traditionally defined statistical errors. However, decision-making for trials that conclude in the Consider zone remains subjective and undefined. In such cases, after qualitatively considering other relevant factors, sponsors must still make a functional Go/No-Go decision on whether to invest further or not. Ultimately, outcomes within the Consider zone are redefined into Go or No-Go decisions in a non-quantifiable manner. In the following, we discuss Lalonde's method in detail.

Lalonde et al. (2007) proposed a multilevel approach based on confidence intervals. The method defines a LRV and a TV, where the TV must surpass the LRV. Comparable to the previous methods aimed at hypothesis testing, decisions at the trial's conclusion hinge on comparing confidence intervals around the observed proportion responding to the LRV and the TV. Specifically, let LLC_{α} and $UL_{1-\beta}$ denote the lower and upper confidence limits of a $(1-\alpha)\times 100\%$ confidence interval and a $(1-\beta)\times 100\%$ confidence interval, respectively. Here α and β correspond to false Go and false No-Go risks, respectively. The decision rules of this method are given in Table 6.7.

Thus, a "Go" decision is predicated on a high level of confidence that the proportion responding meets or exceeds the LRV, with the TV falling within the confidence interval. Conversely, a "Stop" decision is grounded in a high level of certainty that the proportion responding fails to reach the TV threshold. When neither a "Go" nor a "Stop" decision can be made with a high level of certainty, the trial result is categorized as "Consider." This

TABLE 6.7

Decision Rules of the Method

Outcome	Decision
Go	If $LCI_{\alpha} > LRV$ and $UCI_{1-\beta} > TV$
No-Go	If $UCI_{1-\beta} < TV$
Consider	If neither the Go nor No-Go criteria are met

Source: Lalonde et al. (2007).

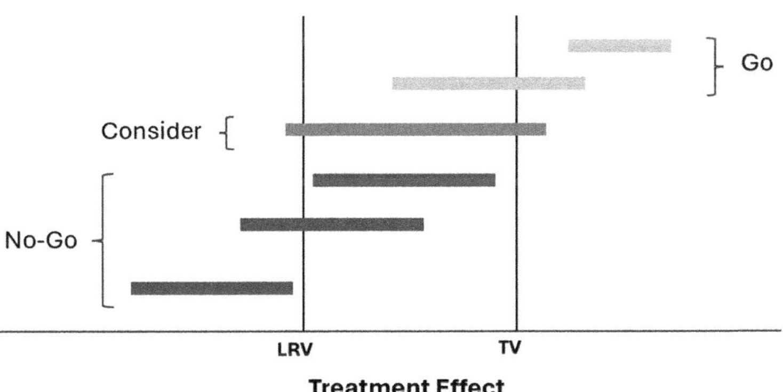

FIGURE 6.3
An illustration of the decision framework by Lalonde et al. (2007).

framework allows for nuanced decision-making based on the available evidence, providing a structured approach to evaluating trial outcomes in scenarios of varying confidence levels.

The two threshold values LRV and TV normally correspond to the minimum and target efficacy values defined in the TPP, respectively. When the trial outcome falls within the "Stop" or "Go" zone, the decision is typically straightforward. However, if the outcome resides within the "Consider" zone, it is imperative to establish decision criteria for transitioning out of this zone. This may entail incorporating secondary endpoints or leveraging competitor data on a similar compound to inform decision-making. By establishing clear criteria for moving out of the "Consider" zone, researchers can effectively navigate uncertain outcomes and make informed decisions regarding the continuation or discontinuation of the trial. This proactive approach ensures that trial outcomes are thoroughly evaluated and that appropriate actions are taken to advance the development process.

Application of the method is illustrated in Figure 6.3.

6.4.2.2.3 Bayesian Multilevel TPP Decision Framework

In the setting of late-stage clinical development, the decision to transition from PoC stage to late-stage trials has signification implications for both patients and sponsors. This necessitates through evaluation of risk and benefit in light of the data from the PoC Phase II trial. Of interest is the estimation of PTS in consecutive phases. Pulkstenis et al. (2017) proposed a Bayesian decision-making framework to facilitate the evaluation of evidence in the context of multilevel TPP.

Compared to the Frequentist approaches discussed previously, this approach has several advantages. Specifically, it addresses the shortcomings of procedures relying solely on *p*-values by directly quantifying the strength of evidence against an established TPP. Furthermore, the Bayesian framework permits proper evidentiary and probabilistic interpretation of various decisions. Lastly, it allows for the integration of prior information into the decision-making process, thus enhancing the robustness of the outcomes (Pulkstenis et al., 2017).

To illustrate the method, consider a PoC oncology trial aimed to demonstrate the activity of a new drug with the primary endpoint of objective response rate. Assume patients are randomized into a treatment arm and control arm. Let $\pi(p_0)$ and $\pi(p_1)$ be two independent prior distributions of let p_0 and p_1, respectively. Let $\theta = p_1 - p_0$,which represents the treatment effect. Denote $X_{n_1} = (x_1, \ldots, x_{n_1})$ and $Y_{m_1} = (y_1, \ldots, x_y)$ be the response data of the treatment and the control, respectively. Further assume that $X_{n_1} \sim \text{binomial}(n_1, p_1)$ and $Y_{m_1} \sim \text{binomial}(m_1, p_0)$ and that beta priors, $\text{Beta}(a_i, b_i)$, $i = 0,1$, are chosen for p_0 and p_1, respectively. Under these distributional assumptions, we can obtain the posterior distributions:

$$p_0 \mid Y_{m_1} = y_{m_1} \sim \text{Beta}\left(a_0 + y_{m_1}, b_0 + m - y_{m_1}\right),$$

$$p_1 \mid X_{n_1} = x_{n_1} \sim \text{Beta}\left(a_1 + x_{n_1}, b_1 + n_1 - x_{n_1}\right).$$

Thus, the posterior probability $\pi\left(\theta \mid X = \left(X_{n_1}, Y_{m_1}\right)\right) = P\left[\theta \mid X_{n_1} = x_{n_1}, Y_{m_1} = y_{n_1}\right]$ is given by

$$F\theta(t) = \begin{cases} \int_{-t}^{1} F_{p1}(t+u) f_{po}(u)du & -1 \le t \le 0 \\ \int_{0}^{1-t} F_{p1}(t+u) f_{po}(u)du + \int_{1-t}^{1} f_{po}(u)du + & 0 \le t \le 1 \end{cases}$$

$$P\left(x_{n_1}, n_1, y_{m1}, m_1, \delta_0\right) =$$

$$\int_0^{1-\delta_0} [1 - F(p+\delta_0; a_1 + x_n, b_1 + n - x_n)] f(p; a_0 + y_{m_1}, b_0 + m - y_{m1}) dp,$$

where the functions $F(\bullet)$ and $f(\bullet)$ are the cumulative probability and density functions of $\text{Beta}(a_1 + x_{n_1}, b_1 + n_1 - x_{n1})$ and $\text{Beta}(a_0 + y_{m_1}, b_0 + m_1 - y_{m1})$, respectively.

Let $\text{LLC}_{\tau_{\text{LRV}}}(X)$ and $\text{UL}_{1-\tau_{\text{TV}}}(X)$ denote the lower and upper confidence limits of a $(1 - \tau_{\text{LRV}} - \tau_{\text{TV}}) \times 100\%$ credible interval of θ, respectively, where τ_{LRV} represents the false Go risk defined as the maximum allowable probability that the minimal case profile is true when a Go decision is made; whereas τ_{TV} denotes the false No-GO risk defined as the maximum allowable probability that the target case is true when a No-Go decision is declared. The posterior probabilities of the minimal, lower, and target PPT case profiles are given in Table 6.8.

The criteria for Go/No-Go decisions are presented in Table 6.9.

As an illustration, Pulkstenis et al. (2017) presented the following example (Figure 6.4).

It is worth noting that there is no direct translation between the decision-making criteria in Table 6.9 and treatment effects that may or may not be statistically significant. In essence, these decisions are directly related to the TPP

TABLE 6.8

Posterior Probabilities of Associated TTP Case Profiles

TPP Case Profile	Parameter Space	Posterior Probability
Minimal	$\text{M}: \theta \le \theta_{\text{LRV}}$	$\Pr_\theta(\text{M}\mid \text{x}) = \int_{\theta\in\text{M}} \pi(\theta\mid \text{x}) d\theta = F_\theta(\theta_{\text{LRV}})$
Lower	$\text{L}: \theta_{\text{LRV}} < \theta < \theta_{\text{TV}}$	$\Pr_\theta(\text{L}\mid \text{x}) = \int_{\theta\in\text{L}} \pi(\theta\mid \text{x}) d\theta = F_\theta(\theta_{\text{TV}}) - F_\theta(\theta_{\text{LRV}})$
Target	$\text{T}: \theta \ge \theta_{\text{TV}}$	$\Pr_\theta(\text{T}\mid \text{x}) = \int_{\theta\in\text{T}} \pi(\theta\mid \text{x}) d\theta = 1 - F_\theta(\theta_{\text{TV}})$

Source: Pulkstenis et al. (2017).

TABLE 6.9

Bayesian Decision Rules for Go/No-Go Decisions

TPP Case Profile	Parameter Space	Posterior Probability
Minimal	$\text{M}: \theta \le \theta_{\text{LRV}}$	$\Pr_\theta(\text{M}\mid \text{x}) = \int_{\theta\in\text{M}} \pi(\theta\mid \text{x}) d\theta = F_\theta(\theta_{\text{LRV}})$
Lower	$\text{L}: \theta_{\text{LRV}} < \theta < \theta_{\text{TV}}$	$\Pr_\theta(\text{L}\mid \text{x}) = \int_{\theta\in\text{L}} \pi(\theta\mid \text{x}) d\theta = F_\theta(\theta_{\text{TV}}) - F_\theta(\theta_{\text{LRV}})$
Target	$\text{T}: \theta \ge \theta_{\text{TV}}$	$\Pr_\theta(\text{T}\mid \text{x}) = \int_{\theta\in\text{T}} \pi(\theta\mid \text{x}) d\theta = 1 - F_\theta(\theta_{\text{TV}})$

Source: Pulkstenis et al. (2017).

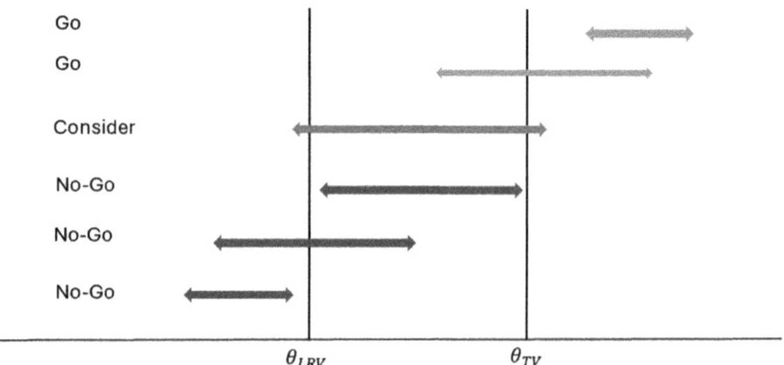

FIGURE 6.4
An example of using Bayesian credible intervals for decision-making. (Adapted from Pulkstenis et al. 2017.)

but de-linked from statistical significance. As previously discussed, careful consideration needs to be given to the selection of the risk levels.

6.4.3 Decision Framework for Multi-Stage Trials

6.4.3.1 Frequentist Approaches

As previously mentioned, positive outcomes of PoC trials may lead to dose optimization Phase IIb trials and Phase III confirmatory trials. When compared to PoC trials, these studies are larger in size and longer in duration. Multi-stage designs have long been used for phase II/III efficacy studies to increase trial efficiency. Early two-stage designs include the work by Gehan (1961) and Simon (1989). Both designs render the sponsor an early termination of the study due to futility. Simon optimal design was constructed to minimize the expected sample size under the null hypothesis. An alternate design called minimax design is also considered by Simon (1989). It is aimed at minimizing the maximum sample size. In either case, Type I and Type II errors are controlled. Using different optimization criteria, many other multi-stage designs have been proposed.

For example, Fleming (1982) proposed a one-sample multiple testing procedure which allows for early termination for futility or efficacy while preserving the size, power and simplicity of the single-stage procedure, Bryant and Day (1995) suggested two-stage design that simultaneously assesses both clinical response and toxicity, Jung et al. (2004) developed a family of two-stage designs that are admissible according to a Bayesian decision-theoretic criterion based on an ethically justifiable loss function. Several group sequential designs were suggested by Pocock (1977, 1982), and O'Brien and Fleming (1979) to allow for repeated testing. Wang and

Tsiatis (1987) and DeMets and Lan (1994) extended the early work by introducing the concept of alpha spending, which rendered greater flexibility in conducting interim analyses.

Although Frequentist multi-stage designs in general increase trial efficiency, they have several limitations. For example, they have the rigid requirement of fixed sample size at each stage of analysis, which is difficult to adhere to particularly in multi-center trials (Green and Dahlberg 1992; Lee and Liu 2008). In addition, Type I and Type II protection is not warranted if the actual interim analysis is not performance strictly in accordance with the original design. As pointed out by Lee and Liu (2008), these designs also lack a formal mechanism to terminate the trial early before the pre-specified sample size is reached. The above-mentioned challenges make it difficult to implement the frequentist multi-stage designs and argue for more flexible alternatives.

6.4.3.2 Bayesian Methods

The challenges of the Frequentist approaches underscore the need for more flexible and adaptive approaches to multi-stage trial design. Incorporating Bayesian methodologies and adaptive trial designs can offer solutions to address these limitations. Bayesian approaches provide a framework for seamlessly integrating prior knowledge, interim data, and evolving trial information, allowing for more informed and efficient decision-making throughout the trial process. Additionally, adaptive designs offer the flexibility to modify trial parameters based on accumulating data, thereby optimizing resource allocation, minimizing ethical concerns, and maximizing the likelihood of successful trial outcomes.

Several Bayesian multi-stage designs have been proposed in literature (Thall and Simon 1994a, 1994b, 1995; Thall et al. 1995; Heitjan 1997; Jung et al. 2004; Lee and Liu 2008). Statistical inferences under these designs are based on either the posterior probability (Thall and Simon 1994a) or predictive probability (Thall and Simon 1994b; Lee and Liu 2008), which incorporate both prior knowledge of parameters of interest, and interim data. Since the inferences are not constricted by the designs, they provide greater flexibility for interim analyses. For example, in fact, the more interim analyses are performed, the smaller the sample size is. Since Bayesian methods depend on prior information from historical data or expert opinion, it is important to use robust methods to elicit prior. Further, interim analysis requires patient response data to be available at the time of analysis, which might not be realistic for some situations in which it takes a long time to assess patient responses. Cai et al. (2014) suggested an approach that treats the delayed response as missing data and copes with them using an imputation method. In the following, we focus our discussion on the Bayesian design based on predictive probability by Lee and Liu (2008) and the Bayesian continuous monitoring design based on posterior probability by Thall and Simon (1994).

6.4.3.2.1 *Bayesian Single-Arm Trials*

For rare diseases or diseases of unmet medical needs, evidence generated from single arm Phase II trial may be adequate to support regulatory approval for the product. Consider the single-arm oncology trial with a binary primary endpoint in Section 6.4.2.1. Assume that throughout the study, the treatment responses out of n patients enrolled follows a binomial distribution, $B(n,p)$ with a prior distribution, $p \sim \text{Beta}(a,b)$.The quantity $\frac{a}{a+b}$ is the mean of the prior distribution; whereas $a+b$ measures how informative the prior is. The above beta distribution can be parameterized in terms of these two quantities. Let $N_{\max}$ be the maximum number of patients to be entered in the study and let X_n the number of responses observed in n patients currently in the study. Consequently, the posterior distribution of the response rate p follows a beta distribution,

$$p \mid X_n = x_n \sim \text{Beta}\left(a+x_n, b+n-x_n\right).$$

6.4.3.2.2 *Go or No-Go Criteria*

One set of "Go or No-Go Criteria" is based on posterior probabilities of efficacy and futility. They are defined as follows:

1. Go criterion is met if $P\left[p > p_1 | x_n\right] > P_E$;
2. No-Go criterion is met if $P\left[p > p_1 | x_n\right] < P_F$;
3. Continue enrollment till one of the above decisions or the maximum sample size N_{Max} is reached,

where P_E and P_F are thresholds of efficacy and futility, respectively.

6.4.3.2.3 *Predictive Probability*

Let $m = N_{\text{Max}} - n$be the number of potential patients to be enrolled in the second stage of the trial and Y the number of patients who respond to the therapy. Hence

$$Y \mid X_n = x_n \sim \text{Beta} - \text{binomial}\left(n,\ a+x_n,\ b+n+1-x_n-y\right).$$

Lee and Jack (2008) suggested the decision of futility and efficacy be made based on the predictive probability of efficacy. The predictive probability measures the strength of a decision, given by

$$PP_E = E\{I(P[p > p_1 | x_n, Y] > P_E)\}$$

$$= \int I\left(P[p > p_1 | x_n, Y] > P_E\right) dP[Y | x_n]$$

$$= \sum_{i=0}^{m} P[y = i \mid x_n] \times I\left(P[p > p_1 | x_n, Y = i] > P_E\right),$$

where $I(\bullet)$ is an indicator function, which assumes 1 or 0 pending on if the event in the parentheses is true or not, and $P[p > p_1 | x_n, Y = i]$ is the posterior probability.

Likewise, the predictive probability of futility can be estimated by

$$PP_F = \sum_{i=0}^{m} P[y = i \mid x_n] \times I\left(P[p > p_0 | x_n, Y = i] > P_F\right).$$

Efficacy and futility criteria are given in the following:

1. Declare efficacy if $PP_E > P_{1E}$;
2. Declare futility if $PP_F < P_{1F}$;
3. Continue enrollment till one of the above decisions or the maximum sample size N_{max} is reached,

where the cut points P_{1E} and P_{1F} are values between 0 and 1, e.g., 95% specified by the investigator.

In practice, the above continuous assessment is carried out after the data of an initial number of patients become available to render reliable estimates of probabilities (Cai et al. 2014). There are other methods suggested by various authors based on predictive probability. For example, Spiegelhalter et al. (2004) suggested a method using prior and current data to calculate predictive probabilities of eventual classical conclusions.

6.4.3.2.4 *Continuous Monitoring of Single-Arm Trials*

Thall and Simon (1994) considered setting of a single-arm study where the aim of the study is to determine if a new therapy is promising relative to the standard of care. The study can be terminated any time for futility or efficacy. The decision is made based on the posterior probability $P(x, n, \delta_0) = P[p_1 > p_0 + \delta_0 | x_n = x]$ where δ_0 is a clinically meaningful difference. It was suggested that an informative prior and a flat or weakly informative prior be used for response rates p_0 and p_1 of the standard care and experimental drug, respectively. Assume that $p_i \sim \text{Beta}(a_i, b_i)$, $i = 0,1$. Note that

$$p_1 \mid X_n = x_n \sim \text{Beta}\left(a_1 + x_n, b_1 + n - x_n\right).$$

The posterior probability p can be obtained by

$$P\left(x_n, n, \delta_0\right) = \int_0^{1-\delta_0} [1 - F\left(p + \delta_0; a_1 + x_n, b_1 + n - x_n\right)] f\left(p; a_0, b_0\right) dp,$$

where the functions $F(\bullet)$ and $f(\bullet)$ are the cumulative probability and density functions of $\text{Beta}\left(a_1 + x_n, b_1 + n - x_n\right)$ and $\text{Beta}\left(a_1, b_1\right)$, respectively.

Define

U_n = the smallest integer x such that $P(x, n, 0) \geq P_{2E}$,
L_n = the largest integer x such that $P\left(x, n, \delta_0\right) \leq P_{2F}$,

where P_{2E} and P_{2F} are pre-specified probabilities between 0 and 1 with P_{2E} being close to 1 and P_{2F} close to zero. It is worth noting that although these two probabilities resemble to Type I error and power in the classical hypothesis testing setting, they bear very different meanings (Thall and Simon 1994). Efficacy and futility decisions are made based on the following decision rules:

1. If $X_n \geq U_n$, declare efficacy;
2. If $X_n \leq L_n$, declare futility;
3. Otherwise, then continue treating another patient.

The trial is deemed to be inconclusive if either of the above conditions is met by the time when the maximum sample size $N_{\max}$ is reached.

6.4.3.2.5 *Comparative Phase II/III Studies*

Single-arm phase II studies rely on historical data of the standard of care. The robustness of the data has an impact on the quality of the study design. In the absence of such data, it is necessary to conduct a randomized trial to demonstrate efficacy of a new drug relative to the standard of care. If the comparator is newly approved for the same indication, it also entails the needs of comparative trials. In these situations, the above Bayesian design with continuous efficacy and futility assessment can be extended to include "Go and No-Go Criteria" based on either posterior or predictive probabilities of efficacy or futility. Suppose that the primary endpoint is binary. Let X_{n_1} and Y_{m_1} be the observes responses from the experiment drug arm and standard of care arm, respectively, which have response rates of p_1 and p_0. The corresponding decisions of efficacy and futility rules based on posterior and predictive probabilities are as follows:

6.4.3.2.5.1 EFFICACY AND FUTILITY BASED ON POSTERIOR PROBABILITY

1. Go: $P\left[p_1 - p_0 > \delta_0 | X_{n_1} = x_{n_1}, Y_{m_1} = y_{n_1}\right] > P_{E_1}$;
2. No-Go: $P\left[p_1 - p_0 > \delta_0 | X_{n_1} = x_{n_1}, Y_{m_1} = y_{n_1}\right] < P_{F_1}$;
4. Continue enrollment till one of the above decisions or the maximum sample size $N_{\max}$ is reached,

where P_1 and P_2 are response rates of the new drug and its comparator, respectively.

6.4.3.2.6 *Efficacy and Futility Criteria Based on Predictive Probability*

1. Declare efficacy if predictive probability $PP\{P\left[p_1 - p_0 > \delta_0 | X_{n_1} = x_{n_1}, Y_{m_1} = y_{n_1}\right] > P_{E_1}\} > P_1$;
2. Declare futility if predictive probability $PP\{P\left[p_1 - p_0 > \delta_0 | X_{n_1} = x_{n_1}, Y_{m_1} = y_{n_1}\right] > P_{F_1}\} > P_2$;
3. Continue enrollment till one of the above decisions or the maximum sample size $N_{\max}$ is reached.

Under the same assumption as in Section 6.4.3.2.1 that $X_{n_1} \sim \text{binomial}\left(n_1, p_1\right)$ and $Y_{m_1} \sim \text{binomial}\left(m_1, p_0\right)$ and priors, $p_i \sim \text{Beta}\left(a_i, b_i\right), i = 0,1$ are chosen for p_0 and p_1, the posterior probability $P\left[p_1 - p_0 > \delta_0 | X_{n_1} = x_{n_1}, Y_{m_1} = y_{n_1}\right]$ can be calculated as follows:

$$P\left(\mathrm{x}_{n_1}, \mathrm{n}_1, \mathrm{y}_{\mathrm{m}1}, \mathrm{m}_1, '_0\right) =$$

$$\int_0^{1-\delta_0} [1 - F\left(p + \delta_0; a_1 + x_n, b_1 + n - x_n\right)] f\left(p; a_0 + y_{m_1}, b_0 + m - y_{m1}\right) dp.$$

Likewise, the predictive probabilities of efficacy and futility can be derived:

$$PP_E = \sum_{i=0}^{n_1} \sum_{j=0}^{m_1} P\left[x = i \middle| x_{n_1}\right] P\left[y = j \middle| y_{m_1}\right] \times I\left(P\left[p_1 - p_0 > \delta_0 | X_{n_1} = x_{n_1}, Y_{m_1} = y_{n_1}\right] > P_{E_1}\right)$$

$$PP_F = \sum_{i=0}^{n_1} \sum_{j=0}^{m_1} P\left[x = i \middle| x_{n_1}\right] P\left[y = j \middle| y_{m_1}\right] \times I\left(P\left[p_1 - p_0 > \delta_0 | X_{n_1} = x_{n_1}, Y_{m_1} = y_{n_1}\right] < P_{F_1}\right).$$

6.4.3.2.6.1 AN EXAMPLE

In this section, we use an example to illustrate the Bayesian continuous monitoring for a phase II oncology trial with treated and control arms with 2:1 allocation, in which a maximum of 60 patients is planned and the first interim look takes place when the responses of the first 24 treated-arm subjects and 12 control-arm subjects are observed. Let p_C and p_T respectively denote the probability of a response in the control and treated arms. Efficacy, futility, or continuation rules are given by

1. Declare efficacy if predictive probability $PP\{P\left[p_T - p_C > 0.2|\text{data}\right] > 0.7\} > 0.9$;
2. Declare futility if predictive probability $PP\{P\left[p_T - p_C > 0.2|\text{data}\right] < 0.1\} > 0.9$;
3. Otherwise continue to full enrollment of 40 total treated-arm and 20 total control-arm patients.

Let x_C and x_T denote the number of responders (out of n_C and n_T) at interim. Assume uninformative priors $p_C, p_T \sim \text{Beta}(½, ½)$. Assume that historical literature/data points to $p_C = 0.2$ and $p_T = 0.45$. This information is used to identify the most likely final set of trial observations.
Note that for i = C, T,

$$p_i \mid \text{data} \sim \text{beta}\left(\frac{1}{2} + x_i, \frac{1}{2} + n_i - x_i\right)$$

Thus,

$$P\left[p_T - p_C > 0.2|\text{data}\right]$$

$$= \int_0^1 \left\{1 - F\left(0.2 + p; \frac{1}{2} + x_T, \frac{1}{2} + n_T - x_T\right)\right\} f\left(p; \frac{1}{2} + x_C, \frac{1}{2} + n_C - x_C\right) dp$$

where $f(.)$ and $F(.)$ denote the beta distribution pdf and cdf.

Conditioned on the interim data, the predictive probability calculations are given in the computer code. Figure 6.5 shows the zones of (x_C, x_T) that are 95% likely under $p_C = 0.2$ and $p_T = 0.45$ and the conditions under which the trial should be stopped early for efficacy/futility or continued.

The above graph provides the investigator with a convenient way to make a decision regarding terminating the trial for futility or efficacy or continue the enrollment at the interim look where the responses of the first 24 treated-arm subjects and 12 control-arm subjects are observed. For example, suppose that 14 out 24 from the treatment arm and 2 out 12 patients from the control

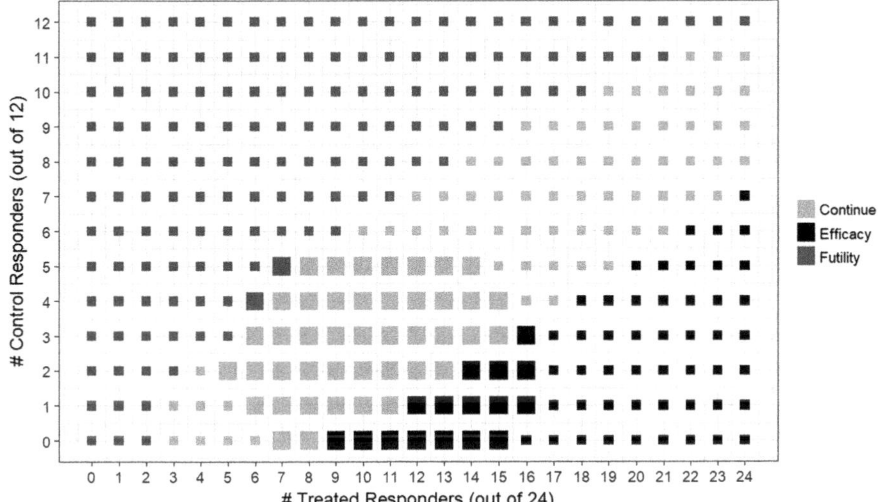

FIGURE 6.5
Conditioned on the interim data (x and y axes), graph shows areas of high probability to stop for efficacy and futility or to otherwise continue the trial. Conditioned on $p_C = 0.2$ and $p_T = 0.45$, the larger box size shows a zone that is 95% likely to be observed. (Adapted from Yang and Novick 2019.)

have positive responses. From the graph, the study should be stopped to declare efficacy.

6.4.4 Assessment of Probability of Technical and Regulatory Success

6.4.4.1 Background

Assessing the Probability of Technical and Regulatory Success (PTRS) is a crucial step in the drug development process, especially considering that nearly 90% of candidate drugs fail to progress. This challenge becomes even more pronounced in late-stage clinical development. As the stakes are high, making an informed decision to advance to Phase III trials requires careful evaluation of numerous factors, including unmet medical need, evidence of the investigational drug's efficacy and safety compared to both approved therapies and drugs in development, patent lifespan, and potential commercial value (Lalonde and Peck 2022).

Weidman and Belsky (2020) emphasized that assessments of technical and regulatory success have distinct objectives and should be conducted separately. The PTS evaluates the likelihood that a drug candidate will meet predefined criteria, such as efficacy, safety, and Chemistry, Manufacturing, and

Controls (CMC), according to its TPP in the final analysis. This ensures that the drug possesses the necessary technical qualities to support its regulatory submission. Conversely, the Probability of Regulatory Success (PRS) assesses the likelihood that, after successful completion of Phase III trials, a program will meet the regulatory requirements for first-cycle approval. It also ensures that the approved product labeling aligns with the base case TPP for key commercial regions.

The PTRS is a composite metric that combines these two probabilities to estimate the overall likelihood that a clinical development program will achieve both technical and regulatory milestones. It is calculated as PTRS = PRS × PTS. PTRS is a critical input for key metrics, such as net present value (NPV), which guide investment decisions and portfolio prioritization. A robust and unbiased estimate of PTRS is essential for sound decision-making, particularly when determining whether to advance a program to Phase III (Lalone and Peck 2022)

6.4.4.2 Estimation of PTS

Several approaches have been proposed for estimating the PTS. One common method relies on industry benchmark success rates for molecules within the same class (Hay et al. 2014; Wong et al. 2019; Holmgren 2014). While this approach has its limitations, such as lack of specificity to the program at hand, it remains useful when internal data is unavailable.

A second method involves expert elicitation, where experts provide estimates of the treatment effect, which are then used to inform PTS calculations (Dallow et al. 2017; O'Hagan 2019). This technique leverages the judgment of individuals with relevant experience but is subject to biases inherent in subjective opinion.

A more sophisticated third approach integrates data from multiple sources through Bayesian analysis, which helps reduce uncertainties that often complicate go/no-go decisions. Various Bayesian methods have been discussed in the literature, including those in Sections 6.4.2 and 6.4.3 (Carroll 2013; Tang 2015; Pulkstenis et al. 2017; Saint-Hilary 2019; Hampson, Björn et al. 2021, 2022; Colin n.d.).

As illustrated in Figure 6.6, these methods generally combine historical data, expert opinion, external competitor data, and insights from the literature to construct a prior distribution representing the anticipated treatment effect. This prior is then updated with real-world data, particularly from internal Phase II outcomes, to form the posterior distribution of the treatment effect. The posterior reflects an updated belief about the treatment's efficacy, informed by both existing knowledge and new evidence. A Monte Carlo simulation is performed using this posterior to generate potential treatment effect outcomes. By comparing these outcomes to a predefined TV, the probability that the treatment effect will exceed this

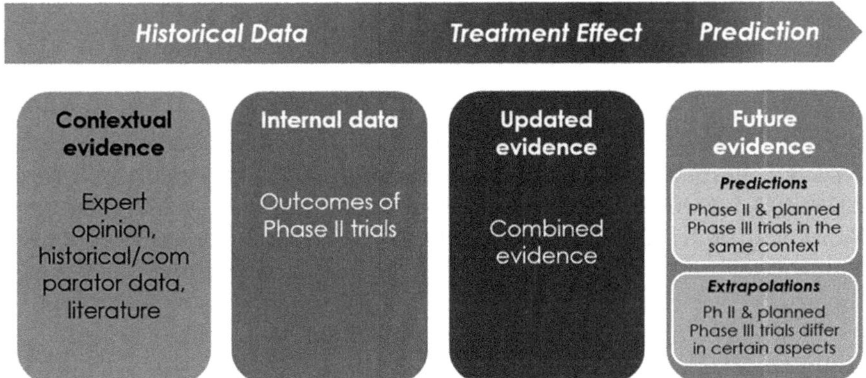

FIGURE 6.6
A general Bayesian approach to PTS assessment. (Adapted from Saint-Hilary 2019.)

threshold is estimated, providing a predictive measure of Phase III success and contributing to the overall PTS assessment. One challenge in using this Bayesian approach is the choice of prior (Carroll 2013). Additionally, the applicability of the method is limited by the requirement that there are no differences in study design between phases. The method also does not consider features of the drug and study design that may potentially impact PTS.

For a detailed description of the Bayesian approach and a practical example, readers may refer to Hampson, Björn et al. (2021).

6.4.4.3 Assessment of PRS

The PRS is influenced by multiple risk factors. Hampson et al. (2022) identified five primary types of risks that can affect the likelihood of regulatory approval for investigational drugs:

1. **Regulatory alignment risks**: These arise when the design or implementation of the Phase III program does not fully align with regulatory expectations.
2. **Unaccounted safety risks**: New safety concerns may emerge, either from within the program or from external sources, that were not previously anticipated.
3. **Unaccounted TPP risks**: Failure to meet key endpoints in the TPP that are critical for marketing approval or market access.
4. **Quality and compliance risks**: Issues related to the quality or compliance of the data or methods used to support regulatory submissions.

5. **Technical development risks**: Challenges such as CMC issues that may arise during development.

In addition to these risk factors, other elements like orphan drug designation can increase the likelihood of regulatory success. The PRS assessment typically employs a semi-quantitative approach (Weidman and Belsky 2020; BIO, Informa Pharma Intelligence, and QLS Advisors 2021; Hampson, Björn et al. 2021, 2022). The process begins with a benchmark PRS, sourced from either internal data or external industry comparisons for drugs in the same class or therapeutic area. Predictive analytics can also be used to derive this benchmark (BIO, Informa Pharma Intelligence, and QLS Advisors 2021).

Subsequent adjustments – either upward or downward – are made to this initial estimate based on specific program risks or opportunities. Table 6.10 outlines potential adjustment factors as described by Weidman and Belsky (2019). By systematically incorporating these factors into the risk assessment, the accuracy of PRS predictions can be improved, enhancing the overall decision-making process.

Figure 6.7 illustrates the adjustments to the PRS of a vaccine for an infectious disease following a successful Phase II efficacy trial. According to

TABLE 6.10

Potential Adjustment to Probability of Regulatory Success Estimates

Adjustment	Description
Upward	Upward adjustment for a product with special designation (such as those in development for "ultrarare" indications
	Medicinal products to address unmet medical need in the treatment of a serious or life-threatening condition
	Need for an advisory committee (US)
	Compliance with health authority advice
	Compliance with health authority data requests/analyses
	Robust clinical data
	The targeted condition is on a health authority list of critical therapies
	Secondary indications, such as a more robustly sound safety profile when compared to a predicate product
Downward	Safety signals cannot be mitigated
	Therapeutic area science is evolving or unclear/not known
	Efficacy is not comparable to current stand of care
	A magnitude of benefit is not expected
	A change in the competitive landscape impacting comparator used
	Compliance/inspection issues
	Non-compliance with HA advice
	Unmitigated technical risk
	Other approvability risks

Source: Weidman and Belsky (2020).

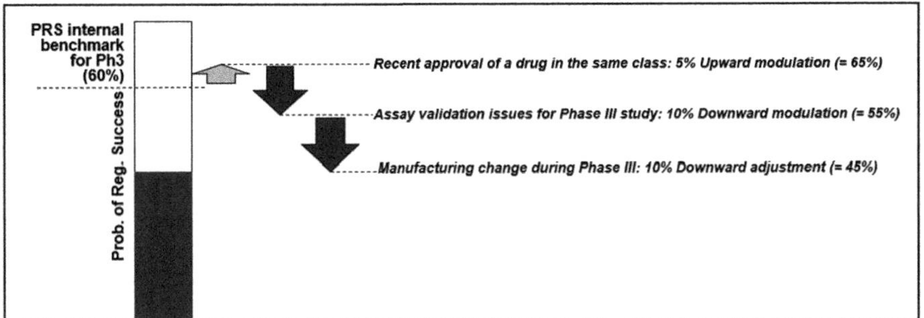

FIGURE 6.7
Adjustments of PRS of a vaccine products that had positive outcomes of Phase II.

the clinical drug development success rate report by BIO, Informa Pharma Intelligence, and QLS Advisors (2021), the success rates for transitioning from Phase III to a Biologics License Application (BLA) and from BLA to approval are 64% and 93.9%, respectively, leading to an initial benchmark PRS of 60%.

An upward adjustment of 5% was made because a vaccine for the same disease was recently approved, increasing confidence in regulatory success. However, despite the Phase III trial using the same primary endpoint as the Phase II study, the assay intended for evaluating this endpoint encountered validation challenges, resulting in a 5% downward adjustment. Additionally, a further 5% reduction was applied due to a planned change in the manufacturing process.

Taking these adjustments into account, the final PRS for the vaccine was estimated to be 55%.

6.5 Applications of AI and RWD in Clinical Trial Decision-Making

6.5.1 Portfolio Strategy

One of the important decisions a pharmaceutical company makes is which indications to target with a specific set of molecules (Anagnostopoulos et al. 2023). The occurs when a company begins a new clinical program or is looking to expand the indications of existing approved product. Traditionally, the input from key opinion leaders, literature reviews, historical and ongoing trials, omics analysis forms the basis for this decision. Since it is challenging to synthesize information from the abovementioned sources, the decision is rarely data-driven, and often rests with the company's board of directors, resulting in subjective and suboptimal selections (Schuhmacher et al. 2021). In contrast, strategies informed by RWD and AI are both objective and comprehensive. RWD and AI can assist uncover new indications and support the selection of most promising to pursue for novel and approved drug products (Harrer et al. 2019; Makady et al. 2017a, 2017b; Wang et al. 2019).

Efforts have been made in using AI tools to assist portfolio strategy. For example, Jung et al. (2020) developed a ML model based on company profile and drug features for recommending and/or predicting drug groups suitable for development by individual pharmaceutical companies (Figure 6.8). The model was shown to have reasonably good performance evidenced by the accuracy and area under the curve (AUC) were 78% and 0.74, respectively, for one of its variants. In particular, the model was applied to predict the success probability of companies developing Coronavirus disease 2019 (COVID-19) vaccines. It was demonstrated that the higher the predicted score from the DDR model, the more progress in the clinical phase of the COVID-19 vaccine development. Although their approach has limitations that should be improved, it makes scientific as well as industrial contributions in that the model can support rational decision-making prior to initiating drug development by considering not only technical aspects but also company-related variables.

Unacceptable safety profiles are one of the major contributors to high attrition rates of clinical trials (Arrowsmith 2011; Hay et al. 2014). When developing portfolio strategies, identifying drugs with significant safety potential is crucial. For example, drug-induced liver injury (DILI) is a significant safety concern in drug development and is frequently cited as a reason for high drug attrition rates and market withdrawals. The FDA recognized the importance of this issue and introduced a guidance document in 2009 to assist in the analysis of liver safety data from clinical trials and address regulatory questions related to liver safety (FDA 2009a). Despite this, both clinical trial data and traditional toxicological studies have proven insufficient in accurately predicting DILI in humans.

To improve the prediction of DILI risk, new tools and approaches are being developed, leveraging novel data sources. Recently, there has been a growing interest in applying ML and AI techniques to this problem. Vall et al. (2021) provided a detailed review of existing AI approaches to predict DILI and elaborate on the challenges that arise from the limited availability of data. Munsaka et al. (2023) presented several examples of the use of these advanced techniques in predicting DILI. Gong et al. (2022) conducted a study to demonstrate the utility of ML methods in identifying undiscovered kinase-adverse event associations for small molecule kinase inhibitors (SMKIs). The method can potentially be used as a tool to prioritize the development of SMKIs target therapies from the safety perspective.

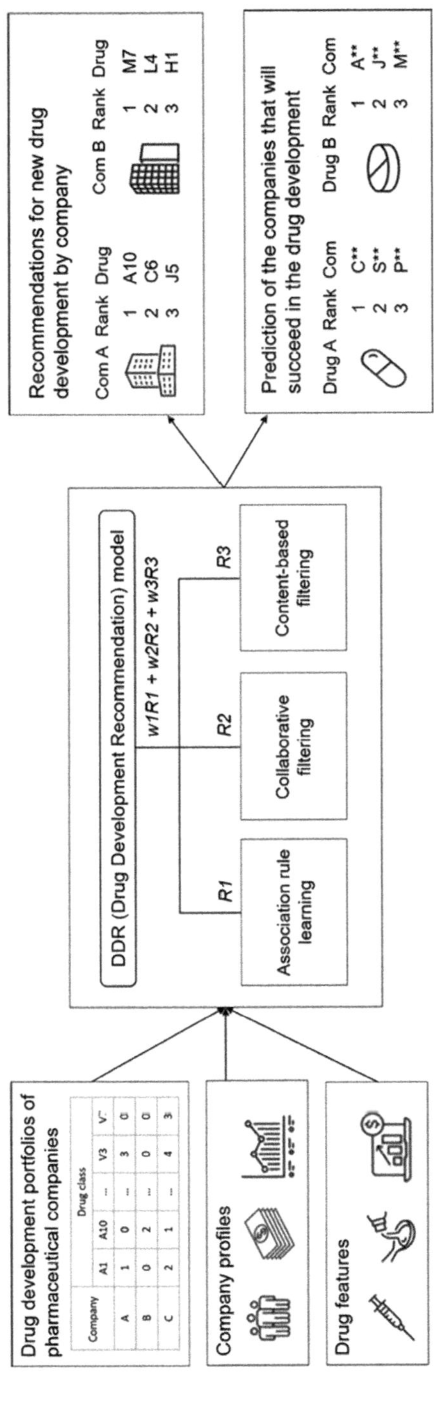

FIGURE 6.8
The framework of the DDR model for new drug development planning. The model consists of three main parts: data input, model building, and output of recommendation and prediction. (Adapted from Jung et al. 2020.)

6.5.2 TPP Development

The development of a TPP for a new drug involves several key decisions, including selections of indication and patient population, target safety and efficacy claims, assessment of the current standard of care, global differences, market size, competitor pricing, and reimbursement environment. Traditionally, these decisions have heavily relied on expert opinions and limited data extracted from scientific literature. However, with the advent of advanced analytics, particularly in RWD analysis, pharmaceutical companies can now leverage vast datasets to gain deeper insights into patient populations with unmet medical needs. By mining and analyzing real-world databases, companies can effectively profile these populations, identifying specific characteristics, demographics, and clinical profiles that may benefit from the proposed intervention. Furthermore, RWD analysis enables the mapping out of treatment pathways for target populations, providing valuable insights into the effectiveness of existing standard-of-care treatments.

The utilization of RWD in TPP development offers several key advantages. Firstly, it provides a more nuanced understanding of the target population, allowing for the identification of subgroups that may derive the greatest benefit from the new intervention. This granularity enables pharmaceutical companies to tailor their development strategies to better address the needs of specific patient cohorts, ultimately enhancing the likelihood of clinical success.

Moreover, RWD analysis facilitates the prediction of clinical and economic outcomes associated with the proposed product based on its defined TPP (Gerlinger et al. 2020). By extrapolating data from real-world sources, companies can estimate the potential impact of the new intervention on patient outcomes, healthcare resource utilization, and overall economic burden. This predictive modeling not only informs internal decision-making processes but also enhances discussions with regulatory agencies and other stakeholders regarding the potential value of the product.

Incorporating RWD into TPP development requires a multidisciplinary approach, bringing together expertise from various domains such as data science, epidemiology, and clinical research. Collaborations with healthcare providers, patient advocacy groups, and other relevant stakeholders are also essential to ensure that the TPP accurately reflects the needs and preferences of the target population.

Furthermore, ongoing data collection and analysis are critical to refining and updating the TPP as new insights emerge throughout the clinical development process. This iterative approach allows for the incorporation of real-world evidence gathered from post-market studies, observational research, and other sources, ensuring that the TPP remains aligned with evolving clinical practice and healthcare trends.

Overall, the integration of advanced analytics and real-world data into the development of TPPs represents a significant advancement in the field of clinical development. By leveraging these tools and methodologies, pharmaceutical companies can enhance their understanding of target populations, optimize their development strategies, and ultimately bring innovative therapies to market more efficiently and effectively.

6.5.3 Clinical Trial Phase Transition and Regulatory Approval

Drug developers often rely on general regulatory approval rates or estimates based on a drug's therapeutic class when managing their portfolios of investigational drugs. However, this approach overlooks critical factors such as the specific characteristics of the drug compound and clinical trial design, leading to significant uncertainties in decision-making related to clinical trial phase transitions and regulatory approvals. Furthermore, this method offers limited insights into the factors that influence pipeline performance.

Recently, there has been growing interest in integrating AI and ML methods to enhance the prediction of clinical development success. These advanced techniques can provide more accurate risk assessments for investigational drugs at various stages of development, reducing uncertainty and improving the efficiency of drug development and portfolio decision-making.

Several research groups have explored the potential of ML models to predict clinical phase transitions and drug approvals based on drug class, indication, and trial design features (DiMasi et al. 2015; Lo et al. 2018; Beinse et al. 2019; Feijoo et al. 2020; Hegge et al. 2020). DiMasi et al. (2015) pioneered a predictive model aimed at forecasting regulatory approval post-Phase II for oncology trials, using data from 98 oncology drugs developed by the top 50 pharmaceutical companies. They employed logistic regression alongside ML techniques such as Random Forests (RFs) for variable selection to identify key factors influencing regulatory success.

Similarly, Lo et al. (2018) conducted a study focused on completed trials to predict drug approval probabilities and identify critical variables using drug development and clinical trial data from 2003 to 2015. Using a variety of ML models – including RF, neural networks, and decision trees – they found that k-nearest neighbor (k-NN) imputation combined with RF delivered the best predictive performance. Their analysis highlighted trial outcomes, trial status, and accrual rates as the most important predictive factors.

Beinse et al. (2019) developed a model to predict oncology drug approval after Phase I trials. In their test set, 73% of drugs predicted to be approved were ultimately approved, while 92% of drugs predicted to be nonapproved remained so. They reported that a predicted approved drug was 16 times more likely to be approved than a predicted nonapproved drug.

Feijoo et al. (2020) introduced a model leveraging supervised machine learning (SML) and Natural Language Processing (NLP) algorithms to predict phase transitions and Limits of Agreement (LoA) for clinical trials. Their model uniquely incorporated a metric for the complexity of eligibility criteria, extracted from the textual data on ClinicalTrials.gov. This predictor, combined with other features, enabled the SML model to predict clinical phase success with 80% accuracy. They identified eligibility criteria complexity and the number of endpoints as key predictors, suggesting the model could offer valuable insights for clinical protocol design, operational feature management, portfolio assessment, and decision-making.

More recently, Aliper et al. (2023) described an AI system that integrates data from multiple sources and uses deep learning models, such as generative graph and text transformers, to predict clinical outcomes. Kavalci and Hartshorn (2023) trained a model on 420,268 clinical trial records and 24 fields extracted from ClinicalTrials.gov, including logistical, administrative, and study design features. Their model achieved an AUC of 0.80 and an accuracy of 0.70.

6.5.4 In Silico Simulation for PTS Assessment

6.5.4.1 Prediction of Phase III Trial Outcomes

The Go/No-Go decisions under the Bayesian methods discussed in Sections 6.4.2–6.4.5 are predicted on the interim data of ongoing trials or the outcomes at the end of Phase II trials. Since it often requires a statistical model to enable the prediction, these methods are consistent with the MIDD approach initiated by the FDA (Madabushi et al. 2022). However, there are several drawbacks. Firstly, the prediction is carried out at the population level, solely based on patient response data without considering patient and disease characteristics and study design features. As a result, it does not provide insight as related to how to enhance the probability of success for the late-stage trials, utilizing enrichment strategy. Secondly, it does not take advantage of data from clinical trials of other drugs for the same indication or drugs in the same class.

AI and ML methods have made inroads into the prediction of trial success at both the population and individual patient levels. These techniques have demonstrated their efficacy in various aspects of clinical development, including early disease detection and prognosis, thereby enhancing the overall success of clinical trials (Sangari and Qu 2020). Qi and Tang (2019) developed a method that predicts Phase III trial outcomes based on aggregation of predicted individual treatment responses from Phase II trial, using a recurrent neural network model. Chekroud et al. (2016) used ML to predict responses to an anti-depressant after 12 weeks. In a separate study, Beacher et al. (2021) applied ML based on patient baseline measurements to predict two-year treatment outcomes of bicalutamide for prostate cancer, utilizing data from three comparable clinical trials. ML methods were also developed to predict drug approvals (Lo et al. 2018; Beinse et al. 2019; Shiah et al. 2021).

Expanding on the role of AI in predicting late-stage clinical trial success, it is crucial to delve into the methodologies and applications driving these advancements. ML algorithms, such as support vector machines, RFs, and deep learning neural networks, have demonstrated remarkable efficacy in analyzing complex datasets to identify patterns and predict outcomes. By leveraging diverse data sources, including genomic, proteomic, and clinical data, these algorithms can predict patient responses to treatment, anticipate adverse events, and optimize trial protocols.

Furthermore, AI facilitates the identification of biomarkers that stratify patient populations, enabling more targeted and personalized therapies. This personalized approach not only enhances patient outcomes but also increases the likelihood of trial success by focusing resources on the most promising avenues of research.

In oncology, AI is revolutionizing clinical trials through the creation of in silico trials, which utilize clinical data to construct simulated cohorts that emulate treatment effectiveness (Kolla et al. 2021). These trials hold immense potential in reducing late-stage development failures by identifying better responders early in the process. However, a significant challenge persists due to the scarcity of high-quality, curated, and comprehensive datasets, which often hinders the full realization of AI applications.

Moreover, AI-powered virtual trials offer a cost-effective and ethically sound alternative to traditional trials, particularly in scenarios where patient recruitment is challenging, or ethical concerns arise. By simulating virtual patient cohorts based on real-world data, these trials accelerate the evaluation of treatment efficacy and safety, reducing the time and resources required for traditional trials.

6.5.4.2 An Example

In this section, we present an example of using digital twins generated through in silico simulation to predict the outcome of a clinical study, FAURA2. FLAURA2 is a Phase III clinical trial by AstraZeneca evaluating the combination of Tagrisso (osimertinib) with chemotherapy against Tagrisso alone for patients with advanced EGFR-mutated non-small cell lung cancer (NSCLC). This example illustrates the potential of AI-assisted modeling and simulation to guide Go/No-Go decision for Phase III trials at the end of PoC studies.

Nova IN SILICO is a pioneering life sciences company specializing in in silico clinical trial simulation through its innovative jinkō platform. This platform was designed to inform and optimize clinical trial designs, leveraging

TABLE 6.11

Summary of Efficacy Outcome of FLAURA 2 Study and Model-Predicted Results

	FLAURA2 Phase III Results (Investigator)	FLAURA2 Phase III Results (BICR)	Mode-Predicted Results
HR	0.62 (0.49, 0.79)	0.62 (0.48, 0.80)	0.602 (0.49, 0.74)
FPS/TTP (Month)	PFS: 25.5	PFS: 29.4	TTP: 25.9
(Osimertinib + Chemo)	(24.7, NC)	(25.1, NC)	(25.1, 27.1)
PFS/TTP (Month)	PFS: 16.7	PFS: 19.9	TPP: 17.3
(Osimertinib Alone)	(14.1,21.3)	(16.6, 25.3)	(16.8, 18.0)

Source: Nova In Silico (2023).

Note: NC = Not calculable.

quantitative systems pharmacology (QSP) models. On September 12, 2023, nova IN SILICO released the results of its forward-looking prediction of outcomes from the FLAURA2 trial. Utilizing solely publicly available data from AstraZeneca's Phase I and Phase II trials, alongside the Phase III trial protocol accessible on ClinicalTrials.gov, nova IN SILICO employed its proprietary NSCLC disease model to inform the jinkō trial simulation.

In total 5,000 virtual patients were simulated, based on demographic profiles resembling those of the 30 lung cancer patients enrolled in the safety run-in trial of FLAURA2. The model used from the simulation accounted for over 65,000 disease model parameters, underscoring the depth and comprehensiveness of their analysis. The results are presented in Table 6.11 along with the outcomes of the FLAURA1 trial.

The findings from the phase III FLAURA2 study demonstrated that combining osimertinib with platinum-based chemotherapy and pemetrexed maintenance therapy significantly reduced the risk of disease progression or death compared to osimertinib alone, as evidenced by a hazard ratio (HR) of 0.62 (95% confidence interval [CI] 0.49–0.79; $p < 0.0001$). The median progression-free survival (PFS) was notably extended to 25.5 months (24.7, not calculable) in the osimertinib plus chemotherapy arm, in contrast to 16.7 months (14.1–21.3) in the single-agent osimertinib arm.

Blinded independent central review (BICR) of PFS results yielded consistent outcomes, with an HR of 0.62 (0.48–0.80) and a median PFS of 29.4 months (25.1, not calculable) for the osimertinib plus chemotherapy arm, compared to 19.9 months (16.6, 25.3) for the single agent osimertinib arm.

The results from the nova IN SILICO simulation study revealed a time to progression (TTP) hazard ratio (HR) of 0.602 (95% prediction interval [PI] 0.49–0.736), with a median TTP of 25.9 months (25.1–27.1) in the experimental arm and a median of 17.3 months (16.8–18.0) in the single-agent comparator arm. TTP served as a surrogate outcome measure for PFS during in silico trial modeling. The predicted results were consistent with the actual study findings.

This simulation of the phase III trial showcases the transformative potential of in silico clinical trials in shaping the future of drug development. Such results, if utilized prior to the commencement of human trials, can facilitate the identification of the most appropriate patient cohorts, refine trial design, and ultimately expedite therapeutic advancements and ultimately realize the collective aspiration that cancer evolves into a highly treatable condition.

As noted by Duruisseaux (VPH Institute 2023), in the near future, in silico clinical trials can play a pivotal role in establishing a "target" Hazard Ratio. This metric will serve as a foundation for constructing the statistical hypothesis of the next generation of clinical trials, heralding a more efficient and precise approach to drug development.

6.5.4.3 Challenges and Considerations

Despite the immense potential of AI in predicting late-stage clinical trial success, several challenges must be addressed to realize its full benefits. These include the need for standardized data formats, interoperable data repositories, and robust data privacy frameworks to ensure the quality, accessibility, and security of data. Additionally, regulatory frameworks must adapt to accommodate the dynamic nature of AI-driven clinical research, balancing innovation with patient safety and ethical considerations.

The integration of AI and ML methods holds tremendous promise in revolutionizing the prediction of late-stage clinical trial success. By harnessing the power of data analytics, predictive modeling, and virtual trials, AI enables more efficient, cost-effective, and patient-centered approaches to drug development. However, addressing technical, regulatory, and ethical challenges is essential to fully realize the transformative potential of AI in clinical research and improve patient outcomes on a global scale.

6.5.5 RWD for Go/No-Go Decision

For rare diseases and many oncology trials, it is often operationally unfeasible and unethical to conduct RCTs. The safety and efficacy of a drug is assessed through single-arm open label studies. There is growing need to generate supportive evidence for the safety and effectiveness of a new drug from other sources such as observational studies. A synthetic control arm derived from historical or contemporaneous populations treated in a real-world setting may serve as a comparator for the experimental drug. The comparative evidence based on the synthetic control can be used to support marketing approval, and access and coverage assessment by health technology assessment agencies. Li et al. (2021) provided an overview of how to adopt external control using RWD in clinical development. Several real-world case examples were discussed.

6.6 Case Examples

We present three case studies. The first example by Yang and Novick (2019) is to illustrate the utility of a Bayesian model to inform dose selection and a "Go" decision. The other two highlight the power of AI and RWD to predict clinical phase transition and trial success.

6.6.1 Estimation of "Go" Probability

6.6.1.1 Background

The drug under evaluation is an anti-inflammatory monoclonal antibody, which was designed to have an extended half-life when compared to the standard antibody. One drug in the same class has been shown to have a desirable safety and efficacy profile and is currently on the market. This competitor drug is administered through IV every 2 weeks. Hence, establishing a dose regiment for the experimental drug with at least "class-matching" efficacy, but on every 4-week schedule can be advantageous. It was also desirable to get to the market quickly due to the changing landscape.

Various design strategies were explored by Yang and Novick (2019), including (1) a fixed Phase II designed followed by two Phase III trials; (2) a Bayesian Phase II design with interim efficacy evaluation followed by two Phase III trials; and (3) a seamless Phase II/III study with Bayesian Phase II design followed by an additional Phase III trial. After weighing benefit and cost accounting from sample size, operational ease, time-to-market, and total cost, it was decided to adopt the second option mentioned previously.

6.6.1.2 Bayesian Phase II Trial

The objectives of this Phase II trial were twofold: (1) detect efficacy signal to trigger early "Go" decision, and (2) identify optimal dose or doses for Phase III trials. A dosing schedule of every 4 weeks was to be explored as the dose regimen of the other drug was every 2 weeks. A Bayesian design with multiple interim looks was evaluated to shorten the duration of Phase II trial. In addition, information from the competitor's clinical program was used to define effect size and Go/No-Go criteria. A brief description of the study design is given in Table 6.12.

TABLE 6.12

Study Design of a Bayesian Phase II Trial

Objective	To demonstrate efficacy and determine an optimal dose regimen
Patient population	Patients with a severe inflammatory disease
Design	Randomized and Placebo Controlled Bayesian Phase II with multiple interim looks
Treatment	Every 4-week infusion of either drug at 100mg, 200mg, 400 mg, 800 mg, or placebo for 16 weeks
Primary endpoint	Flare of the disease at week 16
Maximum study duration	52 weeks
Maximum sample size	200

Source: Yang and Novick (2019).

From the results of the competitor's trial, it was determined that a difference $\delta$of 15% between the treatment arm and placebo is deemed to be a clinically meaningful improvement. Therefore, the Go/No-Go decision was defined based on posterior probability of δ:

1. Go if $P[\delta \geq 15\%] > 80\%$ for any dose;
2. No-Go if $P[\delta \geq 15\%] < 20\%$ for all doses,

where $P[\delta \geq 15\%]$ is the posterior probability for the difference in response rate between a dose of the drug and the placebo.

Patients are assumed to enroll into the trial at a constant rate of 20 subjects per week and are randomized into the five treatment arms with 1:1:1:1:1 allocation. Measurements are taken on each patient every four weeks over the course of 16 weeks, yielding four observations per patient. To ensure the reliability of the interim analysis, the first interim look takes place after 16 weeks, after the first 80 subjects have been enrolled so that 20 patients provide a full four weeks of data, 20 patients provide three weeks of data, and so on. Further interim looks occur every eight weeks until a total of 200 patients provide the full four weeks of data.

Patients are evaluated at each time point and are given binary ratings by investigators with observations $Y_{ij} = 1$ if the disease state showed improvement or $Y_{ij} = 0$ otherwise, with $i = 1$ (placebo), 2, ..., 5 (dose = 800 mg) indexing the doses and $j = 1, 2, \ldots, n_i$ indexing the patients on dose i and with $(n_1 + \cdots + n_5) = 200$.

The dose-response relationship among the treatment groups was described through a random-effect logistic model with $P(Y_{ij} = 1) = p_{ij}$ and

$$\log\left[\frac{p_{ij}}{1-p_{ij}}\right] = \theta_{ij},$$

TABLE 6.13

Specifications of Prior Distributions

Prior	Parameter Specification
$E_0 \sim N(\mu_0, \sigma_0^2)$	$(\mu_0, \sigma_0^2) = (-2, 0.5^2)$, with $\mu_0 = -2$ corresponding to 12% response rate of the placebo observed in the competitor's study.
$E_{max} \sim N(\mu_m, \sigma_m^2)$	$(\mu_m, \sigma_m^2) = (2, 0.5^2)$.
$E_{50} \sim N^+(\mu_{50}, \sigma_{50}^2)$	E_{50} follows a truncated normal distribution with $(\mu_{50}, \sigma_{50}^2) = (200, 50^2)$.
$\sigma_e^2 \sim IG(sh_0, sc_0)$	(sh_0, sc_0)=(2, 0.3). Median value of σ_e is 1.4 with middle 95% of distribution from 0.8 to 3.8.

Source: Yang and Novick (2019).

where θ_{ij} is a random variable following a normal distribution with Emax-model mean given by

$$\theta_{ij} = E_0 + \frac{\text{dose}_i E_{\max}}{\text{dose}_i + E_{50}} + \epsilon_{ij}$$

with $\epsilon_{ij} \sim N(0, \sigma_e^2)$ denoting the subject-to-subject errors.

The posterior probability at each interim look can be estimated using the MCMC method and the priors in Table 6.13. At each interim, the random-effect logistic model was fitted to the data using the JAGS model code given in the following.

To evaluate the Go/No Go probabilities for a particular scenario, a small Monte Carlo study was run with 100 computer-generated datasets. Data were generated so that the respective true probabilities to show improvement are 12%, 20%, 28%, 35%, and 48% for placebo, 100 mg, 200 mg, 400 mg, and 800 mg, respectively. The subject error standard deviation was set to $\sigma_e = 1.25$ with errors assigned around the logit link function to the true probabilities. Finally, the four patient responses were assumed independent. Data were generated according to the following R function.

6.6.1.3 Results

The "Go probabilities" (i.e., $P[\delta \geq 15\%]$) are shown via box plot in Figure 6.9 and summarized in Table 6.14. Given the simulated scenario, the results are quite promising, showing a borderline chance for a "GO" to occur for the 400 mg dose and high likelihood for a "GO" to occur at the highest (800 mg) dose. In fact, the "GO" probability is high throughout the interim looks, which can prompt investigators to halt the trial for efficacy at an early time point.

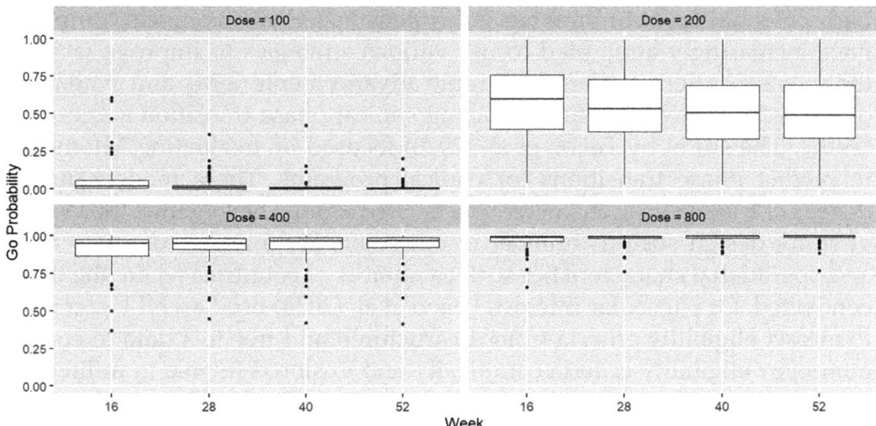

FIGURE 6.9
Box plot of $P[\delta \geq 15\%]$ for each dose at selected interim looks. (Adapted from Yang and Novick 2019.)

TABLE 6.14
Average Value of $P[\delta \geq 15\%]$ and, Parenthetically, Percentage of Runs (out of 100) that $P[\delta \geq 15\%] > 80\%$

	Dose (mg)			
	100	**200**	**400**	**800**
Week 16	3% (0%)	38% (15%)	79% (70%)	93% (93%)
Week 28	1% (0%)	34% (10%)	80% (66%)	94% (96%)
Week 40	1% (0%)	31% (12%)	80% (66%)	95% (97%)
Week 52	1% (0%)	30% (9%)	81% (67%)	95% (98%)

Source: Yang and Novick (2019).

In a final note to this example, given the Emax dose-response model, it is also possible to calculate "GO" posterior probabilities for interpolated doses.

In fact, taking one more step, the smallest dose that produces $P(\delta \geq 15\%) > 80\%$ may be reported for each Monte Carlo run so that the strategy may change to a dose-finding exercise.

6.6.2 Prediction of Clinical Trial Phase Transition

6.6.2.1 Background

As previously discussed, clinical development has faced unprecedented productivity challenges, as evidenced by high attrition in late-stage clinical trials. As a result, the need for making better decisions for transitioning drug candidates from one phase of development to another and invest in high-potential opportunities has only intensified. Pharmaceutical companies have increasingly attempted to use various strategies to improve facilitate the above-mentioned efforts. The rapid advancements of ML and availability of data offers new tools for predicting clinical phase transition success and failure. The study by Feijoo et al. (2020) focused on evaluating ML models to predict phase transitions for clinical programs. These models utilized dozens of trial-specific characteristics as predictors. They include the features of study design such as number of endpoints and number of study arms, and operational characteristics such as number of countries where the trial is conducted. Of particular interest, Feijoo et al. (2020) used an NLP algorithm to extract eligibility criteria from unstructured and free text data to create a metric for eligibility criteria complexity and explored its role in influencing trial outcomes. The ML models reported in this study demonstrate the ability to predict the outcome of a trial phase transition with an average accuracy of 80% within specific therapeutic areas. A model boasting this level of predictive accuracy has the potential to help sponsors to derisk trial failures, at both trial and portfolio levels (Feijoo et al. 2020). At the trial planning stage, one may use to the model as an iterative tool, offering guidance in evaluating various aspects of protocol design and operational characteristics. From a portfolio perspective, the model's predictive capabilities furnish additional information for decision-making in program assessment and resource allocation.

6.6.2.2 Data

For constructing the ML models, Feijoo et al. (2020) utilized clinical trial data sourced from two primary repositories: ClinicalTrials.gov (https://clinicaltrials.gov/) (Tasneem et al. 2012; Zarin et al. 2016) and Biomedtracker (Thomas et al. 2016). ClinicalTrials.gov is a publicly accessible database maintained by the National Library of Medicine at the National Institutes of Health that contains over 268,000 clinical studies. These studies are either privately or publicly funded and contain key information of study protocol and other associated characteristics. The data have been curated by the Clinical Trials Transformation Initiative and made available in a relational database format known as the Aggregate Analysis of ClinicalTrials.gov (AACT) (Tasneem et al. 2012). The version of the AACT database utilized in the study by Feijoo et al. (2020) includes 199,262 unique NCT codes or clinical studies.

In addition to the AACT data, data were sourced from Biomedtracker. It is an independent proprietary database maintained by Informa Business Intelligence Inc that tracks developmental drug through the FDA approval process. This platform provides real-time data and analysis of trial-related events of drugs at all stage of development crucial for financial investment decisions within the pharmaceutical and biotechnology market. The Biomedtracker dataset used by Feijoo et al. (2020) contains information pertaining to 9,526 clinical trials across 42 countries and involving 2,369 lead companies.

Integration of the AACT and Biomedtracker datasets was achieved through a common clinical trial ID, or NCT code. While the AACT dataset primarily furnished descriptive characteristics of each trial, Biomedtracker contributed detailed data not readily available elsewhere. Specifically, Biomedtracker provided information on phase transitions, disease or therapeutic areas, drug classifications, pivotal study details, and endpoint specifications (including primary, secondary, and type). Moreover, Biomedtracker enables the tracking of individual drugs through each phase of the pipeline, up to regulatory approval or failure. This capability facilitates the estimation of phase success rates and the likelihood of overall approval.

This study primarily focuses on trials in Phases II and III due to their elevated risk of failure and their significant contribution to the overall cost of drug development. Following the amalgamation of both datasets using the unique ID NCT code, a final dataset comprising 6,417 unique clinical studies was compiled, encompassing 1,028 unique Phase II clinical trials and 2,919 unique Phase III trials.

6.6.2.3 Methods

Various ML models including RF, neural networks, support vectoring machines, and logistic models were explored. The models were trained for Phase II and Phase III prediction separate and by disease area. The study was focused on the disease areas of oncology, neurology, and cardiology. Within each phase, a model was developed and evaluated for combined data across therapeutic area. The RF models were shown to have better performance when compared with other models.

TABLE 6.15

Summary of Performance Metrics of Random Forest Models by Study Phase and Therapeutics Area

Phase	Therapeutic Area	Accuracy (95% CI)	Sensitivity	Specificity
Phase II	All diseases	0.743 (0.661, 0.774)	0.740	0.746
	Oncology	0.740 (0.634, 0.826)	0.720	0.763
	Neurology	0.750 (0.554, 0.881)	0.750	0.733
	Cardiology	0.852 (0.663, 0.958)	0.933	0.750
Phase III	All diseases	0.672 (0.617, 0.724)	0.692	0.648
	Oncology	0.821 (0.764, 0.924)	0.804	0.842
	Neurology	0.829 (0.651, 0.912)	0.895	0.763
	Cardiology	0.800 (0.563, 0.943)	0.875	0.750

Source: Feijoo et al. (2020).

TABLE 6.16

Summary of AUC by Study Phase and Therapeutic Area

Phase	Therapeutic Area	AUC
Phase II	Oncology	0.790
	Neurology	0.760
	Cardiology	0.840
Phase III	Oncology	0.870
	Neurology	0.820
	Cardiology	0.780

Source: Feijoo et al. (2020).

6.6.2.4 *Results*

The results are presented in Table 6.15. It is evident that predictive power of the RF models is decreased when combing data from all disease areas. For example, for Phase II prediction, there was a reduction in all performance metrics, accuracy, sensitivity, and specificity for the Phase II model relative to the Phase II disease-specific models. The sensitivity and specificity of each RF model appeared comparable to each other, suggesting that the models did not overpredict either success or failure outcomes. The accuracy of the Phase II models was lower than that of Phase III across all disease areas. This implies that it is, in general, more challenging to predict Phase II trial outcomes.

Table 6.16 summarizes the area under the ROC curve by disease, which is the probability for differentiating a positive outcome from negative. The AUC of Phase II trials was estimated to be 0.79, 0.76, and 0.84 for oncology, neurology, and cardiology, respectively. In contrast, the AUC of Phase III trials was 0.87, 0.82, 0.78 for oncology, neurology, and cardiology, respectively.

Feijoo et al. (2020) also evaluated and identified the predictive factors having the most significant influence on phase transition. The results are presented in Figures 6.10–6.12. The most influential predictors are ranked based on the mean decrease in the Gini index (James et al. 2013; Sahni et al. 2018; Grömping 2009). Their analysis revealed that irrespective of the therapeutic area or trial phase, certain factors consistently displayed high importance in predicting a drug's progression toward regulatory approval. Among these influential predictors, metrics such as the complexity of eligibility criteria, number of endpoints, and enrollment figures emerged as particularly crucial for trial transition.

6.6.3 Estimating Probability of Success Using RWD

The following is a case study adapted from a real-life clinical trial, with modifications to ensure confidentiality.

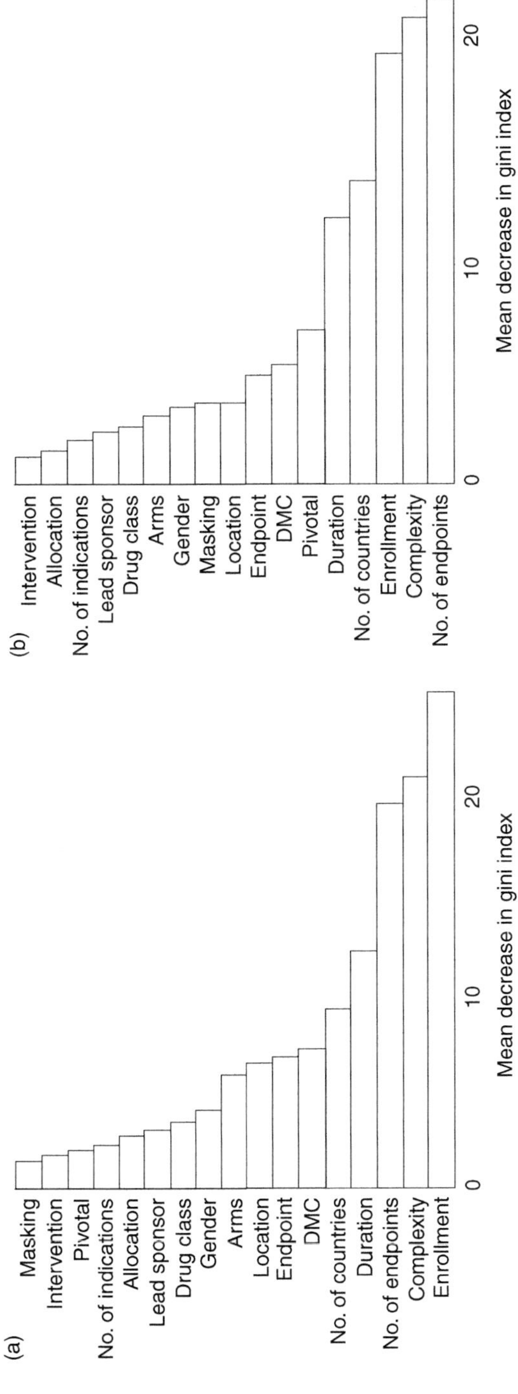

FIGURE 6.10
Random forest (RF) variable importance for (a) Phase II trials and (b) Phase III trials of oncology therapeutic area. (Adapted from Feijo et al. 2020.)

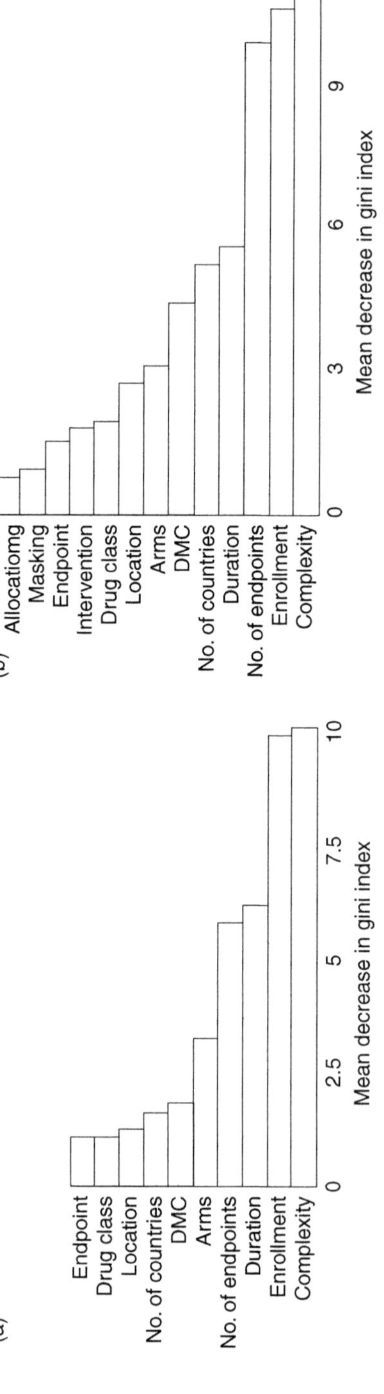

FIGURE 6.11
Random forest (RF) variable importance for (a) Phase II trials and (b) Phase III trials of neurology therapeutic area. (Adapted from Feijo et al. 2020.)

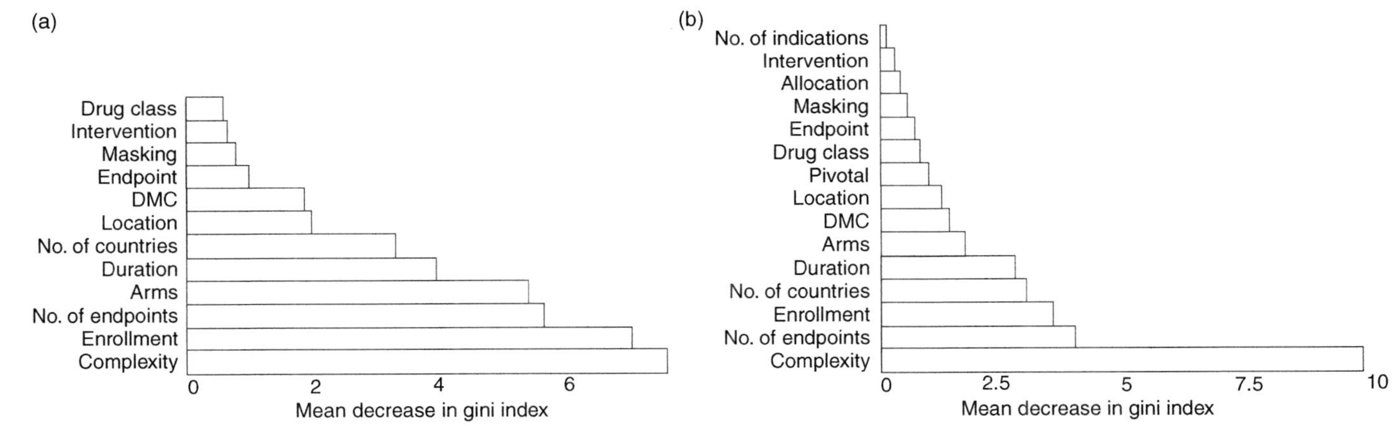

FIGURE 6.12
Random forest (RF) variable importance for (a) Phase II trials and (b) Phase III trials of cardiology therapeutic area. (Adapted from Feijo et al. 2020.)

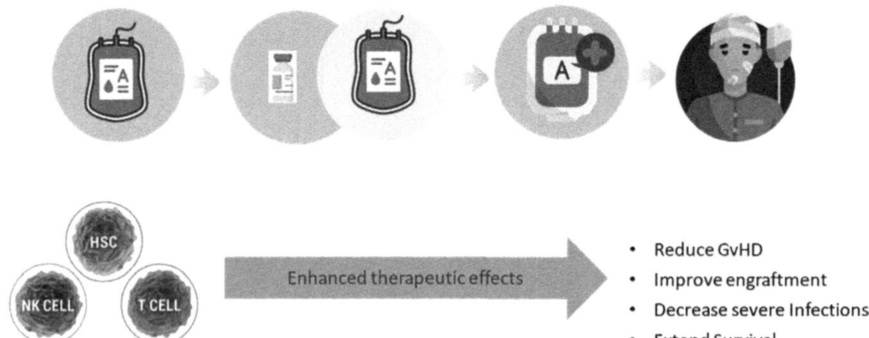

FIGURE 6.13
Process of hematopoietic stem cell transplant.

6.6.3.1 Background

In the context of advanced hematologic malignancies, allogeneic hematopoietic cell transplant (HCT) stands out as a primary therapeutic option, offering potential lifespan extension or even a curative outcome. Typically, peripheral cells such as hematopoietic stem cells (HSCs), natural killer (NK) cells, and T cells are harvested from a matched donor, undergo ex vivo manipulation to enhance therapeutic effects, such as reducing the likelihood of graft-versus-host disease (GvHD) and improving engraftment (Figure 6.13).

A significant concern associated with HCT is GvHD, a complex immune response following transplant characterized by the production and secretion of various cytokines, leading to inflammatory effects and tissue damage. Traditionally, GvHD was categorized into acute and chronic forms based on onset time, with acute GvHD (aGvHD) graded according to severity. Studies have shown markedly worse survival prospects for patients with Grades II–IV aGvHD compared to those with Grade 0 or I aGvHD by Day 100, making it a key efficacy endpoint in HCT trials.

6.6.3.2 Study Design

The study was a Phase II randomized controlled trial with the primary endpoint being Grade II–IV aGvHD at Day 100. The sample size of 56 per arm was determined to achieve 90% power to test the hypothesis of a 43% improvement in aGvHD with the experimental treatment (experimental arm 24% vs. control 42%). This calculation assumed that a two-sided Type I error of 5%. No formal interim analysis was planned (Figure 6.14).

With 70 patients enrolled and anticipating a total enrollment of 112 for the Phase II trial's conclusion, assessing the likelihood of trial success becomes crucial. Based on blinded pooled data, the estimated Grade

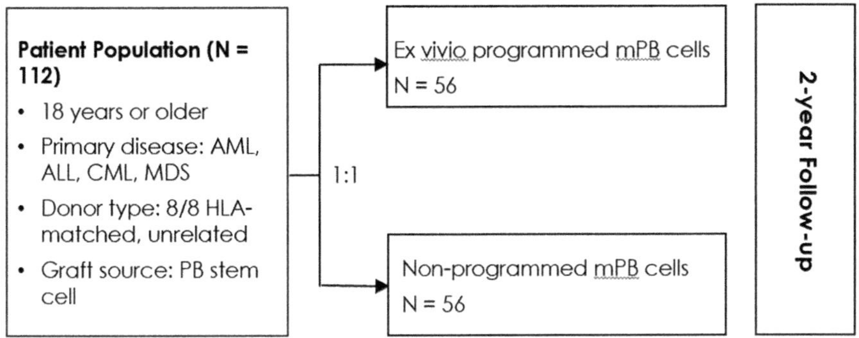

FIGURE 6.14
Diagram of study design.

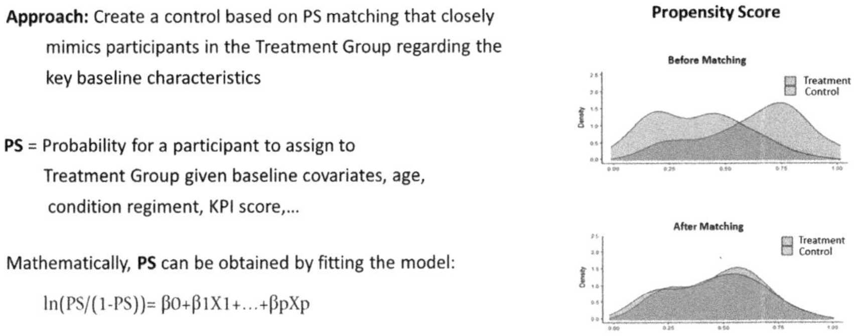

Approach: Create a control based on PS matching that closely mimics participants in the Treatment Group regarding the key baseline characteristics

PS = Probability for a participant to assign to Treatment Group given baseline covariates, age, condition regiment, KPI score,...

Mathematically, **PS** can be obtained by fitting the model:

$$\ln(PS/(1\text{-}PS)) = \beta 0 + \beta 1X1 + \ldots + \beta pXp$$

Propensity Score

Before Matching

Treatment
Control

After Matching

Treatment
Control

FIGURE 6.15
An illustration of creating a synthetic control.

II–IV aGvHD at Day 100 stands at 35%. To ascertain the probability of success, understanding aGvHD rates in both treatment and control arms is necessary.

6.6.3.3 Synthetic Control

Addressing this, a synthetic control is created, matching baseline characteristics of patients randomized to the treatment arm using propensity matching. This method estimates the probability of a patient being assigned to the treatment group given their baseline characteristics, forming a distribution akin to that depicted in the plot at the bottom right of the slide (Figure 6.15).

Key characteristics used for propensity modeling and matched control identification are presented in Figure 6.16, recognized as prognostic factors for this patient population.

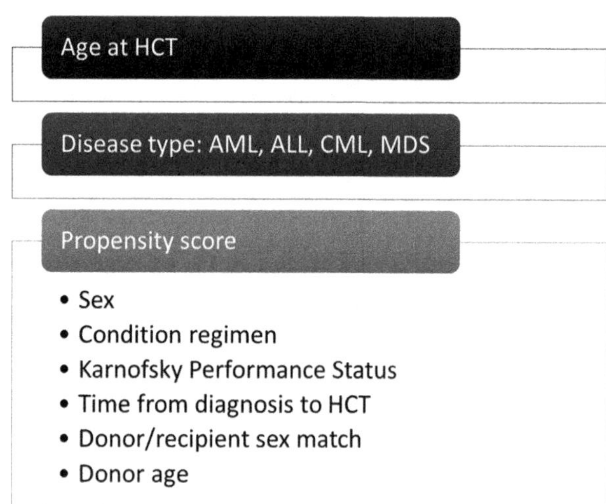

FIGURE 6.16
Prognostic factors used in developing a propensity score model to create a synthetic control.

TABLE 6.17
Summary of Key Prognostic Factors Before Matching

Key Prognostic Factor	Control from RWD (*N* = 450)	In Study (*N* = 70)
Patient Age at HCT, Median (Range), Years	65 (18–76)	55 (22–71)
Sex, *n* (%)		
Male	270 (60)	52 (65)
Female	180 (40)	27 (35)
Conditional Regimen, *n* (%)		
TBI-based	384 (80)	60 (75)
Non TBI-based	166 (20)	20 (25)
Kanofsky Performance Score, *n* (%)		
≥90	234 (52)	51 (64)
<90	216 (48)	29 (36)
Primary Disease, *n* (%)		
AML	153 (34)	41 (51)
ALL	72 (16)	16 (20)
CML	23 (5)	4 (5)
MDS	202 (45)	19 (24)
Donor/Recipient Sex, *n* (%)		
M/M	180 (40)	44 (55)
M/F	113 (25)	18 (23)
F/M	67 (15)	8 (10)
F/F	90 (20)	10 (12)
Donor Age at HCT, Median (Range), Year	29 (18–58)	25 (18–50)
Time from Dx to HCT, Median (Range), Month	6.6 (2–135)	5 (2.5–69)

TABLE 6.18

Summary of Key Prognostic Factors After Matching

Key Prognostic Factor	Control from RWD (*N* = 70)	In Study (*N* = 70)
Patient Age at HCT, Median (Range), Years	56 (18–73)	55 (22–71)
Sex, *n* (%)		
Male	55 (63)	52 (65)
Female	25 (37)	27 (35)
Conditional Regimen, *n* (%)		
TBI-based	59 (74)	60 (75)
Non TBI-based	21 (26)	20 (25)
Kanofsky Performance Score, *n* (%)		
≥90	53 (66)	51 (64)
<90	27 (34)	29 (36)
Primary Disease, *n* (%)		
AML	41 (51)	41 (51)
ALL	16 (20)	16 (20)
CML	4 (5)	4 (5)
MDS	19 (24)	19 (24)
Donor/Recipient Sex, *n* (%)		
M/M	45 (56)	44 (55)
M/F	17 (21)	18 (23)
F/M	8 (10)	8 (10)
F/F	10 (13)	10 (12)
Donor Age at HCT, Median (Range), Year	27 (18–56)	25 (18–50)
Time from Dx to HCT, Median (Range), Month	5.2 (2–78)	5 (2.5–69)

6.6.3.4 Results

6.6.3.4.1 Matching

The real-world data source utilized is the HCT registry from the Center for International Blood and Marrow Transplant Research (CIBMTR), housing data from allogeneic transplants across the United States. Controls from the RWD database before and after matching are displayed in Tables 6.17 and 6.18, respectively. It is evident that the control after matching is comparable to the population enrolled in the study with respect to the key prognostic factors after matching based on propensity scores.

6.6.3.4.2 Probability of Success

With the synthetic control, the estimated Grade II–IV aGvHD at Day 100 stands at 38%. Combining this with the pooled estimate from the study (35%), the aGvHD rate in the treatment arm is calculated at 32%. Considering these estimates, the power to detect differences between treatment and control arms is found to be 12.6%, leading to the decision to terminate the study due to a low probability of success. Actual unblinded results showed the treatment and control aGvHD at Day 100 were 36% and 34%, respectively. The results confirmed the decision previously made was correct.

6.7 Concluding Remarks

In conclusion, decision-making in clinical trials is an intricate process that balances scientific rigor, regulatory requirements, and the pressing need for new therapies. As highlighted in the introduction, the landscape of drug development has seen significant advancements in understanding disease biology and patient selection. However, the high attrition rates in late-stage clinical trials underscore the need for more robust decision-making frameworks.

Throughout this chapter, we have explored various statistical decision frameworks employed at different phases of clinical trials. These frameworks, while valuable, often depend on data from internal studies and traditional model structures, which can limit their effectiveness. The integration of MIDD has marked a significant step forward, providing quantitative tools to enhance decision-making across dose selection, trial design, safety evaluation, and overall drug development strategies.

Moreover, the emergence of AI and ML technologies presents a transformative potential to further refine these processes. By harnessing diverse data sources and leveraging the predictive capabilities of advanced algorithms, AI and ML can address the limitations of conventional methods, offering more accurate and comprehensive insights.

As we move forward, it is imperative to embrace these innovative approaches while ensuring they are rigorously validated and seamlessly integrated into existing frameworks. The ongoing evolution of decision-making in clinical trials will undoubtedly benefit from the synergy between traditional statistical methods and cutting-edge AI technologies. This holistic approach promises to enhance the efficiency and success rates of clinical trials, ultimately accelerating the delivery of new, effective treatments to patients in need.

In summary, the future of clinical trial decision-making lies in the harmonious blend of established methodologies and emerging technologies. By continuing to refine and adopt these strategies, the field of drug development can achieve greater precision, productivity, and impact, paving the way for a new era of medical breakthroughs.

7

AI-Assisted Data Analysis in Clinical Trials

7.1 Introduction

Data analysis is central to clinical trials. The ability to timely integrate and analyze data from clinical trials and real-world sources provides a significant competitive advantage in study design, evidence generation, and sound decision-making. The evidence derived from the analysis of the safety and efficacy of a new drug serves as the backbone of regulatory submissions, steering the approval of new treatments and their subsequent entry to the market. The integration of artificial intelligence (AI) and machine learning (ML) into clinical development has opened new frontiers for improving data analysis, decision-making processes, and overall efficiency. These technologies offer innovative solutions for handling complex datasets, automating routine tasks, and extracting meaningful insights from vast amounts of unstructured data. This chapter explores several key areas where AI and ML are making significant impacts in clinical trials regarding data curation, analysis, interpretation, and reporting.

7.2 Evolving Role of Data Analysis

7.2.1 Current Practice

The use of RCTs for demonstrating drug safety and efficacy has been the gold standard for assessing a drug's safety and efficacy. The evidence generation process of RCTs is consistent with regulatory expectations and has been reinforced through the publication of numerous regulatory guidelines. The current practice of data analysis primarily concentrates on fulfilling regulatory requirements. Data analysis is performed to support study design,

DOI: 10.1201/9781003226086-7

interim go/no-go decisions, phase transitions, and ultimately the assessment of drug safety and efficacy, in support of regulatory submissions. During study design, a meta-analysis is typically conducted, based on the limited published data of the control arm, to support effect size and sample size estimation. For the interim and final analyses, the focus is on the analysis of drug safety and activity. Figure 7.1 depicts a common clinical trial data analysis process, in support of regulatory submissions. It is a multi-stage process involving extracting, transforming, analyzing data, and interpreting and reporting findings.

As discussed in Chapter 2, raw data from clinical sites, local, central, or specialty laboratories, and other clinical trial systems such as IRS are extracted and loaded into a central repository where they are curated and integrated for analysis. These data are mapped to standard datasets according to the Study Data Tabulation Model (SDTM). The SDTM data are further mapped into analysis datasets per the Analysis Data Model (ADaM).

The SDTM is a standardized data model that specifies the structure and content of clinical trial data. Compliance with SDTM is a requirement for data submission to both the FDA and PMDA. This model is employed to organize and present clinical trial data in a uniform format, ensuring that it can be easily understood by regulatory authorities, clinical researchers, and data analysts. SDTM datasets are utilized to summarize patient demographics and baseline information, adverse events (AEs), findings, and other

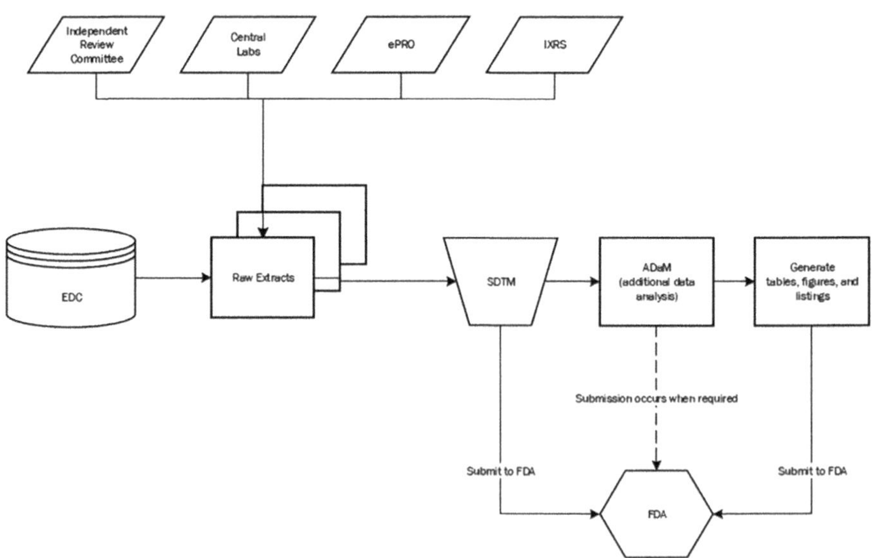

FIGURE 7.1
Process of clinical trial data analysis.

pertinent clinical trial data and are considered source data for regulatory submissions.

In contrast, the ADaM provides the fundamental principles and standards for the creation of analysis datasets and associated metadata. The purpose of ADaM is to provide a framework that facilitates the analysis of clinical trial data while ensuring that reviewers and other recipients can clearly understand the data's lineage from collection through to analysis and results. Both SDTM and ADaM are developed by The Clinical Data Interchange Standards Consortium (CDISC), a global, non-profit organization that develops data standards to streamline clinical research and ensure data quality and interoperability.

Data analysis is carried out in accordance with the methods defined in the Statistical Analysis Plan (SAP). The results of the analysis are presented in tables, figures, and listings (TFLs), which are predefined in the SAP. Finally, the results are summarized and incorporated into the Clinical Study Report (CSR) for regulatory submissions.

7.2.2 Changing Context

In recent years, the high attribution of clinical development and the advent of digital technologies has inspired a shift toward a patient-centric approach for clinical trials (Yang 2023a). There has been a growing interest in using real-world data (RWD) and real-world evidence (RWE) to support clinical trials, regulatory review, and healthcare decisions (Yang and Yu 2021). Regulators, payers, and patients are demanding drug developers to analyze data beyond the confines of clinical trials and generate clinical evidence through a broader perspective accounting for patient heterogenicity, real-world use, and targeted genomic, cultural, and environmental factors that could impact clinical outcomes (Collins and Varmus 2015).

7.2.2.1 New Role of Data Analysis

With the ever-increasing acquisition and access to vast amounts of data from diverse sources including electronic health record (EHR), medical claim, and disease registry, wireless wearable devices, social medica, data analysis in clinical trials has emerged as a cornerstone of innovative study design, synthesis of information of drug safety and efficacy, demonstration of value of a new drug to justify payment, and enablement of personalized medicine. Data analysis also plays a key role in deriving insights from diverse data sources to enhance clinical operations.

Technological advancement has not only broadened the scope of data analysis, but also provided new opportunities for automating and streamlining the largely manual processes of data curation, analysis, and reporting processes and enhancing the accuracy and reliability of analysis results.

7.2.2.2 Expanded Data Analysis Toolbox

Traditional statistical models have been the cornerstone of data analysis in clinical research for decades. However, these models often rely on explicit assumptions of the relationships between the response and covariates and among the covariates. For example, for survival analysis, the Cox proportional hazard regression model is a commonly used method for modeling the relationship between the hazard function and covariates, but it assumes the hazard function is proportional to the baseline hazard. In addition, it is also assumed that the effects of the covariates are additive. Another example concerns the exposure-response analysis where a mathematical relationship needs to be explicitly defined. Owing to rapid advances in digital technologies, more complex and higher volumes of clinical data are collected, often creating datasets which include more covariables than observations. This creates additional challenges for fitting the traditional statistical models to such high-dimensional data. Taking these all together, it is necessary to utilize more sophisticated analytical approaches with minimal model assumptions but great capability to cope with high-dimensional data.

ML algorithms offer several advantages over traditional statistical models in handling and interpreting complex datasets. Since these algorithms are data-driven as opposed to reliance on model assumptions, they can adapt to nonlinear relationships and interactions between variables, offering more flexible and accurate analyses. Even in situations where there are more variables than observations, they can uncover hidden patterns and relationships within data that traditional models might overlook. Techniques such as clustering, decision trees, and neural networks provide robust tools for classification, regression, and prediction tasks. For example, researchers using a random forest model were able to better discriminate between patients who would have better or worse outcomes following cardiac resynchronization therapy compared to using a multivariable logistic regression model (Kalscheur et al. 2018). This demonstrates the ability of random forests to model interactions between features that simpler models cannot capture.

Extensive research has documented the application of ML in disease diagnosis and prognosis. These studies indicate that Random Survival Forests (RSF) exhibit superior or comparable performance to the Cox model in predicting survival outcomes for patients with breast cancer, prostate cancer, and systolic heart failure based on baseline characteristics. Additionally, Artificial Neural Networks (ANN) have been reported to outperform the Cox model in predicting kidney failure survival and breast cancer occurrence. Both ANN and RSF have shown similar predictive performance in head-to-head evaluations of breast cancer survival using microarray data. Gong et al. (2021) conducted a series of comprehensive simulation studies to assess the utility of ML for time-to-event analysis. Their findings suggested high flexibility and reliability of ML-based analyses when compared to the Cox model.

7.3 Applications of AI and Machine Learning in Data Analysis

7.3.1 Study Design

7.3.1.1 Natural History of Disease

The natural history of disease refers to the progression of a disease in an individual over time, in the absence of any treatment or intervention (Centers for Disease Control and Prevention 2021). Understanding disease pathways is critically important for informing the primary study objectives of Phase I studies and defining the key clinical endpoints for assessing the safety and efficacy of a novel therapy. Over the past decade, the study of the natural history of disease has gained significant prominence due to the growing need to understand the etiology of many rare diseases (IQVIA 2020).

For example, Cohen et al. (2020) applied ML and knowledge engineering to a large extract of EHR data to determine whether this combined approach could effectively identify patients not previously tested for acute hepatic porphyria (AHP) who should receive a proper diagnostic workup for AHP. Predictive analytics based on ML and natural language processing (NLP) have been shown to help identify complex clinical patterns from patients' medical histories (Rajkomar et al. 2018; LeCun et al. 2015).

7.3.1.2 Deep Learning for Automating Imaging Classification

Medical imaging is a critical component of clinical research, particularly in fields such as oncology, neurology, and cardiology. The analysis and interpretation of medical images require extensive expertise and time. Deep learning (DL) has the potential to streamline processes by automating imaging classification with high accuracy. Convolutional neural networks (CNNs), a type of DL architecture, are particularly effective for image recognition tasks. CNNs can automatically learn to identify relevant features from raw images, significantly reducing the need for manual feature extraction. These models have been successfully applied to classify various medical conditions from imaging classification through DL enhances diagnostic precision, accelerates research workflows, and allows for the processing of large-scale imaging datasets (Lee and Lee 2020; Weissler et al. 2021).

7.3.1.3 Increasing Effect Size and Improving Statistical Power

Digital markers, also known as digital biomarkers, are objective, quantifiable measures obtained through digital devices such as wearables, smartphones, and sensors. These markers provide real-time data on various physiological and behavioral parameters, offering new opportunities for clinical trials.

The integration of digital markers into clinical studies can significantly increase effect size by providing continuous and precise measurements of patient outcomes. For instance, digital markers can monitor physical activity, heart rate, sleep patterns, and medication adherence, offering a more comprehensive view of patient health and treatment effects. ML algorithms can analyze this data to identify patterns and correlations that traditional methods might miss, thereby enhancing the sensitivity and specificity of clinical endpoints. The use of digital markers not only enriches the dataset but also enables more personalized and timely interventions. In a case study by Zhou and Manser (2020), the authors showed that predictive modeling approaches for the outcomes of amyotrophic lateral sclerosis trials, based on random forest and super learner, were able to explain a moderate amount of variability in longitudinal change. It was further found that including the post-baseline model-predicted results as a covariate in the model for primary analysis may increase statistical power under moderate treatment effects. Guthrie et al. (2019) demonstrated that ML models such as random forest has the potential to transform data from a digital therapeutic into digital biomarkers that predict treatment response in individual participants, thus having potential to improvement treatment outcomes.

7.3.2 Data Analysis

7.3.2.1 AI for Extraction of Endpoints from Unstructured Data

Clinical research often involves large volumes of unstructured data, including clinical notes, patient records, and research articles. Extracting meaningful information from this data is essential for identifying clinical endpoints and generating insights. NLP combined with DL provides powerful tools for this task (Wang et al. 2018). NLP techniques enable the processing and analysis of human language, allowing machines to understand and interpret text. When integrated with DL models, such as recurrent neural networks (RNNs) and transformers, NLP can extract specific endpoints from unstructured data with high accuracy. These models can identify relevant information, such as AEs, treatment outcomes, and patient characteristics, by learning from annotated datasets. The application of NLP and DL in extracting endpoints not only streamlines data processing but also enhances the comprehensiveness and depth of clinical research analysis.

NLP techniques enable the processing and analysis of human language, allowing machines to understand and interpret text. When integrated with DL models, such as RNNs and transformers, NLP can extract specific endpoints from unstructured data with high accuracy. These models can identify relevant information, such as AEs, treatment outcomes, and patient characteristics, by learning from annotated datasets. The application of NLP and DL in extracting endpoints not only streamlines data processing but also enhances the comprehensiveness and depth of clinical research analysis.

Arbour et al. (2021) developed a DL model that uses radiology text reports to estimate gold-standard RECIST-defined outcomes. By utilizing text reports from patients with non-small cell lung cancer (NSCLC) treated with PD-1 checkpoint inhibitors, the model accurately estimated best overall response and progression-free survival. This tool has the potential to determine outcomes at scale, facilitating the analysis of large clinical databases.

7.3.2.2 Automation of SAP and TFLs

NLP, AI, and ML are increasingly being used to automate the creation of SAPs and the generation of mock shells for TFLs. Traditionally, writing an SAP requires significant time and expertise, as it involves translating complex study protocols into detailed statistical methodologies and reporting plans. By leveraging NLP, AI, and ML, these tasks can be expedited. NLP algorithms can parse clinical trial protocols and extract relevant information, while AI can use predefined templates to automatically draft SAPs. ML models, trained on historical trial data and expert-authored SAPs, can generate highly relevant and accurate drafts. Similarly, for TFL mock shells, AI systems can analyze the structure and content of past reports and apply that knowledge to generate shells that adhere to regulatory and reporting standards. This automation not only saves time but also reduces errors, ensuring a more efficient and consistent process for clinical trial reporting.

7.3.2.3 Automation of SDTM and ADaM Mapping

Data integration is a key step toward the maximization of data values. For marketing applications, integrated summaries of safety and efficacy from trials throughout a new drug development is often required. This entails data analysis of the pooled data across clinical trials. Due to the involvement of various vendors and CROs in different stages of clinical development, it is unavoidable that there are variations in metadata including naming of variables for the same type of data such as AEs across studies. In addition, in-license new candidate drug for development often creates incompatibility in data format, resulting significant hurdles when pooled analysis is required to fulfill regulatory requirements or address inquiries from regulatory authorities. Traditionally, these issues are resolved through mapping the data to SDTM format. However, as discussed previously, this is a manual-driven time-consuming process. The advent of AI and ML lends the opportunity to automate this process. Various pilot studies have been conducted to explore the mapping from raw data to SDTM and ADaM datasets. The study by Tomioka (2018) focused on SDTM mapping using a supervised ML approach. The ML model was trained using CDASH and SDTM metadata from CDISC, along with clinical trial data from 20 legacy and ongoing studies spanning Phases I to III. It was shown that the final trained model for the AE domain can generate the mapping specification with an accuracy of 0.94–0.98 (95% CI) and a Cohen's kappa of 0.94 in just a few seconds. The final ensemble model achieved 98% mapping accuracy without any human intervention.

7.3.2.4 Missing Data Imputation

Missing data is a common challenge in clinical research, often leading to biased results and reduced statistical power. Traditional methods of handling missing data, such as mean imputation or complete case analysis, can introduce errors and diminish the quality of findings (Little and Rubin 2002). AI and ML offer advanced techniques for addressing this issue more effectively. ML algorithms, such as k-nearest neighbors (KNN), random forests, and gradient boosting, can predict missing values based on the patterns observed in the available data. These methods leverage the relationships between variables to generate more accurate and reliable imputations. Additionally, DL techniques, such as autoencoders, can be employed to learn complex data representations and provide robust imputation for missing values (Emmanuel et al. 2021). Unlike the traditional imputation methods, these ML algorithms consider multiple causes of data missingness, data-related assumptions, goals, and intended data collection and analytic methods, and impute specific estimates of missing covariate values or averaging over many possible values from a learned distribution to compute other quantities of interest (Weissler et al. 2021). For example, a ML method based on a combination of unsupervised prefilling strategy and supervised learned model for the imputation of missing laboratory test values outperformed baseline and state-of-the-arm models on 13 commonly collected laboratory test variables (Zhang et al. 2020). The use of AI and ML for missing data imputation not only enhances the integrity of the datasets but also improves the validity of the research outcomes.

7.3.2.5 Model Selection for Data Analysis

AI and ML methods play a crucial role in selecting models for clinical data analysis by leveraging their ability to handle complex datasets and optimize predictive performance. These methods facilitate the identification of the most suitable statistical models that best fit clinical data characteristics, thus improving the accuracy and reliability of analyses. Techniques such as cross-validation, regularization methods like Lasso and Ridge regression, and ensemble methods such as random forests and gradient boosting are commonly employed to assess model performance and select the optimal model. AI-driven approaches also integrate feature selection techniques to identify relevant predictors from large datasets, enhancing model interpretability and reducing overfitting. These advancements underscore the transformative impact of AI and ML in enhancing model selection processes, thereby advancing clinical research and healthcare outcomes.

7.3.2.5.1 An Example

Pharmacokinetics (PK) involves the study of how drugs are absorbed, distributed, metabolized, and excreted in the body. Selecting appropriate pharmacokinetic models is crucial for understanding drug behavior and optimizing dosing regimens. These PK models are conventionally based on compartment model structures and are often described by nonlinear mixed effect models (Lee et al. 2018). Using traditional compartment PK models requires a prior understanding of the characteristics of absorption, distribution, metabolism, and elimination (ADME) of a drug. Choosing the right PK model requires expert knowledge and is a time-consuming and labor-intensive process. ML algorithms, such as support vector machines (SVM), random forests, and neural networks, can analyze large PK datasets to identify the most suitable models for different drugs and patient populations. These algorithms can account for various factors, including patient demographics, genetic variations, and disease states, to predict drug kinetics more accurately. The use of ML in PK model selection enhances the precision of drug development and personalized medicine, ensuring that treatments are tailored to individual patient needs. Modern DL technologies, capable of modeling high-level abstractions in data and optimizing a large number of parameters in parallel, offer a promising alternative (Poynton et al. 2009). The application of DL can potentially accelerate the modeling process and enhance predictive accuracy.

The findings of a study reported by Poynton et al. (2009) demonstrated that ensemble models that combined the traditional nonlinear mixed effects models with ML methods such as ANN and SVM improved the accuracy and reduced variance of PK models. More recently, Li et al. (n.d.) proposed a multi-layer fully connected neural network to predict PK profile after a single dose of a solid oral dosage of a drug product. The CNN model has three layers with (Figure 7.2). The inputs of the model include subject ID, sampling time, along with 12 covariates such as gender, race, ethnicity, BMI, weight, and height.

The dataset was divided into training and test sets. The training set was used to develop the DL model, which was then used to make predictions on both the training and test sets. Figure 7.3 presents the predicted PK profiles along with the true PK profiles of Subjects 22 and 8 randomly selected from the training dataset and Subjects 22 and 7 randomly chosen from the test dataset. The shaded areas are range of mean ± standard deviation of drug concentration at across the entire time course. For each subject, there is remarkable similarity between the predicted and true PK profiles.

By leveraging ML algorithms, researchers can gain deeper insights into clinical data, identify novel biomarkers, and improve the overall quality of

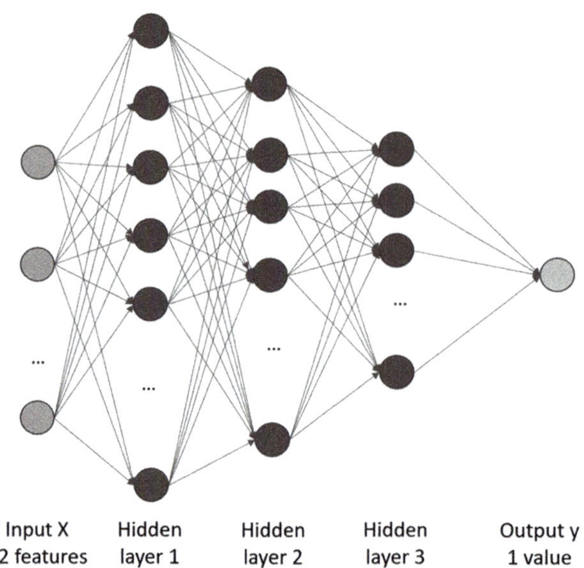

FIGURE 7.2
Structure of CNN for predicting PK profile post a single solid oral dosage form product.

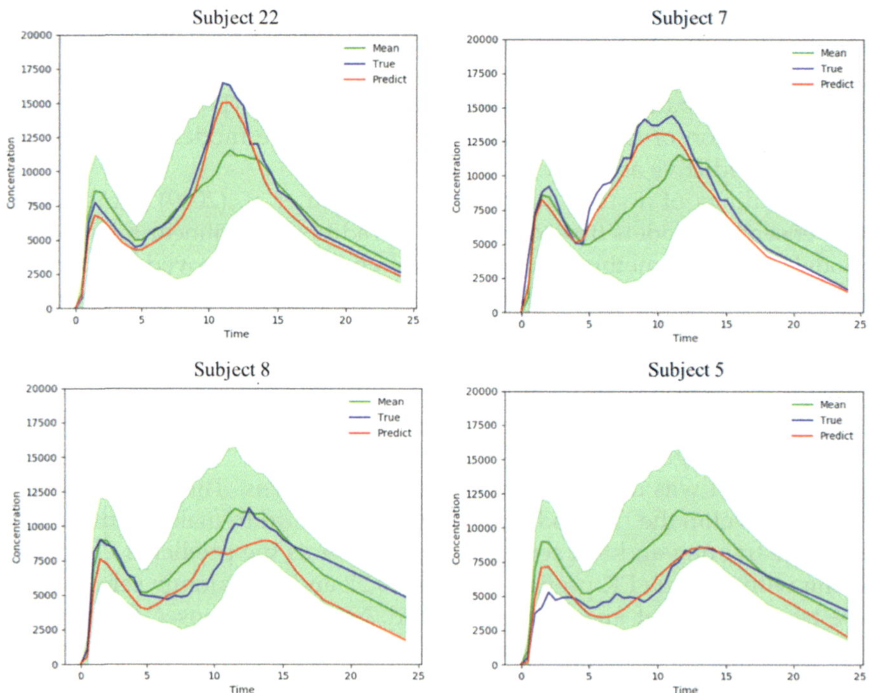

FIGURE 7.3
A sample of PK profiles predicted by the DL model. Subjects 22 and 7 were randomly selected from the training dataset while Subjects 7 and 5 were a random sample from the test dataset.

research findings. This shift toward ML-driven analysis represents a significant advancement in the field of clinical research, enabling more informed decision-making and better patient outcomes.

7.3.2.6 Use of Large Language Models to Process Safety Data

Ensuring patient safety is a paramount concern in clinical research. Large language models (LLMs), such as GPT-4, offer advanced capabilities for processing safety data, including the identification and analysis of AEs. LLMs are trained on vast amounts of text data and can generate human-like responses, making them suitable for understanding complex language patterns and extracting relevant information. In the context of safety data, LLMs can be employed to analyze clinical trial reports, patient feedback, and safety databases to identify potential safety signals. These models can also assist in generating comprehensive safety reports by summarizing findings and highlighting critical information. The use of LLMs for processing safety data enhances the efficiency and accuracy of safety monitoring, contributing to improved patient outcomes and regulatory compliance.

A study was conducted to train and evaluate several DL approaches to individual case safety report processing (Abatemarco et al. 2018). The findings indicated that the cognitive services under development have achieved the minimum evaluative threshold to be considered adequately trained. This demonstrated how ML and NLP techniques can work together to provide accurate outputs, potentially augmenting the ability of pharmacovigilance professionals to process spontaneous ICSRs quickly and accurately.

7.3.3 Knowledge Discovery and Evidence Generation

7.3.3.1 Identification of Treatment Heterogenicity

Patient populations in clinical trials heterogeneous, exhibiting variations in characteristics such as age, sex, disease etiology and severity, comorbid conditions, concurrent medication, and genetic differences (Varadhan and Seeger 2013). These differences among patients can influence how a treatment affects outcomes. RCTs primarily aim at demonstrating safety and efficacy based on average treatment effects (ATEs). This relies on an underlying assumption that a uniform treatment response across the varied patient characteristics. While this assumption may hold true for some treatments, for others, the treatment effect may differ significantly across subgroups. This variation in treatment effects, known as treatment effect heterogeneity, can result from underlying causal mechanisms or artifacts such as chance, bias, or confounding factors. Heterogeneity of treatment effect (HTE) refers to the systematic, explainable variability in treatment effects among individuals within a population. The primary objectives of HTE analysis are to estimate treatment effects in clinically relevant subgroups and to predict whether a specific individual might benefit from a treatment (Goldstein and Rigdon 2019). Subgroup analysis is the most common method for examining HTE.

For RCTs, two types of analyses are usually carried out. The primary analyses are performed to assess the primary objectives of the trial regarding drug safety and efficacy. These analyses are pre-specified in the protocol and the findings serve as the basis for clinical decision-making and regulatory filings. The study is usually sufficiently sized to ensure greater chance of meeting the primary objectives. In contrast, secondary analyses are carried out to understand the HTE, which represents variations of the treatment effect for individuals or among various subgroups. The current practice in secondary analyses usually involves estimating treatment effects of various subgroup, and report the results through a forest plot, which provides a visual representation of the estimated common treatment effect along with associated confidence interval for different groups. The greater similarity of point estimates and more overlapping of conference intervals are, the less heterogeneous the treatment effect is. However, there are several limitations to this approach. Specifically, subgroups used in secondary analyses are defined without considering higher order interaction among factors defining the subgroup and the treatment. In addition, the method does not provide estimation of individual treatment effect. The common remedy for addressing the first issue is to examine the treatment effect using multivariate models that incorporate interactions among covariates that define subgroups. In truth, quantifying individual treatment effect has proved to be more challenging as each patient can often only be exposed to one treatment in a trial.

In statistical literature, a two-step model was proposed to estimate individual treatment effect. This method involves initially constructing separate models for different treatment groups (e.g., treatment versus control). The treatment effect for each individual is then estimated by calculating the difference in predicted responses from these separately built models. Despite its straightforward and intuitive approach, the two-step model has several significant drawbacks. One major issue is that the difference between two independent, accurate models does not necessarily yield an accurate combined model. Additionally, these separate models are often based on ordinary regression techniques, which may not perform well with nonlinear or high-dimensional data.

Alternatively, to evaluate HTE, a regression model can be constructed with pre-specified interactions between treatment and covariates, recognizing that interactions are a key contributor to HTE. However, this approach requires sufficient prior knowledge to identify potential interactions, a challenging and nearly impossible task in the context of high-dimensional data.

Recently, several ML approaches have been developed to estimate HTE. Among these, the decision tree-based HTE method was one of the first to be developed and has gained widespread recognition. The core feature of a decision tree, which involves partitioning the full dataset into subgroups, makes it particularly well-suited for HTE analysis, as it aims to identify subgroups (or individuals) with distinct treatment effects. Goldstein and Rigdon (2019) discussed the use of machine learning to identify heterogeneous effects in randomized clinical trials. Lu, Sadiq et al. (2018) estimated individual treatment effects using random forest methods.

7.3.3.2 Safety Risk Prediction

Exploratory analysis in RCTs is used to assess the association between clinical outcomes and patient characteristics. Since an RCT is rarely adequately powered for exploratory analysis, the findings are often unreliable. The ability to integrate clinical data with data from other sources has greatly increased the utilization of vast amount of study participant data ranging from macro-level physiology and behavior to laboratory, imaging, and "omic" data. ML has the potential to dramatically improve the ability to establish the prognosis of a patient (Obermeyer and Emanuel 2016). Such risk prediction is an important step toward personalized medicine and patient-centric healthcare. A prime example of a successful risk model is the Framingham Heart Score, a risk index for cardiovascular mortality (Berry et al. 2007). Insights into cardiovascular death risk factors identified through this study were credited for a 50% reduction in age-adjusted cardiovascular deaths (Bitton & Gaziano 2010; Berry et al. 2007). ML models applied to predict drug-induced liver injury (DILI) represent another critical area (Munsaka et al. 2022). DILI is a significant concern in drug development, contributing to high attrition rates and drug withdrawals from the market. Recognizing its importance, the FDA introduced guidance in 2009 to guide the analysis of liver safety data in drug submissions (FDA 2009b). Despite advances, current methods from clinical trials and toxicological studies have limitations in predicting DILI in humans, prompting the development of new tools and approaches leveraging AI and ML (Vall et al. 2021).

Efforts have also focused on using ML to predict kinase-AE associations. Gong et al. (2021) employed RSF to assess potential associations between 442 kinases and 2,145 AEs. Their model, trained on patient-level pharmacokinetic and AE data from 4,638 patients across 16 pivotal studies on small molecule kinase inhibitors (SMKIs), offers several benefits: (1) facilitating the discovery of kinase-inhibitor AE pairs, (2) serving as a precision medicine tool to anticipate individual patient safety risks, and (3) aiding in contemporary drug development by comparing SMKI therapies from a safety perspective.

7.3.3.3 Phenotyping

In clinical research, the term phenotype refers to the observable presentation of a disease, such as morphology, biochemical or physiological traits, developmental or behavioral characteristics, without implying any underlying mechanism (Scheuermann et al. 2009). In the context of EHRs, phenotypes are the clinical conditions, characteristics, or features of patients that can be derived solely from EHR data (Yadav et al. 2017). Traditionally, disease phenotyping was carried out through manual review of patients' medical records. However, the scale of EHR data makes this practice challenging.

By using ML and data mining techniques, phenotyping algorithms can identify and characterize diseases, their subtypes, and clusters of comorbidities using EHRs (Kirby et al. 2016), thereby enabling personalized healthcare. Central to this effort is the integration of various modalities of information available in EHRs, along with the systematic handling of missing and sparse data. Early work in this field largely employed unsupervised techniques, but more recent studies have focused on the use of DL algorithms for phenotyping. A comprehensive review of these advancements is provided by Shickel et al. (2018).

7.3.3.4 Evidence-Based Medicine

Analysis of EHR data and derive insights play a critical role in evidence-based medicine, including regulatory approval and label expansion, clinical practice guidelines, and reimbursement policies.

Currently, many clinical trials are conducted in smaller and more specific populations for unmet needs or within niche indications. The specificity of these therapeutic trials has led to an increased acceptance of single-arm trials for therapy registration. While single-arm trials are effective for registration, they are less effective for gaining full approval and reimbursement. In the absence of a standard-of-care (SOC), a synthetic control created from EHR patient data, using rigorous statistical methods, can provide comparative evidence to support product marketing applications (Li et al. 2021).

The aim of clinical guidelines is to guide clinical decisions to achieve optimal patient treatment and care using the most up-to-date information. For example, the American Heart Association/American College of Cardiology (AHA/ACC) recommends using an established algorithm based on risk factors such as hypertension, cholesterol, age, smoking, and diabetes to triage cardiovascular disease (CVD) risk. Evidence-based clinical practice guidelines are the cornerstone of modern medicine (Berry et al. 2007). With the explosion of EHR data, many opportunities arise for creating and revising treatment guidelines, codifying knowledge gained from mining EHR data.

As noted by Berger and Doban (2014), these guidelines will become less population-based and more geared toward individual patients. The guidelines may evolve and be further revised as more is learned about: (1) How did patients fare following the guidelines compared to those who did not? (2) Which patients achieved the best outcomes? (3) Are there clinical characteristics or biomarkers that predict responders and those who experience side effects? These evolving guidelines will continue to integrate insights from real-world EHR data, enhancing personalized patient care.

7.3.4 Results Interpretation and Reporting

7.3.4.1 Automating CSR

Automating the writing of CSRs using NLP and AI presents a transformative opportunity in clinical research. NLP enables machines to interpret and generate human-like text from structured and unstructured data sources, such as clinical trial data, patient records, and regulatory documents (Wang et al. 2019). AI algorithms, including ML models like deep neural networks and transformers, can analyze vast amounts of data to extract key findings, summarize results, and interpret statistical analyses automatically. This capability not only streamlines the CSR writing process but also improves accuracy by reducing human error and ensuring compliance with regulatory standards. By integrating NLP and AI, CSRs can be generated more efficiently, facilitating faster regulatory submissions and enabling researchers to focus more on data interpretation and clinical insights rather than manual report writing. Several technology startups have developed AI-assisted platforms for automating CSR writing (ZYLiQ n.d.).

7.3.4.2 Digitizing Analysis Results

Although integrating and contextualizing analysis results is fundamental in clinical trials, the outputs themselves are frequently underutilized as valuable data sources. Typically, these results are generated as static documents in formats like Word or PDF, which are not inherently machine-readable or easily integrated for future analyses. AI and ML offer promising solutions to digitize and extract insights from these outputs. By applying NLP techniques, such as text extraction and classification, AI systems can parse through these documents, converting them into structured data that is accessible, searchable, and suitable for further computational analysis. This transformation not only enhances data accessibility but also facilitates the integration of analysis results across different studies and trials, promoting more comprehensive meta-analyses and facilitating evidence-based decision-making in clinical research.

7.4 Case Examples

In this section, we present three applications of ML methods and RWD for extracting tumor response endpoints, performing heterogeneous treatment effect (HTE) analysis, and generating clinical evidence through synthetic controls.

7.4.1 Deep Learning to Estimate RECIST from RWD

7.4.1.1 Background

RWD has played a significant role throughout the lifecycle of drug development. Of note, RWD from the extensive number of patients outside of clinical trials presents an important and emerging opportunity for evidence generation. However, since RWD is not collected within the setting of clinical trials, the extraction of treatment-specific outcomes can be challenging. For example, the RWD of cancer patients typically include patient demographics, disease diagnosis, medical history, radiology reports, and mortality that can be readily retrieved. However, the attainment of treat-specific outcomes such as objective response and progression-free survival remains significantly challenging. This is due to the lack of standardized, quantified assessments of response or progression in radiology reports (Arbour et al. 2021). Although sometimes it is possible to extract such information per the standardized RECIST from the RWD, the substantial time and effort required by trained radiologists to review the imaging data retrospectively limits the feasibility of large-scale analyses. In the following, we describe a DL method developed by Arbour et al. (2021) to estimate RECIST using clinical radiology text reports. This tool was trained by gold-standard RECIST reads performed by trained radiologists and has the potential to facilitate the utilization and analysis of large-scale real-world oncology data for clinical trial design and evidence-generation.

7.4.1.2 Data

In total, 453 patients with NSCLC treated with PD-1/PD-L1 inhibitors at Memorial Sloan Kettering Cancer (MSK) and 97 NSCLC patients treated at Massachusetts General Hospital (MGH) were used for the development of the DL model. The 453 patients from MSK were split into a training cohort ($n = 361$) and internal testing cohort ($n = 91$) while the 97 patients from MGH were used as an external testing cohort. In the MSK cohorts, 2,977 clinical imaging reports were available for analysis. These reports were manually reviewed by trained radiologists and objective responses were assessed using the RECIST criteria and reported along with occurrence of progression and date of progression. The RECIST classification was performed prospectively for 199 patients (43%) who were involved in ongoing clinical trials and retrospectively for 24 patients (56%) treated with SOC therapy. These reports and the RECIST reads were used as inputs for model development.

In the MGH cohort, a total of 1,238 clinical imaging reports were included in model testing, along with RECIST reads performed by trained radiologists. Among the 97 patients, the RECIST assessments were performed prospectively and retrospectively for 56 (58%) and 31 (42%) patients, respectively.

The baseline characteristics and the response classifications are summarized in Table 7.1.

TABLE 7.1

Baseline Characteristics

	Training (n = 361)	Internal (n = 92)	External (n = 97)
Age, median [range] – years	67 [32–88]	65 [42–93]	66 [36–85]
Sex – no. (%)			
Female	177 (49%)	49 (53%)	45 (46%)
Male	184 (51%)	43 (47%)	52 (54%)
ECOG PS – no. (%)			
0–1	334 (93%)	86 (93%)	86 (89%)
>1	27 (7%)	6 (7%)	11 (11%)
Smoking history – no. (%)			
Former/Current	320 (89%)	81 (88%)	83 (86%)
Never	41 (11%)	11 (12%)	14 (14%)
Histology – no. (%)			
Adenocarcinoma	272 (75%)	74 (80%)	68 (70%)
Squamous cell lung carcinoma	69 (19%)	12 (13%)	25 (26%)
Other (adenosquamous, LC-NE, NOS)	20 (6%)	6 (7%)	4 (4%)
Line of therapy – no. (%)			
1st line treatment	149 (41%)	43 (47%)	21 (22%)
>1st line treatment	212 (59%)	49 (53%)	76 (78%)
Immunotherapy type – no. (%)			
PD-1 blockade	269 (75%)	70 (76%)	81 (84%)
PD-1 blockade + CTLA-4 blockade	92 (25%)	22 (24%)	16 (16%)
RECIST BOR – no. (%)			
CR	5 (1%)	4 (4%)	4 (4%)
PR	77 (21%)	15 (16%)	25 (26%)
SD	132 (37%)	35 (38%)	28 (29%)
PD	147 (41%)	38 (41%)*	40 (41%)
Clinical Trial – no. (%)	151 (42%)	48 (52%)	56 (58%)
Standard-of-care – no. (%)	210 (58%)	44 (48%)	41 (42%)

* Percentages may not add up to 100 due to rounding.

Source: Arbour et al. (2021).

7.4.1.3 Method

A fully connected deep NLP model was trained for response classification, determination of PD, and date of PD (Figure 7.4). Since the task centered on assessing responses in radiology text reports, the model excluded mortality from consideration. Patients who passed away without experiencing radiologic progression were treated as censored during the training and validation of the progression-free survival analyses.

To train the model, clinical text reports detailing the findings from imaging analysis in the training cohort were categorized as baseline, on-treatment, or

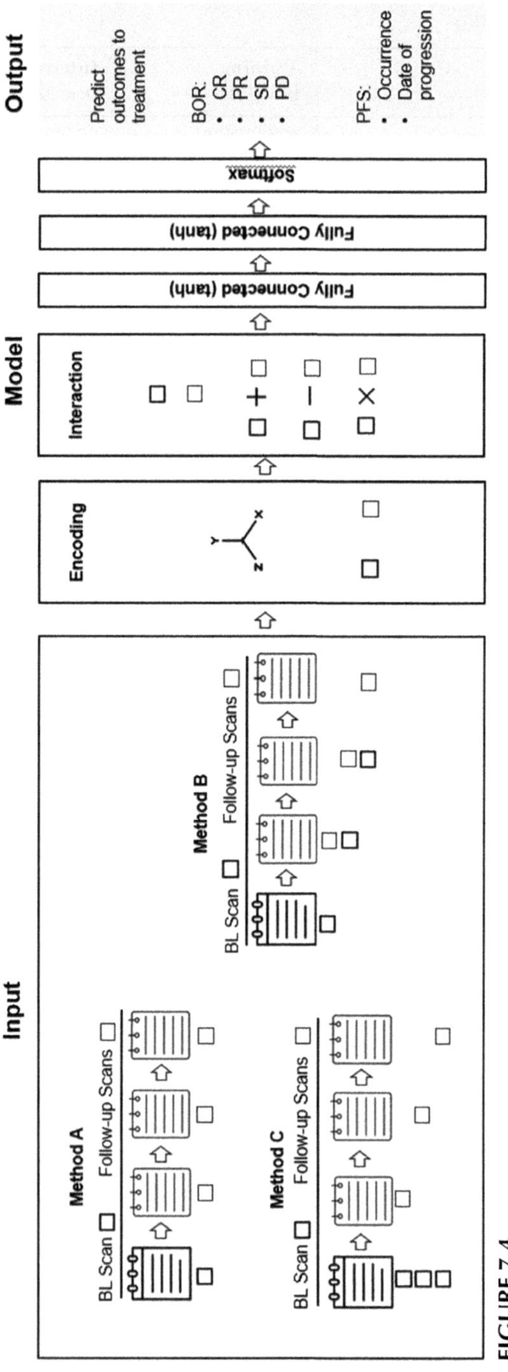

FIGURE 7.4
A fully connected deep learning model. BL = Baseline. (Adapted from Arbour et al. 2021.)

progression. For the test cohorts, reports were only classified as baseline or on-treatment because one of the model's goals was to identify progression.

Three different methods, Methods A, B, and C, for organizing the input were evaluated in the training cohort, reflecting how radiologists review scans to assess response and progression. Method A aggregated text from all follow-up scans which were compared to baseline scans, and Method B compared each scan after the baseline with the one immediately preceding it while Method C compared each scan separately with the original baseline scan. The DL models consisted of an encoding layer, interaction layer, two fully connected layers, and an output layer as briefly described in the following.

7.4.1.3.1 Encoding Layer

This layer processed a pair of tokenized inputs for comparison. Clinical imaging reports are tokenized and embedded into a vector space using pre-trained Global Vectors for Word Representation (GloVe) embeddings.

7.4.1.3.2 Interaction Layer

In this layer, the two vector representations underwent several operations: (1) Concatenation: to capture the information in each vector; (2) Addition: to capture the supplementary relationship between the vectors; (3) Subtraction: to capture the difference between the vectors; and (4) Multiplication: to capture the similarity between the vectors. The outputs of these operations are combined into a single vector, fully representing the interaction between the two inputs.

7.4.1.3.3 Fully Connected Layers

These layers analyze the output from the interaction layer to identify features correlating with specific output classes. The interaction layer's output passes through two fully connected layers, each followed by a tanh activation function.

7.4.1.3.4 Output Layer

The output from the fully connected layers was processed through a softmax activation function to generate a probability score for the predicted class. For the BOR task, there are three output classes: CR/PR, SD, and POD. For the PFS and progression date prediction tasks, there are two output classes: Yes and No. Different networks are used to predict the outputs for different tasks.

7.4.1.3.5 Prediction Methodology

Since different inputs were used for Methods A, B, and C, for Method A, the best objective response (BOR) is the class (CP, PR, SD, or PD) with the highest probability. Likewise, for the prediction of where there was progression of disease at any point, the final prediction is the class with the highest probability. For Methods B and C, because of multiple input scans pairs, a set of output probabilities of classifications were generated for both objective response assessments at different follow-up times. The final prediction is the class with the highest combined probabilities from all outputs for each task.

Finally, for predicting progression, for each patient, the final PFS date is the date of the follow-up scan with the highest probability of over 0.5 in the "Yes" class. If none of the probabilities exceeded 0.5, the patient was categorized as not having progression of disease.

7.4.1.4 Results

Table 7.2 presents the predicted results of BOR of the DL model, based the training cohort for Methods A, B, and C.

The performance metrics, sensitivity, specificity, and overall accuracy, are defined as follows:

$$\text{Sensitivity} = \frac{\text{No. of model predicted patients with CR or PR}}{\text{No. of actual patients with CR or PR}}$$

TABLE 7.2

Summary of Predicted Results of Best Objective Response

		Predicted Label		
Method	**True Label**	**Complete or Partial Response (CR/PR)**	**Stable Disease (SD)**	**Progression of Disease (PD)**
A	CR/PR (n = 82)	83% (n = 68)	15% (n = 12)	2% (n = 2)
	SD (n = 132)	8% (n = 10)	85% (n = 112)	8% (n = 10)
	PD (n = 147)	1% (n = 2)	10% (n = 14)	89% (n = 131)
B	CR/PR (n = 82)	73% (n = 60)	18% (n = 18)	9% (n = 7)
	SD (n = 132)	11% (n = 14)	70% (n = 92)	20% (n = 26)
	PD (n = 147)	3% (n = 5)	16% (n = 24)	80% (n = 118)
C	CR/PR (n = 82)	71% (n = 58)	21% (n = 17)	9% (n = 7)
	SD (n = 132)	11% (n = 15)	61% (n = 80)	28% (n = 37)
	PD (n = 147)	4% (n = 6)	14% (n = 20)	82% (n = 121)

Source: Arbour et al. (2021).

$$\text{Specificity} = \frac{\text{No. of model predicted patients with SD or PD}}{\text{No. of actual patients with SD or PD}}$$

$$\text{Accuracy} = \frac{\text{No. of patients with predicted response matching actual response}}{\text{Total number of patients}}$$

Table 7.3 includes the summary of DL model performance.

From the table, it is evidence that for the training cohort, Methods A demonstrated the best performance across performance metrics when compared to Methods B and C.

Similar analyses were carried out for PFS. Table 7.4 shows the results of Model A in identifying PFS events for the training set. The sensitivity, specificity, and overall accuracy of Method A were estimated to be 91%, 66%, and 86%, respectively. When applying to the training cohort, the AUC values of Methods A, B, and C were 0.79, 0.69, 0.68, respectively. Once again, Method proved to give the best performance.

To determine the date of progression, Method A was unsuitable as it aggregated on-treatment scans instead of treating them as discrete events. Therefore, Arbour et al. (2021) evaluated only Methods B and C, which analyzed each on-treatment scan report individually. The process of pinpointing the date of progression for PFS calculation was streamlined by

TABLE 7.3

Summary of Deep Learning Model Performance

Method	Sensitivity	Specificity	Accuracy	AUC
A	0.83	0.87	0.86	0.90
B	0.73	0.75	0.81	0.84
C	0.71	0.72	0.78	0.81

Source: Arbour et al. (2021).

TABLE 7.4

Summary of Performance of Method A in Predicting PFS Events

	Predicted PD	
True PD	**Yes**	**No**
Yes (n = 284)	91% (n = 259)	9% (n = 25)
No (n = 77)	34% (n = 26)	66% (n = 51)

Source: Arbour et al. (2021).

TABLE 7.5

Rules of Determining PD Date

Rule	Description
1	If both Method A and Method B determine progression, date of progression is defined as date predicted by Method B
2	If Method A determines progression but Method B determines NO progression, last recorded scan date is used as date of progression
3	If Method A determines no progression, the patient is treated as not having progression. Patient is censored at last recorded scan date.

Source: Arbour et al. (2021).

using the follow-up scan date with the highest probability (greater than 0.5) of indicating progression. Method B proved more precise, matching the exact progression date with RECIST in 65% of cases (185/283).

To enhance the model's prediction of PFS in RECIST progression determination from multiple scans, Arbour et al. (2021) adopted an ensemble method combining Methods A and B, using the rules outlined in Table 7.5.

This new approach matched the exact progression date with RECIST in 70% of patients (246/351) and achieved up to 79% accuracy within a two-month window of the RECIST progression date. The overall Spearman correlation between the model by Arbour et al. (2021) and RECIST-determined PFS was 0.83 ($p < 0.001$, two-tailed).

7.4.1.4.1 Model Validation

The model was further validated using the internal and external test cohorts. For the internal cohort, BOR and PFS analyses were performed using Method A and ensemble Method A/B, respectively, resulting in correct classification of 84% CR/PR, 80% SD, and 89% PD (Table 7.6). The corresponding specificity was estimated to be 96%, 88%, and 93%, respectively. For PFS prediction, the model accurately identified progression in 85% of cases. The exact date of progression was correctly determined in 73% of cases and in up to 80% of cases when the model predicted progression within a 2-month window. The same analyses were carried out based on the external cohort from MGH to assess the generalizability of the model. The predicted BOR is summarized in Table 7.6. For CR/PR, SD, and PD classification, the model gave rise to sensitivity of 69%, 43%, and 70% and specificity of 94%, 72%, and 75%, respectively. Each PFS date was correctly predicted for 59% cases, and 82% for cases for which the predicted RECIST PFS data was within 2 months of the exact date. The impact of model-predicted BOR on the overall survival was also evaluated. The hazard ratio between responders (CR/PR) and non-responders (SD/PD) along with a p-value from the log-rank test is presented in Table 7.7. Across all three cohorts, the impact of response classifications by the model on the survival was almost the same as that by the RECIST assessment. For example,

TABLE 7.6

Summary of Predicted BOR by Data Source

		Predicted Label		
Cohort	**True Label**	**Complete or Partial Response (CR/PR)**	**Stable Disease (SD)**	**Progression of Disease (PD)**
MSK	CR/PR (n = 19)	84% (n = 16)	16% (n = 3)	0% (n = 0)
	SD (n = 35)	9% (n = 3)	80% (n = 28)	11% (n = 4)
	PD (n = 38)	0% (n = 0)	11% (n = 4)	89% (n = 34)
MGH	CR/PR (n = 29)	69% (n = 20)	28% (n = 8)	3% (n = 1)
	SD (n = 28)	11% (n = 3)	43% (n = 12)	46% (n = 13)
	PD (n = 40)	3% (n = 1)	28% (n = 11)	70% (n = 28)

Source: Arbour et al. (2021).

TABLE 7.7

Comparison of PFS between Model Predicted Responders and Non-Responders

	Manual Review		Deep Learning Model	
Cohort	**Hazard Ratio**	***P*-Value**	**Hazard Ratio**	***P*-Value**
Training	0.22	< 0.001	0.24	< 0.001
Internal Testing	0.20	< 0.001	0.28	< 0.001
External Testing	0.20	< 0.001	0.30	< 0.001

Source: Arbour et al. (2021).

the hazard ratio between responders and non-responders was 0.22 and 0.24, respectively as determined by the RECIST method and the DL model, each resulting in a p-value less than 0.001.

7.4.2 Machine Learning Method for Heterogeneous Treatment Effect Analysis

7.4.2.1 Background

Treatment effect refers to the impact a treatment or intervention, such as administering an anticancer drug, has on a specific outcome, like the health or disease progression of a patient. This effect is assessed based on counterfactuals, which involve comparing outcomes with and without the

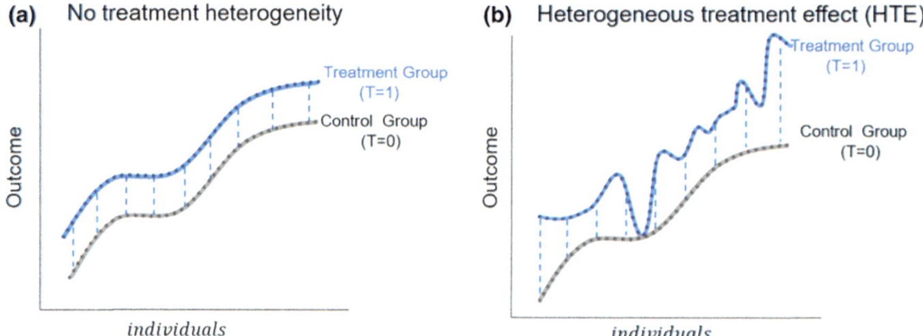

FIGURE 7.5
(a) Homogeneous treatment effect (no treatment heterogeneity): The treatment outcome varies across individuals and between treatment groups, but the treatment effect (i.e., the difference in outcomes depicted by the dotted lines between the two treatment outcome curves) is consistent for every individual. (b) Heterogeneous treatment effect (HTE): The treatment effect varies among individuals. Some individuals benefit more, some less, and some might not benefit at all from the treatment.

treatment. It is important to note that treatment effects are rarely uniform across an entire population. For example, while a new treatment might show similar efficacy to an existing one in the general population, it could be particularly beneficial for a subgroup of patients with specific characteristics. This variability poses challenges in applying ATEs to address individual outcomes, which is crucial in personalized medicine. Understanding these nuances is essential for tailoring treatments to individual patient needs, ensuring more effective and targeted therapeutic interventions (Schork 2015).

Figure 7.5 illustrates the concept of HTE with examples of homogeneous treatment effect (Figure 7.5a) and HTE (Figure 7.5b) among the population. In the case of a homogeneous treatment effect, while the outcomes vary across individuals within each treatment group and differ between treatment groups ($T = 1$ vs. $T = 0$), the treatment effect $Y_i(1) - Y_i(0)$ is the same for every individual and identical to the ATE (Figure 7.5a).

In contrast, with HTE, the population exhibits significant heterogeneity in response to treatments, with some individuals benefiting more (responders), some less, and some not benefiting at all (non-responders) from the treatment (Figure 7.5b). Thus, the ATE provides limited information for individuals, highlighting the need for HTE analysis to understand how treatment effects vary across the entire population.

One commonly used approach to estimate HTE is the two-step method (Gail et al. 1989; Morgan and Winship 2015; Hernán and Robins 2020), which involves building separate regression models for the treatment and control groups. This counterfactual model, comprising the two constructed regression models, is then utilized to estimate the counterfactual differences in individual outcomes, thereby inferring individual treatment effects. Specifically, for an individual with distinct covariate values, each regression model projects outcome values, and the difference between these two outcomes represents the predicted treatment effect.

The two-step method has been conventionally applied in various fields, such as econometrics, social science, epidemiology, and medical science. Despite its intuitive nature and straightforward implementation, this method has limitations. One significant constraint is the reliance on linear regression, which imposes linear relationships unless more complex relationships are explicitly predefined in the model. This can significantly compromise performance in the presence of model misspecification for complex relationships. Additionally, the difference between the two independent "accurate" models does not necessarily result in an accurate HTE estimate, highlighting an intrinsic drawback of the two-step method.

Several ML approaches have been developed to estimate HTE. One notable method is the random forest, which has recently been adapted specifically for HTE analysis through the causal forest method. The causal forest retains the core structure of a traditional random forest, including recursive partitioning, subsampling, and random split selection (Biau and Scornet 2016). However, the tree-splitting criteria are modified to maximize the treatment effect heterogeneity – the difference in estimated treatment effect between daughter nodes (Wager and Athey 2018).

This adaptation is significant because it allows the causal forest to focus on identifying subgroups within the data that exhibit distinct treatment effects. Despite being called a causal forest, this method conducts HTE analysis based on estimated counterfactuals within nodes rather than performing a standard causal inference, which requires specific designs of questions, studies, and analysis (Athey et al. 2019). This distinction is crucial as it highlights the method's utility in estimating treatment effects within the context of available data, rather than establishing causality in a traditional sense (Leacy and Stuart 2014).

Recently, through a simulation study, Gong et al. (2021) evaluated ML-based HTE methods, and demonstrated superior performance compared to conventional HTE methods in two key areas: (1) accurately estimating treatment effects in the presence of nonlinear covariate relationships and (2) identifying influential variables within high-dimensional data with reduced sensitivity to data size and noise level. Details are provided as follows.

7.4.2.2 Evaluation Method

Interaction effects between treatment and subject covariates can induce HTE within a study population, forming the basis of our simulations.

Let $X_i = (x_{i1},\ldots,x_{ik})$ represent the vector of observable covariates for subject i, and $\tau(x) = E\left[Y_i(1) - Y_i(0) X_i = x\right]$ denote the treatment effect for a given covariate set x. HTE scenarios can be simulated using various models (Tian et al. 2014), such that

$$Y_i = f(X_i) + g(X_i)T + \epsilon_i,$$

where $f(\cdot)$ represents the direct impact of covariates on the outcome Y_i, $g(\cdot)$ is a function of covariates X_i that interacting with treatment, and the interaction term $g(X_i)T$ specifies the interaction effect contributing to HTE.

For scenarios with homogeneous treatment effects, outcomes Y can be simulated as:

$$Y_i(T) = f(X_i) + \delta T + \epsilon_i,$$

where each individual's treatment effect is the same as $E[Y_i(1) - Y_i(0)] = \delta$ across all covariates, independent of their values.

7.4.2.3 Simulations

Gong et al. (2021) conducted systematical evaluations to assess the performance of the causal forest and conventional two-step methods for HTE analysis. Four models of increasing HTE complexity were used to simulate the outcomes (Table 7.8): (I) no heterogeneity covariates (no HTE), where all observations exhibit the same treatment effect; (II) a linear relationship between heterogeneity covariates; (III) a nonlinear relationship between heterogeneity covariates; and (IV) a scenario involving high-dimensional data where the number of covariates exceeds the number of individuals/observations. For each simulation model, individual covariates X_i were randomly generated

TABLE 7.8

Models Used for Generating Simulated Treatment with Increasing Levels of Heterogeneity and Complexity

Model	Description	Outcome Model
I	No heterogeneous effect	$Y_i = \beta_0 + \sum_{k=1}^{p} \beta_k x_{ik} + \delta T + \epsilon_i$
II	Linear	$Y_i = \beta_0 + \sum_{k=1}^{p} \beta_k x_{ik} + \left(\gamma_0 + \sum_{k=1}^{p} \gamma_k x_{ik}\right) T + \epsilon_i$
III	Nonlinear + interactive	$Y_i = \beta_0 + \sum_{k=1}^{p} \beta_k x_{ik} + \left[\gamma_0 + \gamma_1 x_{i1}^3 + \gamma_{23}\cos(\gamma_{i2}) x_{i3}\right] T + \epsilon_i$
IV	High-dimensional covariates	$Y_i = \beta_0 + \sum_{k=1}^{p} \beta_k x_{ik} + \left(\gamma_0 + \sum_{k=1}^{p} \gamma_k x_{ik}\right) T + \epsilon_i$

from a mean-zero multivariate normal distribution with covariance matrix $(1-\rho)\boldsymbol{I}_k + \rho 1^T 1$. Treatments $T \in \{0,1\}$ were randomly assigned to the entire population, and random errors $\epsilon \sim N(0, \sigma_0^2)$. No correlations were assumed among covariates ($\rho = 0$) and a noise level of $\sigma_0 = 0.5$ was set. Sample sizes were $n = 2{,}000$ with the number of covariates $k = 10$ for Models I, II, and III, $n = 2{,}000$ with $k = 1{,}000$ for the high-dimensional Model IV.
Model I served as a "baseline" case with no interaction between covariates and treatment, hence no HTE. In Model II, two covariates interacted with the treatment linearly. Model III assumed a scenario where covariates interacted with the treatment in a nonlinear manner.

Model IV was designed to represent a high-dimensional data scenario, where 1,000 covariates were sampled from a multivariate normal distribution with 500 observations. In this model, the 1,000 covariates followed a linear additive relationship with respect to the treatment effect, with covariate coefficients set to zero except for the first five covariates. This setup allowed us to test whether an HTE model could accurately describe such high-dimensional data and identify the important covariates, even when they were sparse (i.e., 5 out of 1,000).

7.4.2.4 Results

This section presents the results of a case example, of which the treatment effect was simulated *using the mode*

$$Y = \beta_0 + \sum_{k=1}^{10} \beta_k x_k + \left(\gamma_0 + \gamma_1 x_1^5 + \gamma_2 e^{x_2} + \gamma_{12} x_1 x_2\right) T + \epsilon,$$

where T assumes a value of either 1 or 0 corresponding to treatment and control, respectively, $\beta_0 = 0.4$, $\beta_2 = 0.5$, $\gamma_0 = 0.8$, $\gamma_1 = \gamma_{12} = 1.6$, and $\gamma_2 = -0.64$, and $\epsilon \sim N(0, 0.5^2)$.

The interaction between the heterogeneity covariates x_1 and x_2 and the treatment was assumed take the nonlinear form

$$\gamma_0 + \gamma_1 x_1^5 + \gamma_2 e^{x_2} + \gamma_{12} x_1 x_2$$

with $\gamma_0 = 0.8, \gamma_1 = \gamma_{12} 1.6$, and $\gamma_2 = -0.64$.

Figure 7.6a illustrates the distribution of true treatment effects as defined by the model, plotted against covariates $x_1$and x_2 (on the x and y axes, respectively). The treatment effects are represented by different shades with the highest treatment effects located in the lower right corner and the most unexpected treatment effects found in the middle of the leftmost region. A nonlinear transition from the highest to the lowest treatment effects is evident in the figure.

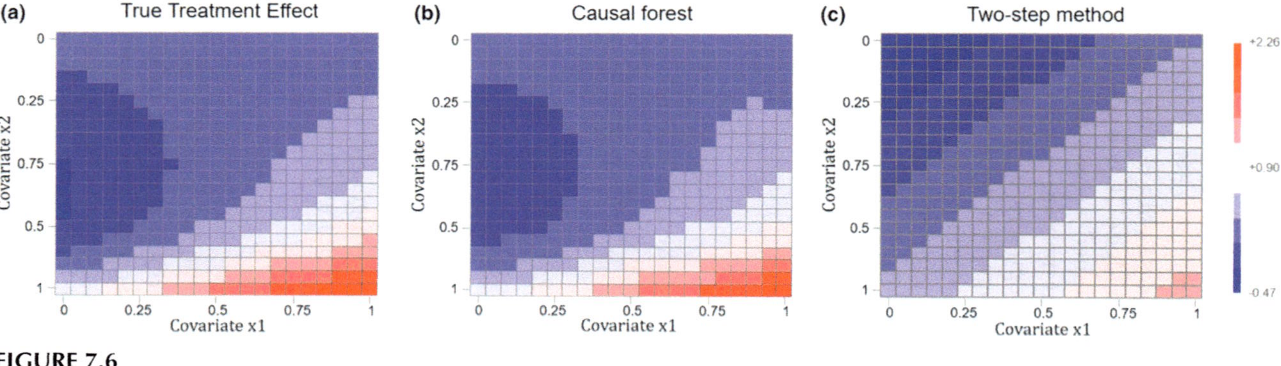

FIGURE 7.6
Treatment effect in a dataset where the relationship between the covariates and the treatment effect is nonlinear. (a) The "true" treatment effect with varying values of covariates x_1 and x_2. (b) Predicted treatment effect using the causal forest method and (c) Predicted treatment effect by the two-step method. (Adapted from Gong et al. 2021.)

Both the causal forest and the two-step methods were applied to the simulated data for HTE analysis. In the two-step method, separate regression models were established for the treatment and control groups. To predict the treatment effect for an individual, their data were input into these two separate models, and the difference between the estimated outcomes was calculated to represent the treatment effect.

For the causal forest method, each tree was grown based on specific splitting criteria. The resulting causal forest model provided the predicted treatment effect for individuals. Figure 7.6b shows the distribution of treatment effects estimated by the causal forest method, which closely matches the true treatment effects. In contrast, Figure 7.6c shows the biased treatment effect estimation with a linear pattern provided by the two-step method, which is limited by its explicitly linear additive regression model. Although this estimation reflects the general trend of the true treatment effects, it fails to capture the nonlinear pattern between the treatment and covariates (Figure 7.6a). These results demonstrate that the causal forest method can detect the underlying HTE even when complex interaction relationships exist between the treatment and covariates.

Figure 7.7 presents the RMSE results calculated from each testing dataset across the four scenarios. The figure shows the mean and standard deviation of RMSE values across 200 simulation replications.

There is no significant difference in prediction performance between the causal forest and the two-step method for Model I (no HTE) and Model II (linear interaction between treatment and covariates). In these scenarios, the two-step method is the correct modeling approach and, as expected, provides accurate predictions. However, the causal forest method performs just as well as the two-step method in these cases.

For Model III (nonlinear or additive interaction between treatment and covariates), the causal forest method offers more accurate treatment effect predictions than the two-step method, as evidenced by significantly lower RMSE ($p < 0.01$). This finding underscores the causal forest's capability for analyzing complex or real-world questions.

The difference in RMSE is most pronounced in Model IV ($p < 0.01$) (Figure 7.7). The two-step method fails to provide a reasonable treatment effect estimation due to parameter identifiability issues, given that the number of observations (500) is less than the number of covariates (1,000). In contrast, the causal forest achieves more than five-fold improvement in RMSE compared to the two-step method.

The variable importance in the causal forest is measured by a weighted sum of the number of times a covariate is split at each depth in the forest. Higher importance values indicate covariates with high predictive power, while lower to zero values indicate variables with low predictive power. Notably, the causal forest is also capable of identifying influential covariates.

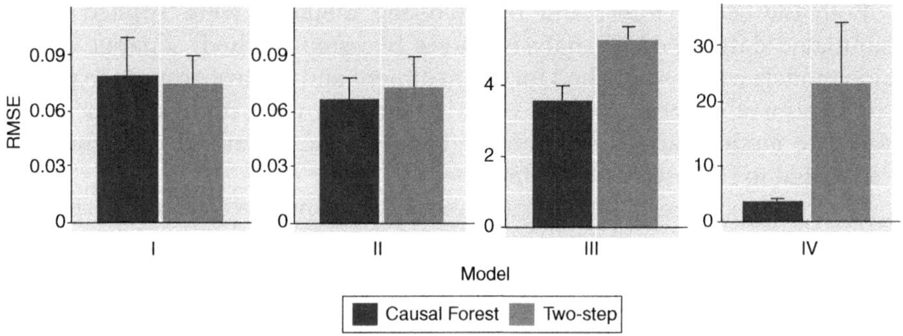

FIGURE 7.7
Comparison of the performance of the causal forest and two-step methods. (Adapted from Gong et al. 2021.)

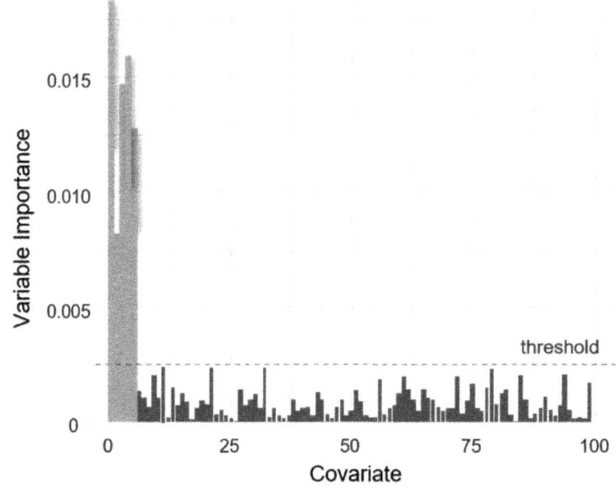

FIGURE 7.8
The variable importance determined by the causal forest for high-dimensional simulated data based on Model IV is shown, with the preset significant covariates highlighted in lighter shade. (Adapted from Gong et al. 2021.)

Note that while causal forests can provide calculated variable importance values, they do not include a criterion to determine if these values are statistically significant or merely due to randomness. To address this, Gong et al. (2021) developed a permutation-based statistical significance test to identify statistically meaningful important covariates (detailed in the Supplemental Information). The results of this significance test indicated that only the first five covariates have statistically significant variable importance values, consistent with the simulation design (Figure 7.8).

7.4.3 RWE Supporting Anakinra for Preventing Respiratory Failure in COVID-19

RCTs are the gold standard for generating evidence in support of a new drug's safety and efficacy assessment. However, there are circumstances where RCTs are either infeasible or unethical. Some trials using single-arm designs have gained FDA approval through expedited programs, especially for drugs and biologics targeting serious conditions and unmet medical needs. Evaluating outcomes in single-arm studies can be challenging due to the lack of an objective reference level. To address this, incorporating an external control arm can be beneficial. These controls can be drawn from historical clinical trials, natural history studies, patient registry data, and other forms of RWD.

In this section, we discussed a use case reported by Kyriazopoulou et al. (2021) in which an open-label, single-arm study, named SAVE (suPAR-guided Anakinra treatment for validation of the risk and early management of SRF by COVID-19), was coupled with an external control arm. The study examined whether early administration of anakinra in patients with lower respiratory tract infection (LRTI) caused by SARS-CoV-2 and suPAR levels equal to or greater than 6 mg/l could mitigate the development of severe respiratory failure (SRF). Anakinra is a recombinant soluble IL-1 receptor antagonist that inhibits both IL-1α and IL-1β. This interim analysis was presented to the European Medicines Agency's Emergency Task Force for COVID-19. The advice provided by the task force informed the design of the pivotal, confirmatory Phase III RCT, known as SAVE-MORE.

7.4.3.1 Background

Patients afflicted with severe infections stemming from the novel coronavirus SARS-CoV-2, commonly referred to as COVID-19, often display elevated levels of pro-inflammatory cytokines circulating throughout their bodies. Although these levels may not reach the same peaks observed in patients with classical acute respiratory distress syndrome (ARDS) and septic shock (Sinha et al. 2020), systemic inflammation remains a prominent hallmark of severe COVID-19.

Emerging evidence suggests that when SRF necessitating mechanical ventilation (MV) arises, two distinct immune phenomena become predominant in the affected individual: (1) macrophage activation syndrome, affecting approximately 25% of patients or (2) a complex immune dysregulation characterized by the downregulation of the human leukocyte antigen DR on circulating monocytes, lymphopenia, and excessive interleukin (IL) production.

The prevailing hypothesis suggests that these immune responses initiate early in patients with LRTI caused by SARS-CoV-2, and gradually intensify, eventually resulting in SRF. This progression is believed to stem from the early release of IL-1 by lung epithelial cells that have been infected by the virus; this IL-1 release further stimulates cytokine production from alveolar macrophages (Conti et al. 2020).

Consequently, it is postulated that initiating anti-IL-1 anti-inflammatory treatment early may mitigate the development of SRF (van de Veerdonk and Netea 2020). However, it is likely that not all patients with LRTI caused by SARS-CoV-2 require early intervention, and the availability of a screening tool to identify those at risk of progressing to SRF would be invaluable. Soluble urokinase plasminogen activator receptor (suPAR) appears to fulfill this role. Recent research, including our own and that of others, has shown that suPAR concentrations exceeding 6 ng/ml anticipate deterioration leading to SRF by as much as 14 days (Azam et al. 2020). The positive predictive value for the early prediction of SRF was notably high at 85.9%. The urokinase plasminogen activator receptor (uPAR) is anchored to the cell membranes of lung endothelial cells. Upon activation of kallikrein, uPAR undergoes cleavage and enters the systemic circulation as its soluble counterpart, suPAR (Pixley et al. 2011).

7.4.3.2 Open-Label Single Arm Study

As described in the previous section, SAVE was an open-label, single-arm study, and non-randomized trial. Key inclusion criteria included: (1) adult patients hospitalized with confirmed infection by SARS-CoV-2 virus by real-time PCR reaction of nasopharyngeal secretions; (2) radiological findings compatible with LRTI; and (3) plasma suPAR level ≥ 6 μg/l using the suPARnostic Quick Triage. Exclusion criteria were: (1) any stage IV malignancy; (2) any do not resuscitate decision; (3) ratio of partial oxygen pressure to the fraction of inspired oxygen (pO_2/FiO_2) less than 150 mmHg; (4) need for MV or non-invasive ventilation under positive pressure (NIV); (5) any primary immunodeficiency; (6) neutropenia (<1,500/mm^3); (7) any intake of corticosteroids at a daily dose ≥0.4 mg/kg prednisone or equivalent the last 15 days; (8) any anti-cytokine biological treatment the last month; and (9) pregnancy or lactation.

Patients who were hospitalized with confirmed COVID infection were screened for eligibility to enroll in SAVE. Among 278 patients evaluated, 132 were enrolled to receive Anakinra in combination with SOC. In parallel, comparators who received SOC were hospitalized at the same time period without being enrolled in the study. In total, 301 patients were screened for study eligibility, and 179 of them met the eligibility criteria (Figure 7.9). Among those participants, 130 were selected for the study, based on a propensity score matching method.

As shown in Table 7.9, significant differences in a variety of baseline characteristics were observed between the 179 parallel comparators and those enrolled in the SAVE study as evidenced by *p*-values less than 0.05.

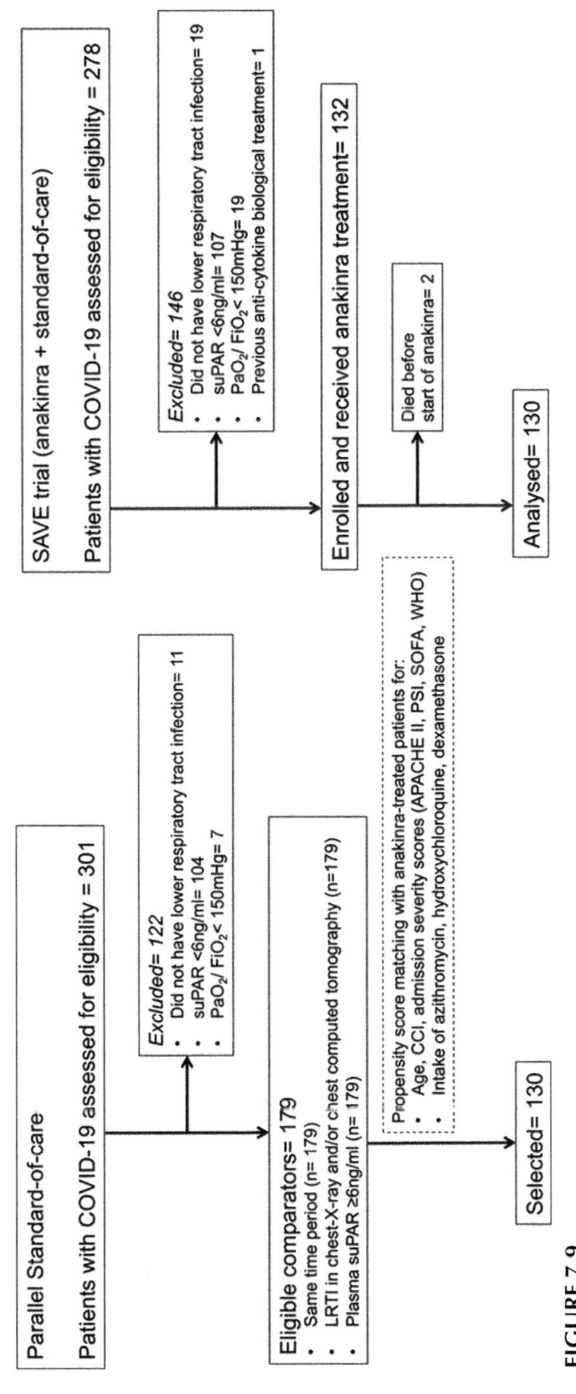

FIGURE 7.9
Trial profile and selection of parallel standard-of-care (SOC) comparators. (Adapted from Kyriazopoulou et al. 2021.)

TABLE 7.9

Baseline Characteristics

	SOC + Anakinra (*N* = 130)	SOC (*N* = 179)	*P*-Value*	SOC after Matching (*N* = 130)	*P*-Value**
Age, years, mean (SD)	63 (14)	66 (14)	0.094	64 (14)	0.839
Male sex, No. (%)	81 (62.3)	116 (64.8)	0.719	84 (64.6)	0.797
Days from onset of symptoms to start of treatment, median (range)	8 (1–23)	6 (4–9)	0.012	7 (1–12)	0.143
Severity indexes, mean (SD)					
Charlson's comorbidity index	3 (2)	3 (2)	0.123	3 (2)	0.927
APACHE II score	7 (3)	8 (5)	0.002	7 (3)	0.957
SOFA score	2 (1)	3 (2)	0.001	2 (1)	0.813
Pneumonia severity index	69 (22)	80 (28)	<0.0001	70 (20)	0.709
WHO classification for COVID-19, No. (%)					
Moderate pneumonia, no oxygen	61 (46.9)	68 (38.0)	0.158	57 (43.8)	0.709
Severe pneumonia	69 (53.1)	111 (62.0)		73 (56.2)	
Comorbidities, No. (%)					
Type 2 diabetes mellitus	41 (31.5)	45 (25.1)	0.248	32 (24.6)	0.270
Chronic heart failure	11 (8.5)	19 (10.6)	0.565	10 (7.7)	1.000
Chronic renal disease	1 (0.8)	12 (6.7)	0.009	4 (3.1)	0.370
Coronary heart disease	10 (7.7)	24 (13.4)	0.141	14 (10.8)	0.521
Arterial hypertension	68 (52.3)	87 (48.6)	0.565	57 (43.8)	0.214
Chronic obstructive pulmonary disease	12 (9.2)	14 (7.8)	0.682	6 (4.6)	0.221
Solid tumor malignancy	8 (6.2)	14 (7.8)	0.658	9 (6.9)	1.000
Cerebrovascular disease	7 (5.4)	13 (7.3)	0.641	4 (3.1)	0.540
Bacterial co-infection, No. (%)	1 (0.8)	5 (2.8)	0.407	3 (2.3)	0.622
Streptococcus pneumoniae	0 (0.0)	2 (1.1)	0.511	2 (1.5)	0.498
Escherichia coli	0 (0.0)	1 (0.6)	1	1 (0.8)	1.000
Laboratory values, median (Q1–Q3)					
White blood cells, cells/mm3	5,400 (4,390, 6,830)	6,215 (4,805, 9,013)	0.005	5,879 (4,760, 7,800)	0.106
Lymphocytes, cells/mm3	950 (722, 1,252)	961 (615, 1,271)	0.307	977 (587, 1,318)	0.382

TABLE 7.9 (Continued)

Baseline Characteristics

	SOC + Anakinra (*N* = 130)	SOC (*N* = 179)	*P*-Value*	SOC after Matching (*N* = 130)	*P*-Value**
Platelets, cells/mm3	184,000 (137,000, 246,000)	202,000 (150,000, 270,000)	0.121	204,400 (150,450, 271,625)	0.107
C-reactive protein, mg/l	47.4 (14.3, 105.5)	64.7 (18.3, 140.0)	0.179	68.8 (19.7, 141.8)	0.117
Procalcitonin, ng/ml	0.14 (0.08, 0.31)	0.23 (0.10, 0.71)	0.149	0.13 (0.08, 0.41)	0.841
Ferritin, ng/ml	0.14 (0.08–0.31)	629.5 (375.0, 1,277.0)	<0.001	607.5 (367.8, 1,196.0)	0.107
Serum soluble uPAR, ng/ml	0.14 (0.08–0.31)	9.2 (7.0, 13.8)	0.344	9.0 (7.0, 11.7)	0.973
PaO2/FiO2, mmHg	0.14 (0.08–0.31)	262.0 (182.0–350.3)	0.09	285.7 (208.5–371.7)	0.917
Concomitant treatment, No. (%)					
Hospitalization at tertiary academic hospitals	127 (97.7)	156 (87.2)	0.191	120 (92.3)	0.167
β-lactamase inhibitors	18 (13.8)	16 (8.9)	0.199	13 (10)	0.444
Third generation cephalosporins	60 (46.2)	66 (36.9)	0.127	43 (33.1)	0.042
Piperacillin/tazobactam	37 (28.7)	61 (34.1)	0.323	42 (32.3)	0.590
Ceftaroline	38 (29.2)	50 (27.9)	0.8	42 (32.3)	0.687
Carbapenem	11 (8.5)	23 (12.8)	0.271	15 (11.5)	0.536
Moxifloxacin/levofloxacin	27 (20.9)	26 (14.5)	0.17	17 (13.2)	0.136
Glycopeptides	2 (1.5)	6 (3.4)	0.475	5 (3.8)	0.447
Azithromycin	96 (73.8)	145 (81.0)	0.164	104 (80.0)	0.303
Remdesivir	8 (6.2)	12 (6.7)	1	11 (8.5)	0.635
Hydroxychloroquine	56 (43.1)	92 (51.4)	0.066	68 (52.3)	0.172
Dexamethasone	52 (40.0)	65 (36.3)	0.553	47 (36.2)	0.610
Predicted probability	0.488 (0.386, 0.571)	–	–	0.488 (0.391, 0.556)	0.629

Source: Kyriazopoulou et al. (2021).

These included days from onset of symptoms to start of treatment, Acute Physiology and Chronic Health Evaluation (APACHE) II score, Sequential Organ Failure Assessment (SOFA) score, pneumonia severity index (PSI), white blood cells, and ferritin of all SOC comparators. These differences had the potential to confound the comparison between the two groups. To address this issue, a propensity matching method was used to select 130 patients from the 179 treated with SOC. There was no difference in baseline characteristics between the patients receiving Anakinra with SOC and patients receiving only SOC treatment.

7.4.3.3 Results

7.4.3.3.1 Efficacy

The primary endpoint was the incidence of SRF by day 14. SRF was defined as any decrease of pO_2/FiO_2 below 150 necessitating MV or NIV. An interim analysis was performed when the 30-day follow-up of the first 130 patients was completed. Comparisons of Anakinra treated patients to propensity-matched controls on SoC were conducted and results reported by Kyriazopoulou et al. (2021).

Anakinra treatment was demonstrated to be superior to SOC with the primary outcome: 22.3% with Anakinra treatment versus 59.2% comparators (hazard ratio, 0.30; 95% CI, (0.20, 0.46)) progressed into SRF (Table 7.10 and Figure 7.9a).

TABLE 7.10

Summary of Primary, Secondary, and Exploratory Outcomes

	Parallel SOC (*N* = 130)	SOC + Anakinra after Propensity Matching (*N* = 130)	OR (95% CI)	*P*-Value
Severe respiratory failure by day 14, N (%)	77 (59.2)	29 (22.3)	0.19 (0.12, 0.34)	<0.0001
Mechanical ventilation (MV) by day 14, *N* (%)	65 (50.0)	25 (19.2)	0.24 (0.14, 0.42)	<0.0001
Non-invasive MV by day 14, *N* (%)	9 (6.9)	6 (4.6)	0.65 (0.23, 1.88)	0.596
14-day mortality, *N* (%)	16 (12.3)	6 (4.6)	0.35 (0.13, 0.91)	0.043
SOFA score on day 7, median (Q1–Q3)	2 (0, 3)	1 (0 to 3)	NA	0.54
SOFA score on day 14, median (Q1–Q3)	2 (0, 9)	1 (0, 3)	NA	0.191
Absolute change of SOFA score by day 7 compared to baseline, median (Q1–Q3)	0 (−1, 1)	0 (−1, 0)	NA	0.356
Absolute change of SOFA score by day 14 compared to baseline, median (Q1–Q3)	0 (−1, 6)	−2 (−1, 0)	NA	0.004
Absolute change of respiratory symptom score by day 7 compared to baseline, median (Q1–Q3)	0 (0, 0.75)	−1 (−3, 0)	NA	0.019
Absolute change of respiratory symptom score by day 14 compared to baseline, median (Q1–Q3)	0 (−0.75, 0)	−2 (−4, −1)	NA	0.016
Ventilator-free days, mean (SD)	18 (11)	25 (8)	NA	<0.0001
30-day mortality, *N* (%)	29 (22.3)	15 (11.5)	0.45 (0.23, 0.90)	0.031
90-day mortality, *N* (%)	40 (30.8)	22 (16.9)	0.46 (0.25, 0.83)	0.013

Source: Kyriazopoulou et al. (2021).

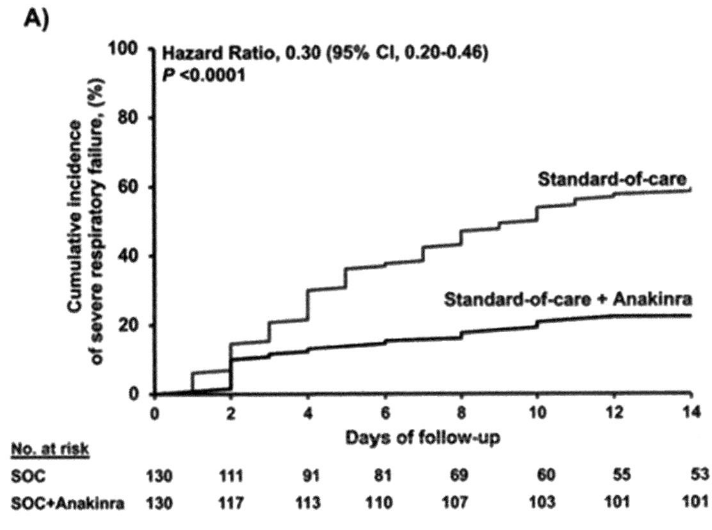

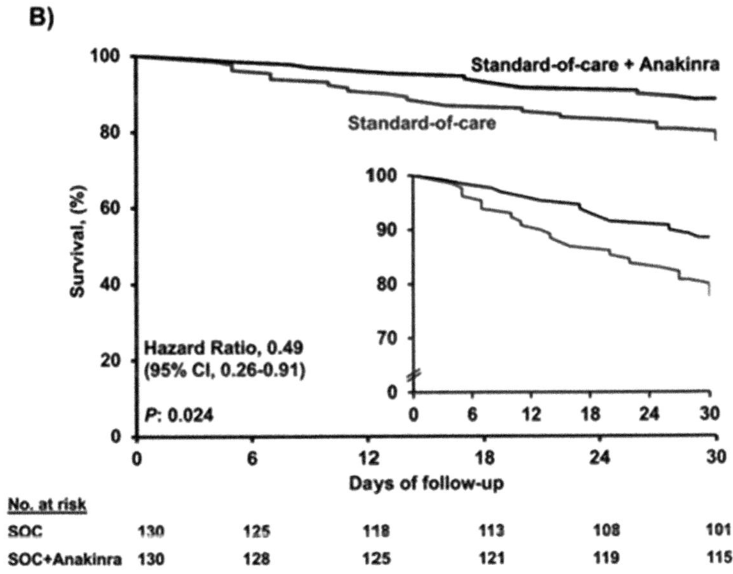

FIGURE 7.10
Outcome of patients treated with Anakinra and SOC. (A) Curves of the cumulative incidence of severe respiratory failure (SRF); and (B) curves of 30-day mortality; the inset shows the curves in enlarged Y-axis. (Adapted from Kyriazopoulou et al. 2021.)

The superiority of Anakinra was further supported by positive outcomes of most of the secondary endpoints, including 30-day mortality (11.5% vs. 22.3%; hazard ratio 0.49 with 95% CI (0.25, 0.9); absolute change of SOFA score by day 14; and absolute change of the respiratory symptoms score by days 7 and 14 (Table 7.10; Figure 7.8b).

Multivariate stepwise Cox regression analysis revealed that anakinra treatment was the sole independent factor associated with reduced 30-day mortality (hazard ratio 0.28; 95% CI, 0.18–0.44; p-value < 0.001) (Table 7.11).

The study examined two main secondary outcomes: the impact of anakinra treatment on circulating inflammatory biomarkers and the function of peripheral blood mononuclear cells (PBMCs). Compared to subjects receiving SOC, those treated with anakinra showed an increase in absolute lymphocyte count and reductions in IL-6, sCD163, and sIL-2R levels (Figure 7.11). The IL-10/IL-6 serum ratio, an indicator of the anti-inflammatory/pro-inflammatory balance in severe COVID-19 (McElvaney et al. 2020), was inversely correlated with the increase in the SOFA score on day 14 in anakinra-treated patients, supporting the anti-inflammatory effect of anakinra. Notably, suPAR levels increased in anakinra-treated patients by day 7, indicating that anakinra provides protection against progression to SRF despite the unfavorable suPAR signal.

Furthermore, the function of PBMCs was modulated in anakinra-treated patients. Specifically, on day 7, patients who did not develop SRF showed higher production of IL-1β and IL-10, suggesting a restoration of the PBMCs' ability to balance anti-inflammatory and pro-inflammatory cytokine production. This was further supported by a positive correlation between the IL-10/IL-1β production ratio in PBMCs on day 7 and the serum IL-10/IL-6 ratio on the same day (Figure 7.12).

7.4.3.3.2 Safety

AEs and serious adverse events (SAEs) recorded during the 14-day study period are listed in Table 7.12. The incidence of these events among SOC-treated comparators is also shown. As indicated in Table 7.12, the incidence of these events was not higher in the anakinra group compared to the SOC group, with the exception of leukopenia, which showed a trend toward being higher in the anakinra group. Additionally, SAEs were fewer among anakinra-treated patients. Overall, Anakinra was shown to have a well-acceptable safety profile at the subcutaneous daily dose of 100 mg.

TABLE 7.11

Summary of Univariate and Multivariate Analyses Assessing Impact of Treatment and Other Factors on 30-Day Mortality

			Univariate analysis		Multivariate analysis	
Variable, no. (%)	**SRF (–) (*N* = 154)**	**SRF (+) (*N* = 106)**	**HR (95% CI)**	***P*-Value**	**HR (95% CI)**	***P*-Value**
Anakinra treatment, *n* (%)	101 (65.6)	29 (27.4)	0.30 (0.20, 0.47)	<0.0001	0.28 (0.18, 0.44)	<0.0001
APACHE II, mean (SD)	6 (3)	8 (3)	1.12 (1.07, 1.19)	<0.0001		
SOFA score, mean (SD)	2 (1)	3 (1)	1.58 (1.38, 1.82)	<0.0001	1.41 (1.21, 1.65)	<0.0001
Pneumonia severity index, mean (SD)	66 (22)	73 (19)	1.01 (1.00, 1.02)	0.017		
Severe COVID-19 by WHO classification, *n* (%)	65 (42.2)	77 (72.6)	2.78 (1.81, 4.27)	<0.0001	1.74 (1.09, 2.79)	0.020
Soluble uPAR, ng/ml, median (Q1–Q3)	8.3 (6.7, 11.0)	10.0(8.0, 13.9)	1.10 (1.04, 1.17)	0.001	1.07 (1.01, 1.14)	0.022
Lymphocytes/mm3, median (Q1–Q3)	1,000 (733, 1,315)	930 (579, 1,197)	0.99 (0.99, 1.00)	0.026		
C-reactive protein, mg/l, median (Q1–Q3)	41.1 (12.7, 95.2)	73.4 (30.6, 153.1)	1.00 (1.00, 1.00)	<0.0001		
Ferritin, ng/ml, median (Q1–Q3)	525.0 (280.0, 804.5)	641.5 (451.3, 1,556.8)	1.00 (1.00, 1.00)	<0.0001		
PaO2/FiO2, mmHg, median (Q1–Q3)	330.8 (231.9, 386.7)	244.4 (161.7, 305.8)	0.99 (0.99, 0.99)	<0.0001		
Treatment with third generation cephalosporin, *n* (%)	73 (47.4)	30 (28.3)	0.53 (0.35–0.82)	0.004		
Treatment with piperacillin/tazobactam, *n* (%)	37 (24.0)	42 (39.6)	1.60 (1.08, 2.38)	0.019		
Treatment with carbapenem, *n* (%)	7 (4.5)	19 (17.9)	2.79 (1.69, 4.60)	<0.0001	2.19 (1.26, 3.83)	0.006
Treatment with glycopeptide, *n* (%)	1 (0.6)	6 (5.7)	2.60 (1.14, 5.95)	0.023		
Treatment with dexamethasone, *n* (%)	46 (29.9)	53 (50.0)	1.73 (1.19, 2.54)	0.005		

Source: Kyriazopoulou et al. (2021).

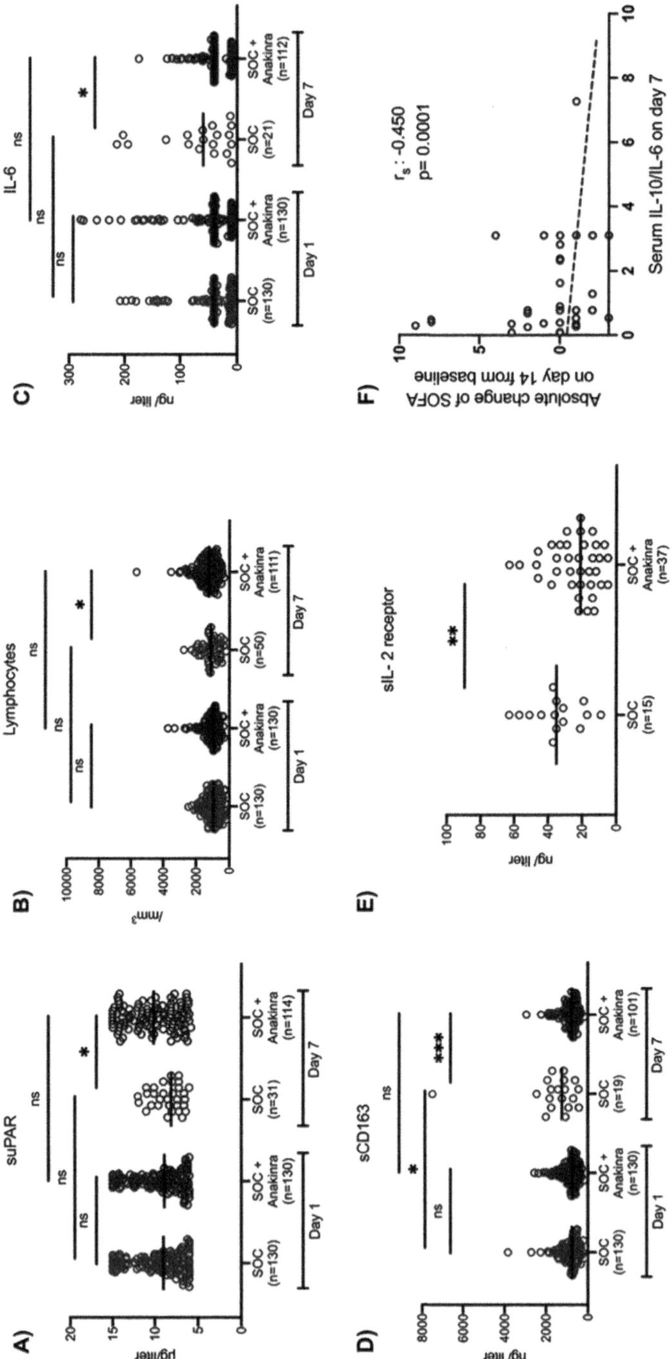

FIGURE 7.11
Effect of anakinra treatment on circulating mediators of patients with lower respiratory tract infection by SARS-CoV-2. (Adapted from Kyriazopoulou et al. 2021.)

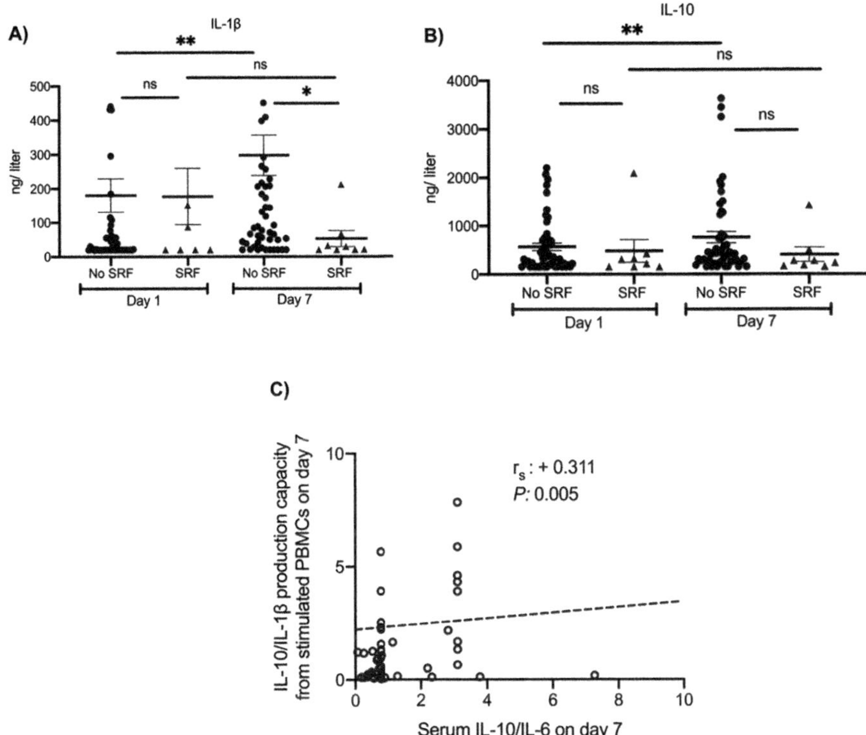

FIGURE 7.12
Effect of anakinra treatment on cytokine production capacity from peripheral blood mononuclear cells (PBMCs). (Adapted from Kyriazopoulou et al. 2021.)

7.5 Concluding Remarks

The integration of AI and ML into data analysis for clinical trials is revolutionizing the landscape of clinical development. These advanced technologies are not only enhancing the ability to manage and analyze complex datasets but also streamlining the entire process from data curation to reporting. The competitive advantage offered by timely and accurate data integration and analysis cannot be overstated, as it underpins the design, evidence generation, and decision-making crucial to successful clinical trials.

AI and ML have demonstrated their potential to transform various aspects of clinical trials. In data curation, AI-driven algorithms are improving the quality and consistency of data, enabling researchers to work with more reliable datasets. The automation of routine tasks, such as data cleaning and normalization, frees up valuable time for researchers to focus on more strategic activities. Moreover, these technologies are adept at handling unstructured

TABLE 7.12

Summary of Adverse Events and Serious Adverse Events

	Parallel SOC (*N* = 130)	SOC + Anakinra (*N* = 130)	OR (95% CI)	*P*-Value
At least one SAE by day 14, *N* (%)	63 (48.5)	32 (24.6)	0.34 (0.21, 0.59)	<0.0001
Extended hospitalization	63 (48.5)	32 (24.6)	0.34 (0.21, 0.59)	<0.0001
Death	16 (12.3)	6 (4.6)	0.35 (0.13, 0.91)	0.043
Shock	56 (43.1)	27 (20.8)	0.34 (0.20, 0.59)	<0.0001
Acute kidney injury	37 (28.5)	15 (11.5)	0.32 (0.17, 0.63)	0.001
Any bacterial infection	30 (23.1)	9 (6.9)	0.22 (0.10, 0.49)	<0.0001
Thromboembolic event	5 (3.8)	2 (1.5)	0.32 (0.06, 1.72)	0.188
Pulmonary edema	0 (0)	1 (0)	NA	1.000
At least one AE by day 14, *N* (%)	89 (68.5)	85 (65.4)	0.87 (0.51, 1.45)	0.693
Gastrointestinal disturbances	9 (6.9)	15 (11.5)	1.74 (0.74, 4.17)	0.284
Electrolyte abnormalities	41 (31.5)	35 (26.9)	0.80 (0.47, 1.37)	0.496
Elevated liver function tests	51 (39.2)	40 (30.8)	0.69 (0.41, 1.15)	0.193
Anemia	26 (20.0)	22 (16.9)	0.82 (0.44, 1.53)	0.632
Leukopenia	3 (2.3)	11 (8.5)	3.91 (1.07, 14.37)	0.051
Thrombopenia	7 (5.4)	9 (6.9)	1.31 (0.47, 3.62)	0.797
Headache	2 (1.5)	4 (3.1)	2.03 (0.37, 11.29)	0.684
Allergic reaction	7 (5.4)	4 (3.1)	0.56 (0.16, 1.95)	0.540
Any heart arrhythmia	22 (16.9)	9 (6.9)	0.37 (0.16, 0.83)	0.020

Source: Kyriazopoulou et al. (2021).

data, extracting valuable insights that were previously difficult or impossible to obtain.

In the realm of data analysis and interpretation, AI and ML are providing innovative tools to uncover patterns and trends within the data, facilitating a deeper understanding of the safety and efficacy of new treatments. These insights are critical for making informed decisions throughout the clinical trial process, from study design to regulatory submissions. The ability of AI to process vast amounts of data at unprecedented speeds also accelerates the timeline of clinical trials, potentially bringing life-saving treatments to market more quickly.

The impact of AI and ML extends to the reporting phase as well. Automated systems can generate comprehensive and accurate reports, ensuring that the evidence presented to regulatory bodies is both robust and reliable. This not only supports the approval process but also builds confidence among stakeholders in the scientific rigor and integrity of the clinical trial.

As highlighted in this chapter, the advancements brought about by AI and ML are reshaping the future of clinical trials. By embracing these technologies, the clinical research community can overcome many of the traditional challenges associated with data analysis, paving the way for more efficient, effective, and ethical trials. The continued evolution of AI and ML will undoubtedly unlock further opportunities for innovation, ultimately improving patient outcomes and advancing medical science.

In conclusion, AI-assisted data analysis is a game-changer in the field of clinical trials. The ability to seamlessly integrate and analyze diverse data sources enhances every stage of the trial process, from initial design to final reporting. As we move forward, the adoption of AI and ML will be crucial in driving the next generation of clinical trials, ensuring that new treatments are developed and delivered with greater speed, accuracy, and confidence.

8

Regulatory Advances in Modernizing Clinical Trials

8.1 Introduction

Clinical trials are essential components of drug development programs for new medicines. Laws and governmental polies often stipulate that a drug's safety and efficacy be established by substantial evidence through adequate and well-controlled investigations to support marketing approval. Given the potential for adverse effects with all drugs, many regulatory agencies integrate a structured benefit-risk assessment as part of the regulatory review process for marketing applications. Recognizing the challenges faced by patients involved in these trials, the extensive time and large patient populations required to complete each phase of development, and the substantial risks and costs associated with failure at any stage, recent years have witnessed a concerted effort to integrate novel and innovative clinical trial designs across all phases of drug development. The convergence of data from diverse sources and digital technologies, including artificial intelligence (AI) and machine learning (ML), has facilitated broad applications of innovative methods for not only clinical trial designs but also execution. This evolution presents regulatory authorities with an unprecedented opportunity to develop guidelines and frameworks that expedite clinical development. This chapter discusses the latest regulatory advancements in modernizing clinical trials, focusing on innovative trial design, the adoption of digital technologies, evidence generation, and regulatory approvals. It also presents several use cases of the FDA's efforts to use AI and ML methods in facilitating regulatory review and decision-making, developing product-specific guidance, and automating administrative work.

DOI: 10.1201/9781003226086-8

8.2 Innovation in Regulatory Agencies

Regulatory agencies are responsible for ensuring the safety, efficacy, and quality of drugs and medical devices by establishing and enforcing laws, rules, and policies that govern the pharmaceutical industry. To effectively and efficiently bring innovative medical interventions to the general public, these regulatory bodies need to adjust their frameworks for product review and approval in response to emerging innovations. For example, the FDA has a long history of innovation and regulation that includes embracing scientific advances, expanding its regulatory scope, and modernizing its approach (Hamburg 2010). A notable example is the launch of the initiative "Innovation/Stagnation: Challenge and Opportunity on the Critical Path to New Medical Products" in 2004 (FDA 2004a). Concerned that the applied sciences essential for medical product development have not progressed at the same rate as the remarkable advances in basic sciences, the FDA stressed the critical necessity to modernize the medical product development process to enhance predictability and reduce costs in product development. The FDA called for a new toolkit for product development, comprising advanced scientific and technical methodologies such as animal or computer-based predictive models, biomarkers for safety and efficacy, and innovative clinical evaluation techniques.

These tools are imperative for enhancing predictability and efficiency throughout the critical path from the initial laboratory concept to the eventual commercial product. Superior development science is required to tackle these challenges and ensure that fundamental discoveries translate into improved medical treatments. It is imperative to invest effort in creating enhanced tools for medical technology development and to establish a knowledge base founded not solely on biomedical research ideas but also on dependable insights into the pathway to patient care.

The past two decades have witnessed significant advancements in technology that allow for generating deeper insights into diseases and patients. For example, the next generation sequencing and comprehensive profiling have improved our ability to differentiate cancers by their genetic mutations (Hirakawa et al. 2018). This has spurred endeavors toward "precision oncology" where therapies are chosen to precisely target cancers based on their genetic mutations (Park et al. 2019). The recent proliferation of big data sourced from diverse channels, coupled with emerging developments in AI and ML technologies further enable drug developers to tailor therapies according to the patient profile. It allows them to gain deeper insights into diseases, harness more efficient tools, and expedite the review and approval processes for drug products (Askin et al. 2023). While these technologies have shown significant utility in trial design and evidence generation, their applications extend beyond clinical activities. As previously discussed in Chapters 2–4, applications of these digital technologies can effectively expedite and streamline various activities regarding clinical trial execution. Digital technology heralds a transformative era in drug product development and the broader healthcare ecosystem, an overarching goal to expedite patient access to new medications while enhancing the efficiency and success rates of clinical trials. The rise of big data and digital technologies also challenges the FDA to modernize its approach to new innovations and create more modern platforms that are better for advancing clinical development.

The FDA and other Regulatory agencies are progressively endorsing the adoption of sophisticated trial methodologies, and digital technologies, evident in the proliferation of guidelines and reflection papers addressing this trend. These advances hold immense promise for all stakeholders involved, including developers, patients, prescribing physicians, insurers, and regulators. Despite the considerable potential of innovative methodologies and technologies, their full realization necessitates the evolution of current regulatory frameworks. There exists a pressing need for clear guidance concerning the acceptability of data, the validity of novel approaches and methodologies, such as ML algorithms, for processing and analyzing this wealth of information, and the delineation of pathways for product approval within these emerging paradigms.

With this as the backdrop, the following sections describe in further detail recent regulatory developments aimed at harnessing the potential of novel clinical design methods such as adaptive design (Bhatt and Mehta 2016; FDA 2019b) and master protocol (Woodcock and LaVange 2017; FDA 2023a), and modern technologies including AI and ML (EMA 2020; FDA 2020b, 2021f, 2023e). Several use cases are provided, showcasing the adoption of AI and analytics by the FDA in guidance development, workload planning, evidence synthesis, and regulatory review.

8.3 Clinical Trial Designs

The FDA has played a pivotal role in fostering and championing innovations in clinical trial design. The inception of the Critical Path Initiative in 2004 marked a significant initial stride, while the passage of the 21st Century Cures Act further underscored the agency's commitment to facilitating more streamlined advancements in treatments (FDA 2016a). In recent times, the FDA has proactively developed a series of guidelines aimed at assisting the industry in tackling challenges inherent in contemporary oncology drug development. Notably, one of these recent guidelines concentrates on master protocols, while another zeroes in on expansion cohorts within first-in-human clinical trials. Presently, both guidelines are in draft form. Demonstrating a commitment to expeditiously providing valuable guidance, the FDA prioritizes the production of informative guidelines to support the efficient development of effective new therapies and aid sponsors in navigating the approval process.

8.3.1 Enrichment Trial

8.3.1.1 Background

Clinical trials are not intended to prove treatment effectiveness in a random population. Instead, sponsors use various strategies to select subsets where drug effects are more easily seen. Examples include choosing patients with stable diseases, high treatment compliance, or specific characteristics suggesting treatment response. These strategies, called enrichment, use patient characteristics to boost the likelihood of detecting drug effects. Enrichment encompasses demographic, pathophysiologic, historical, genetic, clinical, and psychological factors. It may involve analyzing subsets within a broader population, enhancing the study's ability to detect drug effects. These principles can apply to safety assessments as well.

In 2019, the FDA issued a draft guidance on enrichment strategies for clinical trials (FDA 2019a). An enrichment design is defined as "the prospective use of any patient characteristic to select a study population in which detection of a drug effect (if one in fact is present) is more likely than it would be in an unselected population" (FDA 2019a). Enrichment strategies have become increasingly important in the context of personalized medicine, where the goal is to tailor treatments to individual patient profiles for maximum efficacy and safety (Rothwell 2005; Temple and Enas 2009).

Furthermore, innovative statistical methodologies and adaptive trial designs have been developed to support enrichment strategies, improving the precision and efficiency of clinical trials (Chow and Chang 2008; Thall, Simon, and Estey1995). These advancements help in identifying the right patient populations and making informed decisions throughout the drug development process.

8.3.1.2 Enrichment Strategies

Table 8.1 presents three basic kinds of enrichment addressed in the FDA guidance. The guidance outlines key enrichment strategies across various categories, detailing study design options with their respective advantages and disadvantages and addressing result interpretation in enrichment studies. These strategies are primarily discussed within the framework of randomized controlled trials, typically influencing patient selection pre-randomization.

TABLE 8.1
Enrichment Strategies

Strategy	Description
Strategies to Decrease Variability	These include choosing patients with baseline measurements of a disease or a biomarker characterizing the disease in a narrow range (decreased interpatient variability) and excluding patients whose disease or symptoms improve spontaneously or whose measurements are highly variable (decreased intrapatient variability). The decreased variability provided by these strategies would increase study power.
Prognostic Enrichment Strategies	These include choosing patients with a greater likelihood of having a disease-related endpoint event (for event-driven studies) or a substantial worsening in condition (for continuous measurement endpoints). These strategies would increase the absolute effect difference between groups but would not be expected to alter relative effect.
Predictive Enrichment Strategies	These include choosing patients who are more likely to respond to the drug treatment than other patients with the condition being treated. Such selection can lead to a larger effect size (both absolute and relative) and can permit use of a smaller study population. Selection of patients could be based on a specific aspect of a patient's physiology, a biomarker, or a disease characteristic that is related in some manner to the study drug's mechanism. Patient selection could also be empiric (e.g., the patient has previously appeared to respond to a drug in the same class).

8.3.1.2.1 Decreasing Variability

Decreasing variability is an effective way to enhance study power. This can be accomplished by implementing well-thought-out study inclusion and exclusion criteria aimed at reducing patent heterogeneity, ensuring adherence to the study protocol both by the site and patient, among other measures. Prognostic enrichment is often utilized to increase trial efficiency. Although prognostic enrichment, in general, does not impact relative risk reduction, it increases the number of events in a shorter period, resulting in a smaller sample size.

8.3.1.2.2 Prognostic Enrichment

Prognostic enrichment strategies have been successfully utilized in registrational trials (FDA 2019a; McMurray et al. 2014; Pedersen et al. 2004; Fisher et al. 1998). For instance, studies on adjuvant therapy using tamoxifen revealed not only a delay in the progression of metastases in breast cancer patients but also a decreased risk of contralateral tumors (new primary tumors) in this high-risk group (identified as high risk due to prior breast cancer diagnosis). Subsequently, tamoxifen was investigated in a cohort of 13,000 high-risk women (identified using the Gail model) with no history of breast cancer, who were monitored for four years in the National Surgical Adjuvant Breast and Bowel Project (NSABP) P-1 study (Fisher et al. 1998). The findings demonstrated a 44% relative reduction in the risk of invasive breast cancer, leading to FDA approval of tamoxifen for reducing breast cancer risk in high-risk individuals identified by the Gail model calculator. Conversely, conducting a study in patients at lower risk would necessitate a significantly larger sample size.

It is worth noting that clinical trials aimed at demonstrating a decrease or delay in the occurrence or recurrence of cancer are notably more likely to succeed in patients identified as high risk for such events, often through genetic or other identifiable characteristics. For example, a study focusing on patients with a risk level of 25% of that of the NSABP P-1 study population would require approximately 20,000 patients to detect an effect comparable to that observed in the NSABP P-1 study with 90% statistical power.

8.3.1.2.3 Predictive Enrichment

By contrast, predictive enrichment holds significant potential in bringing novel medicines to the market. By identifying patients more likely to respond to the treatment, predictive enrichment strategies enhance the probability of trial success. For predictive enrichment, the FDA guidance (FDA 2019a) details three study design options with their respective advantages and disadvantages as shown in Figure 8.1.

In recent years, several predictive enrichment strategies have been employed in the pivotal trials of oncology drugs, leading to marketing approvals.

One notable example is the drug Herceptin, which demonstrated a significant survival benefit of five months for patients with metastatic breast cancer exhibiting high HER2-neu expression, constituting roughly 25% of all breast cancer cases (Slamon et al. 2001). However, the overall survival improvement was less than two months across all patients. By focusing part of the trial on the subset of patients with high HER2/neu-expressing tumors, the drug's approval was supported, despite significant cardiotoxicity (Piccart-Gebhart et al. 2005). This strategic approach underscored the challenge of gaining approval in an unselected population, where the observed benefits would have been smaller and only a fraction of patients would have benefited.

Similarly, the recently approved drug Kalydeco targets the genetic defect responsible for cystic fibrosis but is effective in only about 4% of patients with the disease who possess a specific genetic abnormality (Ramsey et al. 2011). Although this percentage seems low, the drug's impact within this subset of patients is profound. A study conducted on an unselected population, where the majority would see little benefit, likely would have yielded different results.

Moreover, advancements in hepatitis C treatments offer another dimension of personalized medicine. While newer treatments demonstrate faster and

(A) Biomarker-Stratified Design

Assess Biomarker
Stratify
Biomarker positive
Randomize
Treatment A
Treatment B
Biomarker negative
Randomize
Treatment A
Treatment B

(B) Enrichment Design

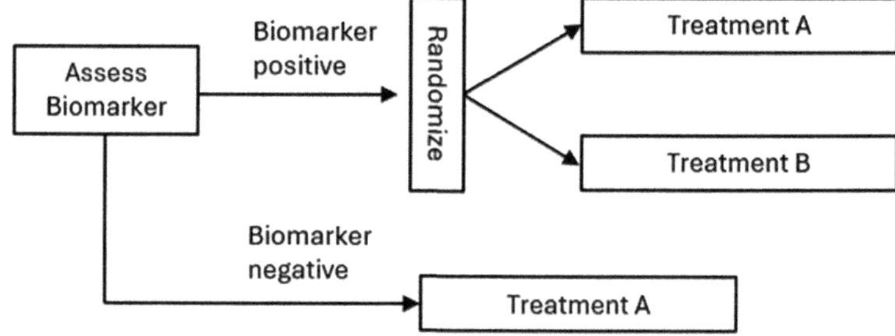

(C) Biomarker-Strategy Design

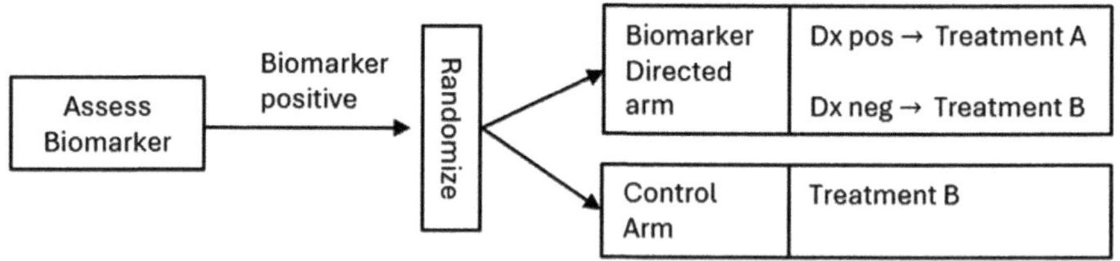

FIGURE 8.1
Types of enrichment designs. (Adapted from Freidlin et al. 2010.)

more effective results, they are specifically tailored to patients with hepatitis C Type 1, showing diminished efficacy in Types 2 and 3 (Jacobson et al. 2011). However, ongoing drug development endeavors aim to address these variations.

8.3.1.2.4 Biomarkers in Enrichment Trials

Advances in genomics, particularly in tumor sequencing, have enabled detailed classification of tumors using molecular biology (Kumar-Sinha and Chinaiyan 2018). These advancements allow for the identification of specific surface or genetic characteristics of cancer cells that can predict the efficacy of certain drugs in treating particular tumors. As highlighted by the EFPIA Clinical Trial Design Task Force (EFPIA 2020), prognostic and predictive enrichment strategies significantly depend on the efficacy of the screening test and its feasibility for real-world application post-approval. If the test isn't already in widespread clinical use, the development of a companion diagnostic may be warranted. A companion diagnostic is defined as a test crucial for ensuring the safe and effective utilization of a corresponding drug or biological product.

To facilitate these efforts, the FDA has published several guidelines, including "Principles for Co-development of an In Vitro Companion Diagnostic Device with a Therapeutic Product" (FDA 2016c), "Biomarker Qualification: Evidentiary Framework Guidance for Industry and FDA Staff" (FDA 2018d), and "Developing and Labeling In Vitro Companion Diagnostic Devices for a Specific Group or Class of Oncology Therapeutic Product" (FDA 2019e). Similar progress has been made by the EMA, evidenced by its publication of a reflection paper titled "Concept paper on the development and lifecycle of personalised medicines and companion diagnostics that measure predictive biomarkers which help to assess the most likely response to a particular treatment" (EMA 2017).

Marketing approval of in vitro diagnostic medical devices was previously governed by Directive 98/79/EC on in Vitro Diagnostic Medical Devices. As of May 25, 2017, two pivotal EU regulations – Regulation (EU) 2017/745 governing medical devices (MDR) and Regulation (EU) 2017/746 overseeing in vitro diagnostic medical devices (IVDR) – have taken effect, supplanting the previous trio of EU Directives. Within the updated MDR framework, the scope of "medical device" has been broadened to encompass software utilized for diagnosis, prevention, monitoring, prediction, prognosis, treatment, or alleviation of disease. The MDR employs a risk-based methodology, categorizing medical devices into different classes based on their intended purpose and inherent risks to ascertain the appropriate conformity assessment procedure. Each medical device is assigned to one of four risk categories: Class I, Class IIa, Class IIb, and Class III, indicating low, medium, medium/high, and high risk, respectively. Software that operates a device or influences its usage is classified within the same category as the device itself, as outlined in Section 3.3, Annex VIII of the MDR. Furthermore, the new MDR introduces novel regulations governing the classification of software, which are summarized in Table 8.2.

TABLE 8.2

Rule 11 in Annex VIII for the Classification of Software

Intended Use	Inherent Risk	Risk Category
Provide information used to make decisions with diagnosis or therapeutic purpose	Decision may result in death or an irreversible deterioration of a person's state of health	III
	Decision may cause a serious deterioration of a person's state of health or surgical intervention	IIb
	None of above	IIa
Monitor physiological processes	Monitoring vital physiological parameters, where the nature of variations of those parameters is such that it could result in immediate danger of the patient	IIb
	Not for monitoring vital physiological parameters	IIa
Other software		I

TABLE 8.3

Impact of Prevalence of Marker-Positive Patients on Sample Size

Prevalence of Maker-Positive Patients	Treatment Effect in Maker-Positive Patients (% of Marker-Positive Response)	
	0% Sample Size Ratio	50% Sample Size Ratio
100%	1	1
75%	1.3	1.3
50%	4	1.8
25%	16	2.6

Source: FDA (2019a).

It is worth noting that, unlike the medical device regulations in the United States, medical devices are regulated by the Member States of the European Union rather than by the EMA.

8.3.1.2.5 Impact of Prevalence of Marker-Positive Patients

The extent to which an enrichment strategy can decrease the required sample size for a study hinges on multiple factors, including the prevalence, sensitivity, and specificity of the enrichment marker and the relative efficacy of the drug in marker-positive versus marker-negative populations. The FDA guidance presents an example (Table 8.3), illustrating how sample size ratios – indicating the number of subjects required in an unselected population versus solely in the marker-positive population – vary with differing prevalence rates of marker-positive patients and varying treatment effects in marker-negative patients (with treatment effect in marker-negative patients set at either 0% or 50% of that in marker-positive patients). It is important to note that Table 8.3 operates under the assumption of 100% accurate classification of patients into positive and negative groups.

Generally, a lower prevalence of marker-positive patients in the unselected population, coupled with a smaller treatment effect in marker-negative patients, leads to a greater reduction in sample size for a study of marker-positive patients compared to an unselected population study. For instance, when only 25% of a population is marker-positive and no treatment effect is anticipated in the 75% of marker-negative patients, the required sample size for an unselected population study would be 16 times larger than that for a study exclusively focused on marker-positive patients (Simon and Maitournam 2004). In the context of COPD clinical trials, Gold, Yang, et al. (2014) developed a clinical trial simulator to explore the impact of the prevalence of marker-positive patients, along with other design parameters, on COPD patient outcomes to inform COPD trial planning with biomarkers. Please refer to Chapter 5 for details.

8.3.1.3 Machine Learning-Based Enrichment

Clinical biomarkers encompass a diverse array of markers that may include demographic, physiological, molecular, histological, or radiographic characteristics or measurements believed to be associated with various aspects of normal or abnormal biological functions or processes. To enhance predictiveness of these biomarkers, they are often used in combination. A number of studies have been conducted to improve physicians' prognostic accuracy for patients with advanced or metastatic cancer (Ploquin et al. 2012). A prognostic score called RMH was among the first prognostic indices to predict life expectancy in oncology trials, based on baseline laboratory values including lactate dehydrogenase (LDH), albumin, and number of metastatic sites (Arkenau et al. 2008). GRiM (Bigot et al. 2017) and LIPI (Mezquita et al. 2018) are two other scores developed for similar purposes in cancer trials involving immune checkpoint inhibitors. More recently, Yu et al. (2020) developed a prognostic index based on ML methods, which demonstrated improved performance in identifying patients at risk of early mortality compared to the other three aforementioned methods.

Since nonlinear relationships between predictive biomarkers and patient outcomes, as well as inter-correlation or dependency among biomarkers, exist, it is advantageous to use ML models to capture these complex relationships in enriching clinical studies. In recent years, ML-driven discovery of therapeutic indices has been an area of intense research (Huang and Chiu 2023).

8.3.2 Adaptive Design

8.3.2.1 Background

Designing a study relies on a set of assumptions, such as treatment effect, background variability, and other factors. However, at the initial design stage, our knowledge about these assumptions is often imprecise. As a result, the sample size determined based on these initial assumptions might prove inadequate. This will increase the Type II error, or chance of erroneous claim of ineffectiveness. One way to mitigate the risk is to incorporate an adaptive feature into the design, enabling adjustments of the trial features such as randomization ratio, sample size, and treatment arm during the study based on emerging data from both the ongoing trial and/or external sources. interim analyses.

The application of adaptive design methods to modify ongoing clinical trials based on accrued data has been a longstanding practice in clinical research. This concept traces back to the 1970s, with the introduction of adaptive randomization procedures and a class of designs for sequential clinical trials. As a result of this historical foundation, most adaptive design methods in clinical research are commonly known as adaptive randomization, group sequential designs, and sample size re-estimation at interim stages to achieve desired statistical power.

In 1990, an enhanced Phase I design known as the Continuous Re-Assessment Method (CRM) was introduced by O'Quigley et al. (1990). This method offered improved control over observed toxicity levels during the trial, flexibility in defining the target maximum tolerated toxicity, and more reliable estimates of the maximum tolerated dose (MTD). The introduction of RM subsequently led to the development of a broader range of CRM designs (O'Quigley 1992; O'Quigley and Chevret 1991; O'Quigley and Zhen 1996; Piantadosi and Liu 1996; Heyd and Carlin 1999). More recent efforts by many working groups, including the Pharmaceutical Research and Manufacturers of America (PhRMA) and Biotechnology Industry Organization (BIO), have further popularized the use of adaptive clinical designs.

8.3.2.2 Types of Adaptive Designs

There are a range of adaptive designs. Commonly known adaptive design methods in clinical trials can be categorized based on the type of adaptation or modification intended to make. They include: (1) adaptive group sequential designs; (2) sample size re-estimation; (3) adaptive enrichment; (4) drop-the-loser design; (5) adaptive randomization; (6) adaptive endpoint selection; and (7) designs with more than one adaptation.

The pros and cons of each of the above adaptive designs have been extensively reported in the literature (Chow and Chang 2008; Chang 2008; EFPIA 2020). It is not uncommon to utilize multiple adaptations to address different objectives in complex clinical trials. In recent years, there has been a noticeable uptick in the utilization of adaptive trial designs (Freidlin and Simon 2005). While most instances of adaptive designs occur in the earlier stages of clinical development, particularly in phase II trials, there are significantly fewer examples observed in confirmatory trials. A survey by Bothwell et al.

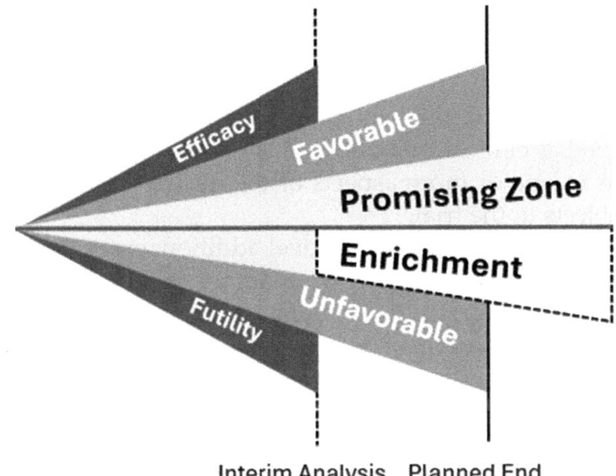

FIGURE 8.2
Adaptive design with one interim look. (Adapted from Cytel website.)

(2018) indicated that recent growth in the use of adaptive designs globally has been noted. The most common adaptations include seamless Phase II/III (57%), group sequential (21%), biomarker adaptive (20%), and adaptive dose-finding designs (16%). Approximately one-third (32%) of trials included an independent DMC, while 6% reported blinded interim analysis. Among adaptive trials, 9% were considered for FDA product approval, and 12% for EMA product approval. Regulators had varied experiences with adaptive trials, often involving extensive correspondence between sponsors and regulators, sometimes resulting in required revisions or alterations to research designs. EPIA reported that among the various disease areas, adaptive designs have predominantly been employed in oncology settings (EFPIA 2020) (Figure 8.2).

8.3.2.3 Regulatory Developments

The concept of adaptive design was first introduced in ICH E9 Statistical Principles in Clinical in 1990 Trials (ICH 1998). The guideline states that trials could be modified as determined by Independent Data Monitoring Committees.

In February 2010, the FDA released a draft guidance, Guidance for Industry: Adaptive Design Clinical Trials for Drugs and Biologics, with a primary focus on adaptive designs for confirmatory trials (FDA 2019c). In late 2016, the U.S. Congress enacted the 21st Century Cures Act, directing the FDA to revise its guidance regarding adaptive designs for sponsors of investigational drugs and biological products. This legislation characterizes adaptive designs as "modern" and "novel" methodologies. In September

2018, FDA issued a new draft Guidance for Industry on Adaptive Designs for Clinical Trials of Drugs and Biologics and published the final version in November 2019 (2019c).

In both the draft and final guidance, the FDA states that "an *adaptive design* is defined as clinical trial design that allows for prospectively planned modifications to one or more aspects of the design based on accumulating data from subjects in the trial."

The 2018 document introduced a novel addition: examples of five studies showcasing the potential benefits of adaptive designs. These adaptations include: increasing sample size to guarantee adequate power via a pooled, non-comparative evaluation of interim data (Bolland et al. 1998); interim analyses incorporating efficacy-based stopping rules (McMurray et al. 2014); adaptive dose selection (Chen et al. 2015); an adaptive platform trial that evaluates multiple treatments against a common control and employs pre-planned adaptations to identify promising therapies (Sydes et al. 2012); and a Bayesian adaptive design study (PREVAIL II Writing Group et al. 2016; Dodd et al. 2016).

For adaptation based on comparative data, the guidance states that stopping or adaptation rules can be specified on a variety of different scales, such as the estimate of treatment effect, fixed sample *p*-value, conditional probability of trial success, Bayesian posterior probability that the drug is effective, or Bayesian predictive probability of trial success. The choice of scale is relatively unimportant as long as the operating characteristics of the designs are adequately evaluated. The guidance emphasizes in multiple instances the crucial role of clinical trial simulations, underscoring their particular significance in the planning and design of adaptive trials. It also reiterates the importance of having robust and comprehensive plans established before the initiation of the study. These plans should detail the processes designed to regulate access to information and systematically document such access throughout the trial so as to maintain trial integrity.

In 2007, EMA published a reflection paper on methodological issues in confirmatory clinical trials planned with an adaptive design. Subsequently, in 2012, EMA issued guidelines specifically addressing the use of adaptive designs in clinical trials for patients with acute coronary syndrome. The PMDA has provided guidelines similar to the FDA and EMA, focusing on adaptive designs (PMDA 2007).

Taken together, regulatory agencies globally recognize the potential benefits of adaptive design in clinical trials and provide guidelines to ensure these designs maintain scientific rigor and reliability. These guidelines cover statistical principles, methodological considerations, ethical standards, and practical recommendations for conducting adaptive trials. The ICH, EMA, FDA, and PMDA emphasize the importance of maintaining trial integrity, minimizing bias, and ensuring the reliability of results through careful planning, appropriate statistical methodologies, and thorough documentation.

8.3.3 Bayesian Design

8.3.3.1 Background

Clinical development is an empirical problem-solving process, characterized by trial and error. Knowledge gleaned from the previous study is often used to guide the next experiment. The sequential learning nature of experimentation and reliance on knowledge from various clinical development stages calls for statistical methods that can synthesize information from disparate sources to aid better decision-making. Despite the controlled nature, clinical trials also encounter many uncertainties such as varying medical practice at different sites and unknown patient or disease characteristics which are unaccounted for in the trial inclusion or exclusion criteria. Robust evaluation of drug safety and efficacy lies in the researcher's ability to synthesize information from various sources including different stages of clinical development and incorporate known variability in the inference of the clinical research questions. It also requires statistical methods to be adaptive in the sense that inference can be updated based on new information. Bayesian statistics provide a flexible framework for continuous update of learning based on new information. In addition, Bayesian analysis allows for the incorporation of prior knowledge, either in terms of expert opinion or historical data, in its statistical inferences (Yang and Novick 2019).

In the past two decades, significant advances of Bayesian applications in clinical trial and analysis have been made (Spiegelhalter et al. 2004). There has been also greater regulatory acceptance of Bayesian approaches to study design and analysis as evidenced by the publication of the FDA guidance on the use of Bayesian statistics in medical device clinical trials (FDA 2010) and acceptance by regulatory agencies of Bayesian methods for both early and late phases of drug development. Bayesian adaptive study design and analysis have been widely used for Phase I dose-finding studies (Chevret 2006) and Phase II efficacy assessment (Berry and Stangl 1996; Spiegelhalter et al. 2004). In the following section, we discuss regulatory advances in the use of Bayesian methods in drug development and regulatory approval.

8.3.3.2 Guidance for Use of Bayesian Methods

The Federal Food, Drug, and Cosmetic Act mandates that a drug's effectiveness be supported by "substantial evidence," defined as adequate and well-controlled investigations by qualified experts. FDA's regulatory review ensures this standard is met, integrating a structured benefit-risk assessment (Ionan et al. 2023).

While the preponderance of statistical methods used for evidence generation in support of regulatory filings are from traditional frequentist approaches, there have been growing applications of the Bayesian approach in drug development in recent years. In the ICH E9 Statistical Principles for Clinical Trials, it is stated "Because the predominant approaches to the design and analysis of clinical trials have been based on frequentist statistical methods, the guidance largely refers to the use of frequentist methods when discussing hypothesis testing and/or confidence intervals. This should not be taken to imply that other approaches are not appropriate: the use of Bayesian and other approaches may be considered when the reasons for their use are clear and when the resulting conclusions are sufficiently robust." The guidance presents opportunities for Bayesian methods to contribute to the totality of evidence in support of regulatory filings.

On February 5, 2010, the FDA finalized and published the *Guidance for the Use of Bayesian Statistics in Medical Device Clinical Trials* (FDA 2010). It clearly states that "The information from a current trial is augmented and the precision may be increased by the incorporation of prior information in a Bayesian analysis. The Bayesian analysis brings to bear the extra, relevant, prior information, which can help FDA in marketing approval decision-making. The guidance emphasizes the importance of pre-specification and agreement on both the prior information and the model, which includes amount of borrowing of information from previous studies. The document specifically discusses the utility of Bayesian hierarchical models to combine results from multiple studies to obtain estimates of safety and effectiveness parameters.

Although the guidance is intended for medical device clinical trials, many of the principles and methods are application for drug clinical development. However, unlike the medical device setting where the mechanism of action for many medical devices is well understood and availability of prior information from earlier version of the device, more care is needed when applying the Bayesian methods for clinical trials. In recent years, the FDA has been increasingly promoting innovative trial designs including the applications of Bayesian methods. The FDA guidelines, *Guidance for Industry: Adaptive Designs for Clinical Trials of Drugs and Biologics* (2019) and Guidance for Industry: Interacting with the FDA on Complex Innovative Trial Designs for Drugs and Biological Products (2020), outlines the FDA's recommendation on the Bayesian trial designs and synthesis of information from the current and external studies.

FDA's commitment to advancing appropriate use of Bayesian methods is further evident in its acceptance of several Bayesian designs such as BOIN as "fit-for-purpose" (Shord et al. 2023), co-sponsoring several workshop on Bayesian innovation, publications of successful use cases (Travis et al. 2023; Ionan et al. 2023), and its intent to a draft guidance on the Use of Bayesian Methodology in Clinical Trials of Drugs and Biologics by the end of 2025 as a part of the Prescription Drug User Fee Act VII commitment (Ionan et al. 2023).

8.3.3.3 Areas of Emerging Applications

A recent article published by FDA reviewers highlights a broad array of applications of Bayesian in various clinical contexts (Ionan et al. 2023).

Specifically, Bayesian approaches find relevance across clinical contexts and study types. Early drug and biological product development benefits from Bayesian model-based design for dose selection (O'Quigley 1992; O'Quigley and Chevret 1991; O'Quigley and Zhen 1996; Piantadosi and Liu 1996; Heyd and Carlin 1999). Bayesian methods extend to noninferiority trials, offering tools for design and analysis. In fact, such potential applications are recognized in the FDA guidance on noninferiority trials (FDA 2016b). Examples of applying Bayesian methods to determine the noninferiority margin and analyze noninferiority data can be found in Rothwell et al. (2020), Gambo et al. (2014) and Price and Scott (2021).

Adaptive clinical trials leverage Bayesian frameworks for dynamic study modifications. Applications in this context encompasses: (1) Utilization of predictive statistical modeling, potentially integrating external information, to dictate the timing and decision rules for interim analyses; (2) Incorporation of presumed dose-response relationships to guide dose escalation and selection; (3) Explicit incorporation of information from external sources (such as previous trials, natural history studies, and registries) through informative prior distributions to enhance trial efficiency; and (4) Employing posterior probability distributions to establish criteria for trial success.

Conducting pediatric clinical trials is often challenging due to technical difficulties, concerns, and, sometimes, small patient pool. Informative Bayesian methods prove beneficial in pediatric trials, facilitating extrapolation of efficacy from other populations. Rare disease trials, in particular, can benefit, using prior information integration to address limited data challenges.

Bayesian subgroup analysis enhances precision by pooling data across subgroups, offering more reliable treatment effect estimates. Bayesian hierarchical models for subgroup analysis have garnered significant attention in literature and have been highlighted in CDER impact stories as an innovative statistical approach, offering more reliable treatment outcome information to patients and clinicians (Ionan et al. 2023; FDA 2019g). The fundamental concept involves analyzing pertinent subgroups collectively rather than individually. This is achieved through Bayesian methods by connecting individual subgroups via a prior distribution that encapsulates underlying treatment effects in each subgroup. Leveraging information from other subgroups enhances the overall accuracy of estimates across the board, a particularly valuable trait when analyzing data from smaller subgroups.

Ionan et al. (2023) presented several use cases highlighting the successful applications of Bayesian methods in CDER- and CBER-reviewed clinical trials. Section 8.6 presents one case example by Travis et al. (2023) where data from adult trials with a comparable design to the pediatric study are utilized to assess the effectiveness of the drug for the pediatric indication.

8.3.4 Master Protocol

8.3.4.1 Background

There is a growing interest in streamlining late-stage drug development, through timely generation of evidence demonstrating drug safety and efficacy. However, traditional methods of generating such evidence through single or dual intervention clinical trials have become increasingly costly and difficult. This has left many critical clinical questions unanswered. Precision medicine trials aiming to evaluate targeted therapies face challenges in recruiting patients with rare genetic subtypes of diseases. Additionally, there is growing interest in mechanism-based trials, where eligibility criteria extend beyond traditional disease definitions. The overarching need is to address more questions efficiently and swiftly (Woodcock and LaVange 2017).

In response to these challenges, there is a clear need to use trial designs that enable the assessments of multiple drugs and/or diseases concurrently within a single protocol. Such trials, commonly referred to as *master protocols,* including umbrella, basket, and platform trials, offer a more streamlined approach, sharing design components and operational aspects for improved coordination compared to independently conducted trials. These master protocols may facilitate direct comparisons of competing therapies or simultaneously evaluate different therapies against respective controls. They may utilize existing infrastructure or establish new trial networks specific to the protocol.

Given the complexity and regulatory implications of these trials, it is imperative to ensure their meticulous design and execution to safeguard human subjects and produce data meeting regulatory standards for demonstrating safety and effectiveness of each investigational drug. Through pre-trial discussions among sponsors, trial conductors, and governance parties are crucial to address data use, publication rights, and regulatory submission timing concerns before trial commencement.

8.3.4.2 FDA Guidance

In 2018, the FDA issued the draft guidance, Guidance for Industry: Master Protocols: Efficient Clinical Trial Design Strategies to Expedite Development of Oncology Drugs and Biologics Guidance for Industry. The guidance was subsequently finalized in March 2022 (FDA 2023a).

In both documents, a master protocol is defined as "a protocol designed with multiple sub-studies, which may have different objectives and involve coordinated efforts to evaluate one or more investigational drugs in one or more disease subtypes within the overall trial structure." In general, the FDA advises sponsors to establish the Recommended Phase 2 Dose (RP2D) for investigational drugs before utilizing a master protocol. However, individual drug sub-studies within the master protocol may include an initial dose-finding phase, particularly in pediatric subjects, if adequate adult data inform a starting dose and the drug offers potential clinical benefit to pediatric patients (21 CFR 50.52). A master protocol can also serve exploratory or marketing application purposes and may concurrently evaluate different drugs against their respective controls or a common control. sponsors have the flexibility to design the master protocol with a fixed or adaptive approach, allowing modification or termination of individual sub-studies within the protocol.

8.3.4.3 Basket, Umbrella, and Platform Trial Designs

8.3.4.3.1 Basket Trial Design

A *basket trial* refers to a master protocol structured to evaluate either a single investigational drug or a drug combination across various populations delineated by different cancers, disease stages within a specific cancer, histology, prior therapy counts, genetic or other biomarkers, or demographic traits is typically termed a basket trial (as depicted in Figure 8.3). In the case of a basket trial assessing an investigational drug combination, there might be a dose-finding or safety lead-in segment aimed at pinpointing safe doses of the combination before advancing to an activity-estimating phase.

8.3.4.3.2 Umbrella Trial Design

A master protocol aimed at assessing multiple investigational drugs, either as individual agents or in combination, within a single disease population is often termed an *umbrella trial* (Figure 8.4). Sub-studies within umbrella trials may encompass dose-finding or safety lead-in components to ascertain safe doses of investigational drug combinations prior to initiating an activity-estimating phase. As reiterated earlier, sponsors must confirm that the Recommended Phase 2 Dose (RP2D) for each investigational drug has been determined prior to their inclusion in a master protocol.

8.3.4.3.3 Platform Trial Design

A distinctive iteration of the master protocol is the *platform trial,* which assesses multiple interventions, all within the same study framework (Figure 8.5). Figure 8.5 presents an example of platform trial with one SoC and three treatment arms being entered into the study at different time points. Based on the outcomes of interim analyses. Adaptive randomization was used to assign more participants to treatment(s) with greater probability of success.

Arguably, platform trials differ from the traditional clinical trial designs. The key differences are highlighted in Table 8.4. What sets platform trials apart is their scale in terms of the number of drugs that can be evaluated and their utilization of a common control group. They also have the capacity to modify the therapies being investigated during active enrollment, facilitated by Bayesian

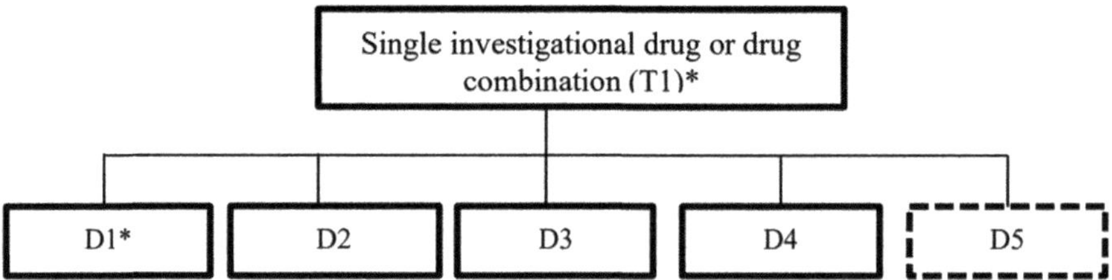

FIGURE 8.3
Schematic representation of a master protocol with basket trial design. T = investigational drug; D = protocol-defined subpopulation in multiple disease subtypes; D5 = dashed lines indicate potential amendments to include additional subpopulations. (Adapted from FDA 2023a.)

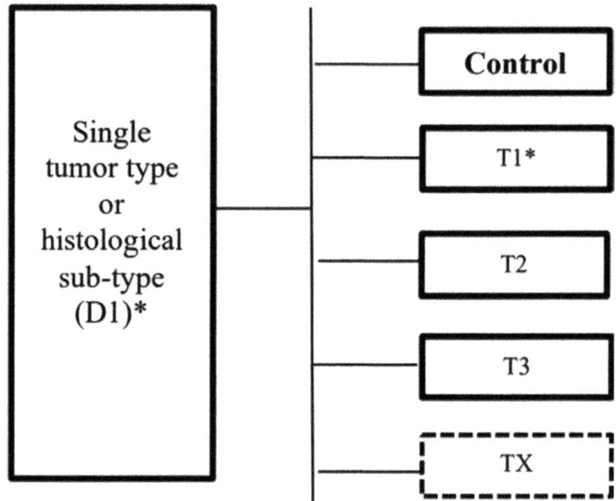

FIGURE 8.4
Schematic representation of a master protocol with umbrella trial design. T = investigational drug or investigational drug combination; D = protocol defined subpopulation in single disease subtypes; TX = dashed lines indicate potential amendments to include future treatment arms. (Adapted from FDA 2023a.)

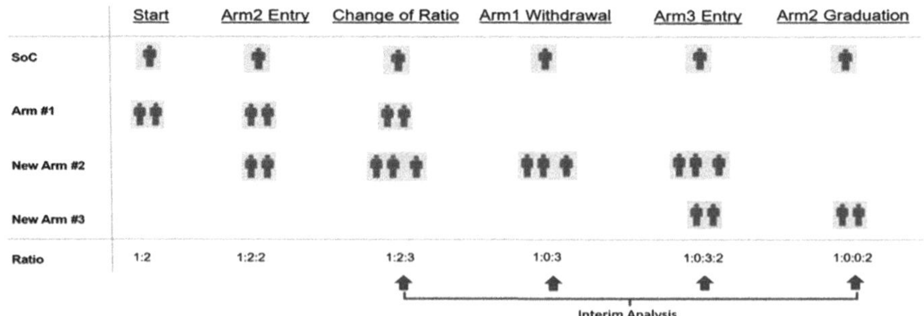

FIGURE 8.5
An example of a platform trial featuring adaptive randomization.

TABLE 8.4

Differences between Traditional and Platform Trial Designs

Characteristic	Traditional Design	Platform Design
Scope	Single agent	Multiple agents
Population	Homogeneous	Heterogeneous
Duration	Finite	Long-term
Number of Treatment Groups	Pre-specified	Multiple / may change over time
Randomization	Fixed	Adaptive / Dynamics
Decision Rules	Entire trial may be stopped early for success of futility	Individual arms may be removed from trial due to efficacy or futility

interim analysis. This methodological flexibility not only accelerates the development of new therapies but also ensures a more ethical and effective use of resources in clinical research (Saville and Berry 2016). By dynamically reallocating patients toward better-performing treatments, adaptive randomization facilitates a more responsive and patient-centric approach to clinical trials, which is increasingly crucial in the rapidly evolving landscape of medical research (Collignon et al. 2020).

8.3.4.4 Applications of Master Protocols

Master protocols have proven successful in numerous clinical trials, demonstrating their potential to streamline the drug development process and enhance therapeutic outcomes. One notable example is the BRAF-V600 study, which focused on patients with tumors harboring the BRAF-V600 mutation across various cancer types (Hyman et al. 2015). This trial highlighted the efficiency of targeting specific genetic mutations regardless of the cancer's origin, providing significant insights into personalized medicine. The NCI-MATCH trial, led by the National Cancer Institute, further exemplifies the utility of master protocols by matching patients with rare cancer types to targeted therapies based on genetic abnormalities identified in their tumors (Flaherty et al. 2020). Similarly, the LUNG-MAP trial has been pivotal in advancing treatments for squamous cell lung cancer by using a master protocol to test multiple drugs simultaneously, allowing for more rapid and flexible investigation of potential therapies (Papadimitrakopoulou, V. 2018). The I-SPY2 study described by Barker et al. (2009) has revolutionized breast cancer treatment by employing an adaptive trial design to evaluate the effectiveness of novel agents in the neoadjuvant setting, dynamically assigning patients to different treatments based on real-time results. The AcSé Trial in France is another example, evaluating off-label use of targeted therapies in patients with rare mutations, thereby expanding treatment options for these populations (Massard et al. 2017). The ALCHEMIST trial (Adjuvant Lung Cancer Enrichment Marker Identification and Sequencing Trials) conducted by United States National Cancer Institute (NCI 2021) aims to identify early-stage lung cancer patients with specific genetic alterations and match them with targeted therapies post-surgery. Lastly, the FOCUS4 trial in colorectal cancer uses a modular approach to test multiple therapies in parallel, adapting treatment strategies based on the genetic profile of the tumors (Kaplan et al. 2017). These studies illustrate the capacity of master protocols to enhance clinical trial efficiency and effectiveness, fostering more rapid and precise therapeutic advancements.

8.3.4.5 Challenges

While master protocols offer significant advantages in enhancing the efficiency and effectiveness of clinical trials, they also present several challenges. One primary challenge is the complexity of trial design and implementation, which requires sophisticated statistical methodologies and advanced planning to ensure that the study remains scientifically robust and valid throughout its duration (Berry et al. 2015; Woodcock and LaVange 2017). This complexity extends to regulatory approval, as different therapeutic agents and patient subgroups may necessitate varied approval processes and timelines (Berry et al. 2015). Additionally, logistical issues arise from the need to coordinate multiple stakeholders, including pharmaceutical companies, regulatory bodies, and clinical sites, each with their specific requirements and expectations. Managing and interpreting data from diverse sources can also be daunting, requiring robust data management systems and analytical tools to ensure data integrity and reliability. Furthermore, there are ethical considerations related to patient consent and safety, especially when introducing new therapies or modifying treatment arms based on interim results. Finally, funding and resource allocation can be challenging, as master protocols often require substantial financial and infrastructural investment. Despite these challenges, ongoing advancements in trial design and regulatory frameworks continue to support the successful implementation of master protocols (Berry et al. 2015).

8.3.5 Dose Optimization

8.3.5.1 Background

Dose-finding trials, encompassing dose-escalation and dose-expansion phases, have traditionally focused on determining the MTD for oncology drugs. This approach, rooted in the era of cytotoxic chemotherapy, relied on observing steep dose-response relationships, the lack of drug target specificity, and the readiness of patients and physicians to endure significant toxicity to combat life-threatening diseases (Ratain et al. 1993). The MTD was established through incremental dose increases in small patient cohorts, monitoring for severe or life-threatening dose-limiting toxicities (DLTs). Subsequent trials often adopted the MTD or a closely related dose without further optimization.

However, the landscape has shifted with the advent of modern oncology drugs, particularly targeted therapies like kinase inhibitors and monoclonal antibodies. These therapies, designed to interact with specific molecular pathways underlying oncologic diseases, exhibit dose-response dynamics distinct from traditional chemotherapy. Lower doses than the MTD may offer comparable efficacy with reduced toxicity. Moreover, achieving the MTD may prove unattainable in certain scenarios, as patients may receive targeted therapies over extended periods, potentially leading to prolonged, albeit lower-grade, symptomatic toxicities that pose ongoing challenges.

Despite these nuances, registration trials – aimed at evaluating safety and efficacy to support marketing applications – often adhere to the MTD paradigm or the highest dose explored in dose-escalation trials if the MTD is undefined. This approach risks recommending dosages poorly tolerated by patients, compromising their functioning and quality of life and potentially hindering treatment adherence and clinical outcomes. Moreover, patients experiencing adverse reactions may encounter difficulties tolerating subsequent treatments, particularly if toxicities overlap (Parulekar and Eisenhauer 2004).

The conventional MTD paradigm often overlooks valuable data on low-grade toxicities, dosage adjustments, drug activity, dose-exposure relationships, and specific patient populations. Dose-finding trials that systematically explore a range of doses, selecting those for further investigation based on comprehensive clinical data and understanding of dose-response relationships, offer a more nuanced approach to identifying optimal dosing regimens.

Given the persistent high unmet medical need in advanced cancers, expedited access to safe and effective therapies remains paramount. Some oncology development programs employ a seamless approach, facilitating swift transitions from initial dose-finding to registration trials to streamline development (Blumenthal et al. 2021).

8.3.5.2 Regulatory Initiatives

8.3.5.2.1 Project Optimus

In 2021, the FDA's Oncology Center of Excellence launched Project Optimus, aiming to spearhead a proactive shift in the dosing landscape within oncology. Specifically, Project Optimus seeks to educate, innovate, and collaborate with companies, academia, professional societies, international regulatory authorities, and patients to move forward with a dose-finding and dose optimization paradigm across oncology that emphasizes selecting a dose or doses that maximize not only the efficacy of a drug but also its safety and tolerability (FDA 2021c). The initiative's specific goals include: (1) Communicating expectations for dose-finding and dose optimization through guidance, workshops, and other public meetings; (2) Providing opportunities for and encouraging drug developers to meet with FDA Oncology Review Divisions early in their development programs, well before conducting trials intended for registration, to discuss dose-finding and dose optimization; and (3) Developing strategies for dose finding and dose optimization that leverage nonclinical and clinical data in dose selection, including randomized evaluations of a range of doses in trials (FDA and AACR 2024). Emphasis will be placed on performing these studies as early and efficiently as possible in the development program to bring promising new therapies to patients.

This initiative unites scientists specializing in clinical pharmacology, medical oncology, biostatistics, pharmacology/toxicology, and related fields both within and outside the FDA. Its primary objective is to redefine the approach to dose selection for oncology drugs, with a dual focus on enhancing efficacy while prioritizing safety and tolerability (FDA 2021c).

8.3.5.3 FDA Guidance on Dose Optimization

In 2023, FDA released its draft guidance on optimizing the dosage of human prescription drugs and biological products for the treatment of oncologic disease (FDA 2023d). The guidance represents the shift of the agency away from the MTD paradigm in oncology drug dosing, advocating for the evaluation of dose ranges to understand exposure-response relationships, aligning with practices in other therapeutic areas. The guidance emphasizes that doses chosen at each development stage must be supported by relevant clinical and nonclinical data. Sponsors are encouraged to discuss dose optimization plans in FDA meetings, and expedited pathways do not excuse lack of optimization. The FDA may issue a clinical hold if a selected dose presents unreasonable safety risks. It is recommended to optimize dosages before approval, as delaying until after approval risks exposing a large number of patients to suboptimal dosages. Failure to optimize doses for marketing applications could lead to approval failure or necessitate additional clinical trials to demonstrate safety and efficacy. Furthermore, conducting comparative trials post-approval can pose logistical challenges once a drug is already in use for a specific indication.

Ever since the release of the guidance, FDA has organized or cosponsored several workshops (FDA-ASCO; FDA and AACR 2024) discuss research and clinical challenges for dose optimization and highlight strategies to improve dose optimization for anticancer agents.

8.3.5.4 Innovative Designs for Dose Optimization

Shord et al. (2023) discussed various alternative trial designs and analytical methods that leverage nonclinical data and clinical observations beyond short-term safety data for dose optimization. These include utilization of controlled backfill, strategically planned expansion cohorts, and randomized parallel dosage comparisons to provide additional clinical data, increasing confidence in selected dosages for registration trials. With advancements in methodology and regulatory experience with model-informed drug development (MIDD), the FDA is increasingly interested in implementing model-informed clinical trial designs such as algorithm-based Bayesian designs. They methods enable more informed decisions regarding dose selection due to the ability to incorporate nonclinical and external information of drugs in the same class (Shord et al. 2023)

Acknowledging the challenges in identifying optimized dosages before drug approval, seamless, adaptive designs are being explored to address key questions around dosage optimization, safety, and efficacy determination (FDA and AACR 2024). These approaches may involve comparing multiple dosages as part of a registration trial after an expedited proof-of-concept.

A clearly delineated plan for establishing dose- and exposure-response relationships is crucial, including pre-specifying approaches to evaluating the association between relative dose intensity and clinical outcomes. Historical post hoc dosage refinement has limitations, typically evaluating only one dosage in registration trials, limiting the robustness of inferred optimized dosages. Thus, evaluating dose- and exposure-response relationships is essential to inform dosages for efficacy trials (FDA and AACR 2024). Such a relationship is often assessed through either randomized dose-controlled trials (RDCTs) or randomized concentration-controlled trials (RCCTs).

The concept of RCCT was first formalized by Peck in the late 1980s (Peck 1990) and subsequently popularized in early 1990s (Sanathanan et al. 1991; Sanathanan and Peck 1991). RCCTs represent a powerful alternative to RDCTs when understanding drug concentration-response relationships is a primary objective. Unlike RDCTs, where subjects are randomized to predefined doses, RCCTs involve randomizing participants to specific target drug concentration ranges, achieved through adaptive, individualized dosing. This approach minimizes variability arising from pharmacokinetics, thereby enabling a more precise assessment of the concentration-response relationship. Early simulation studies demonstrated the sample size efficiency of RCCTs compared to RDCTs, sparking interest in their application (Sanathanan and Peck 1993). Over time, RCCTs have been utilized in various therapeutic areas, including oncology, neurology, and transplant medicine, highlighting their versatility and impact (Grahnén and Karlsson 2001; Kraiczi et al. 2003). While their implementation demands additional resources and expertise, their potential to enhance the precision and reliability of clinical trial outcomes makes them a valuable tool in dose optimization.

8.4 FDA Complex Innovative Trial Design Pilot Meeting Program

The FDA introduced the Complex Innovative Trial Design (CID) Pilot Meeting Program through a notice in the Federal Register on August 30, 2018 (FDA 2018c). This initiative aimed to achieve two main objectives: (1) Facilitate the advancement and adoption of complex adaptive, Bayesian, and other innovative clinical trial designs and (2) Foster innovation by providing a platform for the FDA to publicly discuss trial designs developed through the pilot meeting program as case studies, even before the drug under study has received FDA approval.

The CID pilot meeting program was particularly geared toward pioneering clinical trial designs that necessitated simulations to estimate operating characteristics. It offers sponsors whose meeting requests are granted the opportunity for increased interaction with FDA staff to discuss their proposed CID approach. To promote innovation in this area, trial designs developed through the meeting program may be presented by FDA (e.g., in a guidance or public workshop) as case studies, including trial designs for medical products that have not yet been approved by FDA. Notably, several examples of CID case studies conducted under this program have been published (Price 2021; FDA and AACR 2024). More recently, the FDA has extended the program under the new name "Complex Innovative Trial Design Meeting Program" (FDA 2022a), affirming the agency's commitment to support the goal of facilitating and advancing the use of complex adaptive, Bayesian and other novel clinical designs.

In December 2020, the FDA published guidance for industry titled "Interacting with the FDA on Complex Innovative Trial Designs for Drugs and Biological Products." The guideline provides a comprehensive framework for sponsors seeking to engage with the FDA regarding the development and implementation of complex innovative trial designs (CIDs). This guidance outlines best practices for early and ongoing communication with the FDA to facilitate the development of CIDs, which include adaptive designs, Bayesian approaches, and other advanced statistical methods. The guidance emphasizes the importance of detailed pre-submission meetings to discuss trial objectives, design elements, simulation plans, and analysis strategies. Additionally, it highlights the necessity of including robust simulation studies to demonstrate the operating characteristics of the proposed designs. The FDA encourages sponsors to seek feedback through the CID Pilot Meeting Program, which offers a collaborative platform to discuss innovative methodologies and ensure that trial designs are scientifically sound and aligned with regulatory expectations. By fostering open dialogue and providing detailed recommendations, the FDA aims to enhance the development and regulatory review of CIDs, ultimately accelerating the availability of effective and safe therapies for patients.

8.5 AI and RWD for Innovative Trial Designs

8.5.1 Applications in Clinical Trials

With big data and AI being recognized as pathways toward modernizing clinical development, a range of opportunities has been identified and explored, as discussed in previous chapters. This progress is supported by the widespread adoption of EHR systems and other digital technologies, such as wireless wearable devices and sensors, along with the expansion of clinical trials generating large volumes of clinical, molecular, and imaging data (Askin et al. 2023).

AI-powered solutions can address many common pitfalls of current RCTs, such as manual processes, poor patient selection, lack of predictability of trial outcomes, over-reliance on concurrent control for evidence generation, and limited insights for personalized treatment. Conducting clinical trials involves vast amounts of manual effort and administrative work. Robotic process automation and ML can help streamline and automate many of these tasks, reducing manual effort and associated human errors (Weissler et al. 2021).

AI methods have been used to detect asymptomatic cases in clinical trials to reduce patient heterogeneity. Real-world data (RWD) is increasingly mined for prognostic and predictive enrichment to enhance trial design, particularly in oncology. Combining genomic data with clinical outcomes can uncover biomarkers for personalized drug development.

Predicting clinical outcomes is central to precision medicine, informing clinical trial design and phase transitions. ML-based analytics, coupled with in silico simulations, have been broadly explored to predict patient outcomes and trial success.

For rare diseases and many oncology trials, conducting RCTs is often infeasible and unethical. Drug safety and efficacy are evaluated through single-arm open-label studies. Additional evidence from observational studies is increasingly necessary. A synthetic control arm, based on historical or contemporaneous populations, can serve as a comparator. This comparative evidence supports marketing approval and health technology assessments (Li et al. 2021).

Pragmatic clinical trials (PCTs), with less constrained designs and broader populations, bridge RWD and RCT evidence. PCTs maintain patient randomization principles in controlled settings with inclusive populations, not strictly adhering to protocols. Properly designed PCTs inform regulatory and payer decisions. Examples like the Salford Lung Study (SLS) demonstrate successful PCT implementation (Vestbo et al. 2016; Woodcock et al. 2017). The SLS assessed the effectiveness and safety of fluticasone furoate in COPD patients using EHR data. Accepted outcomes fulfill post-approval commitments. Real-world evidence (RWE) in chronic myeloid leukemia (CML) aids early treatment milestones and patient perspectives, enhancing treatment optimization. RWD and RWE contribute to clinical decisions and healthcare insights.

8.5.2 Regulatory Developments

AI is set to transform not only clinical development but also other stakeholders, including regulatory agencies. This progress presents a unique opportunity to revamp the current regulatory framework governing how new medicines are evaluated and approved for marketing. Working closely with various stakeholders, regulatory authorities have begun embracing the potential benefits of AI, advanced analytics, and data from diverse sources in key decision-making processes, fostering the use of AI-enabled solutions to accelerate clinical development and regulatory approval processes.

8.5.2.1 FDA Perspective

In 2023, the FDA, through its CDER, CBER, CDRH, and DHCoE, published a discussion paper titled "Using Artificial Intelligence & Machine Learning in the Development of Drug & Biological Products" to initiate dialogue with stakeholders (FDA 2023e). By facilitating this discussion, the FDA aims to enhance public health protection and foster innovation. Rapid technological advancements in data collection and computing offer transformative potential in drug development, accompanied by unique opportunities and challenges. Collaboration with domestic and international partners is essential to harnessing these innovations for public benefit. Stakeholders, including developers, regulators, and academics, are exploring AI and ML integration throughout the drug development process. This paper serves as an initial step to promote mutual learning and engagement, emphasizing the broad scope of AI/ML applications. While not regulatory guidance, it encourages feedback to inform future regulatory considerations. This inclusive approach seeks to familiarize stakeholders with these technologies and solicit input on their opportunities and challenges, ultimately shaping the regulatory landscape in drug development and medical devices.

Prior to this publication, the FDA Digital Health Center of Excellence (DHCoE) provided guidance on the development and certification of SaMD, including a pre-certification pilot program for software (FDA 2017b, 2021a). These guidelines are applicable to AI and ML solutions utilized within the context of clinical trials if classified as SaMD. Additionally, the FDA has published Good Machine Learning Practices for Medical Devices to address challenges such as dataset representativeness and ongoing model assessment.

In December 2018, the FDA released a new strategic framework to advance the use of RWE to support the development of drugs and biologics (FDA 2018a, 2018b), which was followed by the publication of several guidelines related to RWE (FDA 2019b, 2019d). In January 2025, FDA published draft guidance on the use of AI to produce information or data intended to support regulatory decision-making regarding safety, effectiveness, or quality for drugs. The guidance outlines a risk-based framework for AI model development and validation (FDA 2025).

8.5.2.2 EU Regulatory Development

Similar to the FDA, initiatives surrounding AI are underway across various levels in the European Union (EU). The European Commission (EC) is spearheading the development of a comprehensive framework and governance model for AI, emphasizing excellence and trust as foundational principles (EC 2020, 2021). Key priorities within this initiative include the establishment of robust infrastructure, cultivation of expertise, and adherence to fundamental principles such as accuracy, supervision, security, privacy, transparency, diversity, and non-discrimination. These principles serve to safeguard ethical considerations and uphold EU fundamental rights.

At the forefront of EU regulatory efforts is the proposal for the Artificial Intelligence Act, aimed at regulating high-risk AI applications, including Software as a Medical Device (SaMD). This proposal entails conformity assessment procedures covering risk management, data governance, automatic record-keeping, human interface design, and cybersecurity requirements. However, the implementation of this regulation poses challenges, particularly in the creation of an EU-shared database and ensuring full access to datasets in both pre- and post-marketing phases.

Historically, The European Medicines Agency (EMA) has promoted the use of RWD to complement and enhance evidence collected in RCTs. In recent years, it has launched several initiatives regarding applications of RWD and AI. For example, in the document titled "EMA Regulatory Science to 2025 Strategic Reflection," it outlines the strategy for RWD and AI. Stakeholders within the pharmaceutical sector are prioritizing the development of regulatory frameworks and guidelines for AI validation and assessment, fostering collaboration with academia and expert centers. While harmonized guidance from EMA is pending, several Member States have taken national initiatives.

For instance, the Italian Regulatory Agency (AIFA) has issued guidelines addressing the incorporation of AI and ML in clinical trials, emphasizing the safety of participants and the validity of collected clinical data.

8.5.2.3 Global Development

On a global scale, the International Coalition of Medicines Regulatory Authorities (ICMRA) has recognized AI as one of the top innovation topics challenging current regulation (ICMRA 2021). Concerns include the acceptance of AI-generated data and the evolution of algorithms, necessitating prior scientific advice. Sponsors planning to integrate AI tools into clinical trials are encouraged to disclose algorithms for evaluation and demonstrate advantages over traditional methods.

Furthermore, challenges highlighted by the ICMRA encompass trial implementation complexities, including informed consent procedures and the usability of AI tools by participants and investigators. Building AI expertise within regulatory agencies, ethical committees, and data and safety monitoring boards is deemed essential to support evaluations effectively.

Lastly, the World Health Organization (WHO) underscores ethical use and governance as primary concerns regarding AI's healthcare applications (WHO 2020). Ethical challenges in the context of clinical trials revolve around the management of confidential data, assurance of inclusiveness to mitigate bias and inequality, and access and control over algorithms governed by private commercial interests. WHO recommends adherence to established principles in constructing regulatory frameworks and healthcare practices that integrate AI seamlessly.

8.5.3 Applications of AI and RWD in FDA Filings

Recent regulatory developments in policies and guidance governing the utilization of big data, AI, and ML in drug research and development have ushered in unprecedented opportunities for sponsors. AI is increasingly being utilized in regulatory filings. As delineated in a comprehensive article authored by FDA reviewers (Liu et al. 2022), the Center for Drug Evaluation and Research (CDER) has encountered and assessed a multitude of submissions incorporating various ML components from 2016 to 2021 (Figure 8.6).

The authors showed that there has been a steady increase in submissions that encompass AI/ML components between 2016 and 2021. Of note is a sharp increase in 2017 with a total of 132 submissions (Figure 8.6A). Figure 8.6 B breaks down the total number of submissions by development stage. It is evident that most of the AI/ML applications are seen in clinical development with a total number of filings of 140. The distributions of these submissions are shown in Figure 8.6 C. Notably, there is broader adoption of AI/ML tools in the oncology area followed by psychiatry, gastroenterology, neurology and so on.

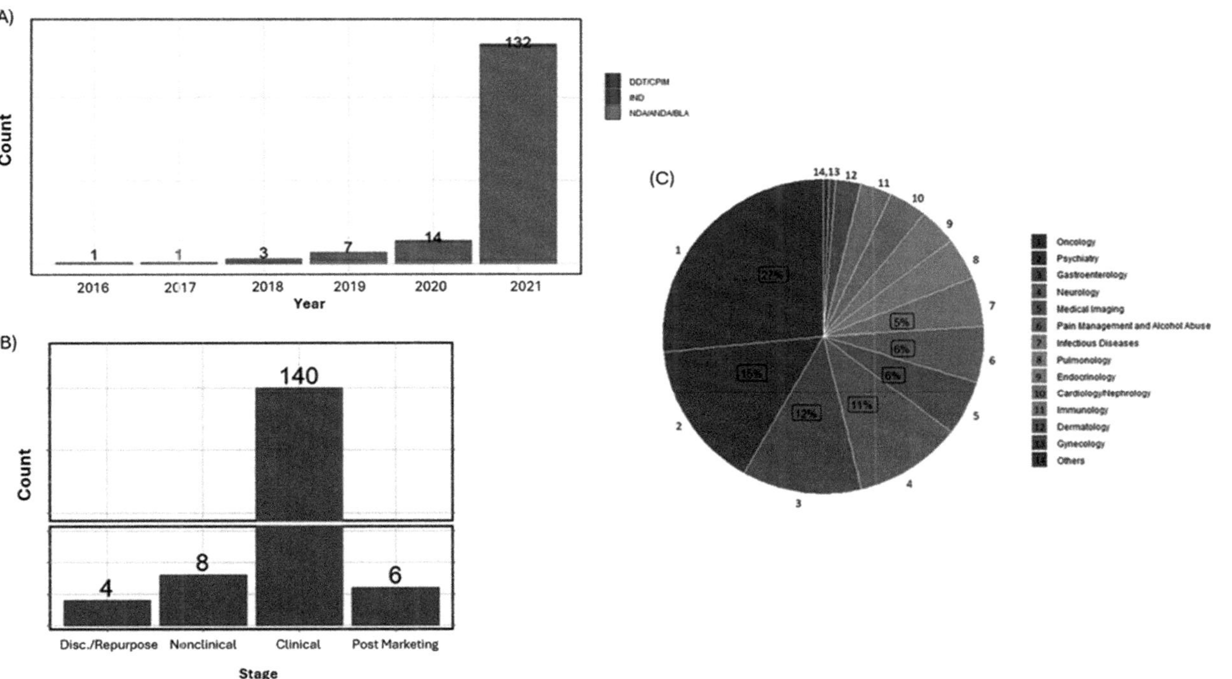

FIGURE 8.6

Regulatory submissions between 2016 and 2021 to the Center for Drug Evaluation and Research (CDER) at the FDA that included artificial intelligence or machine learning components. (A) Distribution of applications by year; (B) distribution of applications by development stage; and (C) Distribution of applications by therapeutic area. IND = Investigational New Drug applications (IND); NDA = New Drug Applications (NDA); ANDA = Abbreviated New Drug Application (ANDAs); BLA = Biologic License Applications (BLA); CIPM = Critical Path Innovation Meeting (CPIM); and DDT = Drug Development Tools. (Adapted from Liu et al. 2022.)

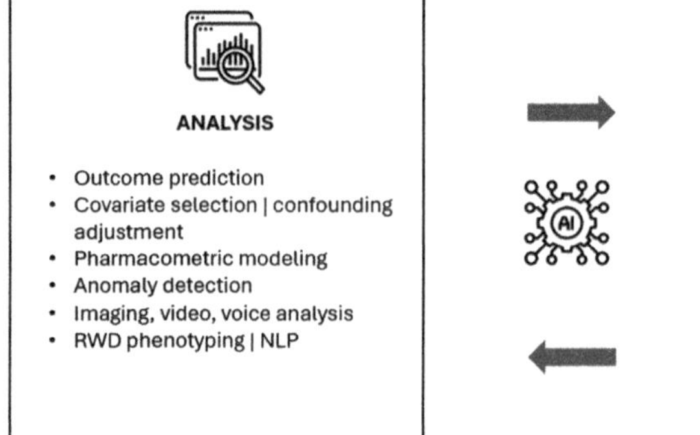

FIGURE 8.7
AI/ML-assisted analyses supporting various objectives in regulatory submissions. (Adapted from Liu et al. 2022.)

These submissions encompass a diverse array of AI/ML-assisted analyses for various purposes as shown in Figure 8.7. These examples underscore the burgeoning role of ML and AI in augmenting traditional approaches to clinical evidence generation and regulatory decision-making. Such innovations not only enhance the efficiency and accuracy of drug development processes but also broaden the scope of regulatory considerations in evaluating novel therapeutic interventions.

8.6 Case Studies

In this section, we present four examples of innovative approaches employed by the FDA in regulatory review, approval, work planning, and guidance development. The first concerns borrowing information from adult trials to generate evidence in support of regulatory approval for a pediatric indication of systemic lupus erythematosus (SLE). The second pertains to the development of an alternative patient selection criterion based on ML. The third is to forecast generic applications to assist strategic workload and research at the FDA. The last example is regarding the use of a large language model for developing product specific guidance.

8.6.1 sNDA for Belimumab in Pediatric Patients

8.6.1.1 Background

In 2018, the FDA received a supplemental New Drug Application (NDA) for belimumab, a B-lymphocyte stimulator (BLyS)-specific inhibitor, aimed at treating pediatric patients aged 5 to 17 with active, auto-antibody-positive SLE who were on standard therapy. Given the rarity of this disease in children, conducting a large-scale pediatric study posed significant challenges. Despite this, a study over five years successfully enrolled and randomized 93 participants.

The study's primary goal was to determine the proportion of patients who met the SLE Responder Index (SRI) Response criteria at Week 52. According to the predefined primary analysis in Table 8.5, there was no statistically significant difference between belimumab and placebo, as the 95% confidence interval of the odds ratio included the value of 1.

Faced with the challenge of conducting another study in the pediatric population, the review team explored reanalysis of the primary endpoint using an informative Bayesian method.

8.6.1.2 Bayesian Borrowing

As a perquisite of this reanalysis, a prior distribution needed to be determined. The relevant data in this case were the treatment arms from two adult SLE studies that compared two dose levels (1 mg/kg and 10 mg/kg) against a placebo. The study team believed that the disease characteristics and patient responses would be similar between adults and children, given the analogous underlying pathophysiology and management of SLE in both groups. Additionally, BLyS, the target of belimumab, plays a similar role in both populations, and systemic belimumab exposures were comparable. The results at 10 mg/kg from the adult studies are detailed in Table 8.6.

The odds ratio estimates for these two studies were synthesized by the FDA team, based on a random effects model, resulting in a single probability distribution of treatment effect, $\theta_1 \sim N(0.51,\ 0.16)$. This distribution was then combined with a vague non-informative prior $\theta_2 \sim N(0.51,\ 0.16)$ such that the treatment effect is given by

$$\theta_0 \sim (1-w)N(0,\ 8.27) + w\,N(0.51, 0.16),$$

TABLE 8.5

Primary Efficacy Analysis of SRI Response Rate at Week 52 from Pediatric Study

	Placebo $N = 40$	**Belimumab** $N = 53$
Response, % (n)	44% (17)	53% (28)
Observed difference	–	9.20%
Odds ratio (95% CI)	–	1.5 (0.6, 3.5)

Source: Ionan et al. (2023).

Note: One participant in the placebo arm did not have baseline SELENA SLEDAI assessment and, therefore, did not contribute to SRI analyses.

TABLE 8.6

Primary Efficacy Analysis of SRI Response Rate at Week 52 from Two Adult Studies

	Adult Study 1		Adult Study 2	
	Placebo $N = 275$	**Belimumab** $N = 237$	**Placebo** $N = 287$	**Belimumab** $N = 290$
Response, % (n)	34% (93)	43% (118)	44% (125)	58% (167)
Observed difference	–	9%	–	14%
Odds ratio (95% CI)	–	1.5 (1.1, 2.1)	–	1.8.(1.3, 2.6)

Source: Ionan et al. (2023).

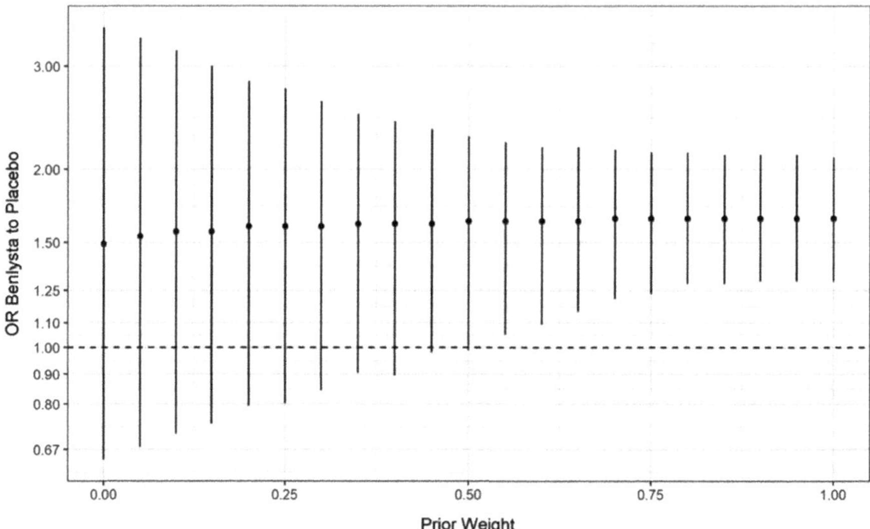

FIGURE 8.8
Posterior mean (points) and 95% credibility intervals (lines) of the odds ratio of SRI response. (Source: Ionan et al. 2023.)

where the vague non-informative prior was assumed to be centered at zero, reflecting no effect, with a single patient's worth of information per arm (Travis et al. 2023).

The parameter w is relative weight allotted to each distribution, with $w = 0$ and 1 corresponding to an independent analysis of the pediatric trial and a pooled analysis of the pediatric and adult studies, respectively. The Bayesian analysis was carried out for a range of weights from no borrowing to full borrowing.

8.6.1.3 Results

The results of the Bayesian analysis are shown in Figure 8.8. The Bayesian analysis revealed that a prior distribution weight of at least 0.55 resulted in posterior probabilities of positive treatment effects exceeding 97.5%. The review team deemed this weight of 0.55 reasonable, thus providing support, alongside additional evidence, for belimumab's approval in the pediatric SLE population.

8.6.2 Emergency Use Authorization for Kineret (Anakinra)

8.6.2.1 Background

On February 4, 2020, the Secretary of Health and Human Services determined, pursuant to section 564 of the Federal Food, Drug, and Cosmetic (FD&C) Act, that a public health emergency existed due to the outbreak of a novel coronavirus, SARS-CoV-2, which causes COVID-19. Following this determination, on March 27, 2020, the Secretary declared that circumstances existed justifying the authorization of emergency use of drugs and biologics during the COVID-19 outbreak, as per section 564 of the FD&C Act, subject to the terms of any authorization issued under that section.

Subsequently, on May 10, 2022, the FDA approved Olumiant (baricitinib) for the treatment of COVID-19 in hospitalized adults with pneumonia requiring supplemental oxygen, non-invasive or invasive mechanical ventilation, or extracorporeal membrane oxygenation (ECMO). Three months later, on August 24, 2022, the FDA approved a supplemental new drug application (sNDA) for Veklury® (remdesivir) to treat COVID-19 in people with mild, moderate, and severe hepatic impairment who are at risk of progressing to severe respiratory failure (SRF). Both approvals were under the Emergency Use Authorization (EUA).

Kineret (Anakinra) is an interleukin-1 receptor antagonist initially approved for the treatment of rheumatoid arthritis and other inflammatory conditions. During the COVID-19 pandemic, the manufacturer submitted a Biologics License Application (BLA) to the FDA, requesting an EUA for Kineret in the treatment of hospitalized adults who tested positive for SARS-CoV-2. Anakinra was evaluated in an open-label, single-arm study named SAVE and a Phase III randomized controlled study named SAVE-MORE, with the latter being the primary source of evidence of effectiveness for the EUA submission.

Patients in the SAVE and SAVE-MORE trials were selected based on a biomarker cut-off, suPAR $\geq$ 6 ng/mL, to enrich the trials with patients at risk for progressing to severe respiratory failure. At this time, the suPAR test is not commercially available in the United States. Given this enrichment strategy and the inconsistent results in other studies involving a broader population, it is unclear if the results seen in the SAVE-MORE trial can be generalized to the broader scope of the EUA as proposed by the sponsor (i.e., a population defined regardless of suPAR level).

To identify a comparable population to that in the SAVE-MORE trial, the FDA review team worked with the sponsor to explore ways to select patients most likely to have suPAR ≥ 6 ng/mL based on commonly measured patient characteristics. Two ML methods were developed to effectively predict whether a patient had suPAR ≥ 6 ng/mL based on baseline characteristics. To operationalize this method in clinical practice, an alternate scoring rule, named "SCORE 2," was developed and recommended in the fact sheet for healthcare providers: Emergency Use Authorization for Kineret (FDA 2022b).

8.6.2.2 Clinical Studies

Anakinra was evaluated in the studies SAVE and SAVE-MORE for the treatment of COVID-19, which are briefly described in the following.

SAVE Study: The SAVE study was an open-label, single-arm trial conducted to assess the safety and efficacy of anakinra in combination with the standard of care (SoC) in patients aged 18 or above who were hospitalized with a confirmed SARS-CoV-2 infection, lower respiratory tract infection, and plasma suPAR ≥ 6 ng/mL. The blood protein suPAR (soluble urokinase plasminogen activator receptor) was used in patient selection, as previous investigations suggested that measuring suPAR can identify pneumonia patients at the highest risk for developing respiratory failure. The primary efficacy endpoint was the incidence of severe respiratory failure (SRF) by day 14. SRF was defined as any decrease of pO_2/FiO_2 below 150 necessitating mechanical ventilation (MV) or non-invasive ventilation (NIV).

The results of the first 130 participants from the SAVE trial were published, comparing anakinra-treated patients to propensity-matched controls on SoC (Kyriazopoulou et al. 2021). Key findings include: (1) 22.3% of anakinra-treated patients and 59.2% of comparators (hazard ratio, 0.30; 95% CI, 0.20–0.46) progressed to SRF; (2) 30-day mortality was 11.5% for anakinra-treated patients and 22.3% for comparators (hazard ratio, 0.49; 95% CI, 0.25–0.97); and (3) Anakinra treatment was associated with a decrease in circulating interleukin (IL) sCD163 and sIL2-R; the IL-10/IL-6 ratio on day 7 was inversely associated with SOFA score; patients were allocated to less severe WHO-CPS strata.

The SAVE-MORE study was a multi-center, prospective, randomized, double-blinded, placebo-controlled Phase III trial. It compared 100 mg of anakinra in combination with SoC to a placebo with SoC. Study visits included days 1–10, 14, 28, 60, and 90, with WHO Clinical Progression Ordinal Scale (CPS) status collected at these time points (Table 8.7). The study was conducted in hospitalized adults (≥ 18 years) with confirmed SARS-CoV-2 infection by real-time PCR of nasopharyngeal secretions, radiological findings compatible with lower respiratory tract infection, and suPAR level ≥ 6 ng/mL.

TABLE 8.7

WHO Clinical Progression Scale

Patient Stage	Descriptor	Score
Uninfected	Uninfected; no viral RND detected	0
Ambulatory mild disease	Asymptomatic; viral RND detected	1
	Symptomatic; independent	2
	Symptomatic; assistance needed	3
Hospitalized; moderate disease	Hospitalized; no oxygen needed	4
	Hospitalized; oxygen by mask or nasal progs	5
Hospitalized; severe disease	Hospitalized; oxygen by NIV or high flow	6
	Intubation and mechanical ventilation $pO_2/FiO_2 \geq 150$	7
	Mechanical ventilation $pO_2/FiO_2 < 150$ ($SpO_2/FiO_2 < 200$)	8
	Mechanical ventilation $pO_2/FiO_2 < 150$ and vasopressors, dialysis	9
Dead	Dead	10

Note: FiO_2 = Fraction of inspired oxygen; NIV = Noninvasive ventilation; pO_2 = Arterial oxygen pressure; RNA = Ribonucleic acid; SpO_2: Oxygen saturation; WHO: World Health Organization.

8.6.2.3 Results

The primary endpoint was the distribution of the 11-point WHO-CPS at Day 28. Two different analyses, pre-specified in the statistical analysis plan (SAP), were performed for the primary endpoint: univariate (no covariate adjustment) and multivariate proportional odds models. These models estimated the odds ratio (OR) for higher WHO-CPS (disease severity) comparing anakinra to placebo. The multivariate model adjusted for variables collected at the time of screening and used in stratified randomization: dexamethasone use, moderate/severe disease, BMI, and country. For binary endpoints, univariate and multivariate logistic regression models estimated the odds ratio (OR) comparing anakinra to placebo.

Several secondary and supportive endpoints, including mortality by Day 28, were also evaluated. The results of the primary endpoint and mortality by Day 28 are presented in Table 8.8.

Although the results had several limitations, such as a lack of multiplicity control across secondary endpoints and multiple analyses for the same endpoints, the FDA team concluded that the efficacy results from the SAVE-MORE trial provided evidence that anakinra may be effective in the high-suPAR selected population.

Safety analysis was also carried out using data from SAVE-MORE, SAVE, and post-marketing safety data from Europe. The overall incidence of treatment-emergent adverse events (TEAEs) was similar in patients in the anakinra + SoC arm (343 patients, 84.7%) compared to those in the placebo + SoC arm (161 patients, 85.2%). The proportion of patients with

TABLE 8.8
Summary of Efficacy Results

Endpoint	Placebo N (%)	Anakinra N (%)	Risk Difference (95 CI)	Univariate Analysis	Multivariate Analysis
				OR (95% CI)	OR (95% CI)
WHO-CPS at Day 28	$N = 189$	$N = 405$	–	0.36 (0.26, 0.49)	0.37 (0.26, 0.50)
				HR (95% CI)	HR (95% CI)
Mortality by Day 28	13 (6.9%)	13 (3.2%)	−3.7% (−7.7%, 0.3%)	0.45 (0.21, 0.98)	0.46 (0.22, 1.04)
Mortality by Day 60	18 (9.7%)	21 (5.3%)	−4.4% (−9.2%, 0.4%)	0.52 (0.28, 0.98)	0.56 (0.30, 1.04)
SRF by Day 28	62 (32.8%)	88 (21.2%)	−11.6% (−19.4%, −3.8%)	0.61 (0.44, 0.85)	0.66 (0.48, 0.92)

Source: FDA Anakinra EUA review (FDA 2022b).

treatment-related TEAEs, serious TEAEs, TEAEs leading to study drug discontinuation, and TEAEs leading to death was slightly lower in the anakinra + SoC arm.

Based on the efficacy and safety assessment, anakinra was granted an EUA for the treatment of COVID-19 in hospitalized adults with positive results of direct SARS-CoV-2 viral testing, pneumonia requiring supplemental oxygen (low or high-flow oxygen), who are at risk of progression to severe respiratory failure and are likely to have an elevated suPAR.

8.6.2.4 Patient Selection

As discussed in Section 8.6.2.1, the review team collaborated with the Sponsor to explore methods of identifying patients most likely to exhibit suPAR ≥ 6 ng/mL based on commonly measured patient characteristics. On May 11, 2022, the FDA review team issued an Information Request (IR) to the Sponsor, strongly urging them to consider developing suPAR as a companion diagnostic tool to pinpoint patients who would benefit most from anakinra treatment. Additionally, the IR advised exploring alternative combinations of clinical characteristics and laboratory tests to reliably identify the population with suPAR ≥ 6 ng/mL if developing suPAR as a companion diagnostic was not feasible.

8.6.2.4.1 Machine Learning Algorithm-Based Methods

In response to this Information Request (IR), the Sponsor analyzed various combinations of baseline variables that might predict suPAR ≥ 6 ng/mL using data from the SAVE-MORE trial. They proposed a potential scoring rule, dubbed "SURROGATE," which incorporated factors such as age (≥75 years), severe pneumonia based on WHO criteria, current or previous smoking status, Sequential Organ Failure Assessment (SOFA) score ≥3, hemoglobin ≤10.5 g/dL, blood urea ≥50 mg/dL, a medical history of renal disease, and a medical history of ischemic stroke. According to the Sponsor, patients meeting at least two of these criteria would be deemed positive for the SURROGATE score and highly likely to have suPAR ≥ 6 ng/mL.

The FDA review team thoroughly assessed the Sponsor's proposal for the SURROGATE scoring system. Additional analyses were conducted using data from the SAVE-MORE trial to potentially refine the scoring rule. Two distinct AI/ML algorithms, namely artificial neural network and elastic net regression, were employed to predict whether a patient exhibited suPAR levels ≥ 6 ng/mL based on their baseline characteristics (for detailed analyses, refer to Appendix XXVI). Both algorithms yielded consistent results, indicating the inclusion of an additional criterion: "Neutrophil-to-Lymphocyte Ratio (NLR) ≥7." This biomarker is indicative of inflammation and immune response.

From a biological standpoint, the addition of this NLR-based criterion is rational, considering suPAR's association with inflammation and the immune system, while none of the components of the initial SURROGATE score directly relate to these factors. Moreover, this addition aligns logically with the mechanism of action of anakinra, which pertains to immune modulation. The AI/ML algorithms determined the NLR cutoff of 7, a value supported by several literature papers suggesting its relevance in predicting severity/mortality risk in COVID-19 cases (Li et al. 2020). This provides biological plausibility for incorporating this criterion.

8.6.2.4.2 Rule-Based Approach

It important to note that while a continuous variable might have provided enhanced predictive performance, for the purpose of this scoring system, a dichotomized approach was deemed preferable to facilitate its application by physicians in identifying a similar patient population to those studied.

Consequently, the FDA review team devised an alternative scoring rule, termed "SCORE 2," to identify patients with suPAR levels ≥ 6 ng/mL. SCORE 2 includes the following considerations:

1. Age ≥75 years
2. Severe pneumonia by WHO criteria
3. Current or previous smoking status
4. SOFA score ≥3
5. NLR ≥7
6. Hemoglobin ≤10.5 g/dL
7. Medical history of ischemic stroke
8. Blood urea ≥50 mg/dL and/or medical history of renal disease

Patients meeting at least three of these eight criteria were considered positive for SCORE 2 and highly likely to have suPAR ≥ 6 ng/mL. Compared to SURROGATE, SCORE 2 had higher positive predictive values and higher specificity (lower false positive rate), although it had lower sensitivity.

Based on the biological plausibility and predictive performance, SCORE 2 was proposed to the Sponsor by the FDA review team for validation. The Sponsor concurred with the FDA review team on the development of SCORE 2 and the addition of the criterion based on NLR (a biomarker for inflammation and immune status), considering the nature of suPAR and the mechanism of the drug (modulating the immune system).

8.6.2.5 Performance Evaluation

The SCORE 2 rule was trained on data from the SAVE-MORE study and externally validated using data from the SAVE trial. Table 8.9 summarizes the predictive performance of SCORE 2 in both the training dataset (SAVE-MORE trial) and the external validation dataset (SAVE trial). SCORE 2 demonstrated good performance in both datasets, showing high positive predictive value, high specificity, and a low false positive rate. This suggests that patients with suPAR <6 ng/mL are very unlikely to be incorrectly selected by SCORE 2 and given anakinra treatment, thereby avoiding the risk of missing the opportunity to be treated with other medications.

As a trade-off, SCORE 2 has relatively low sensitivity, meaning some patients with suPAR ≥ 6 ng/mL will not be identified by SCORE 2. Despite this limitation, including information on SCORE 2 in the fact sheet may help guide physicians in identifying patients similar to those studied in SAVE-MORE.

As suPAR was intended to identify patients at risk for progressing to severe respiratory failure (SRF), the FDA review team used data from the SAVE-MORE trial to conduct an exploratory analysis evaluating whether SCORE 2 can help identify such patients. The review team compared the prognosis for score-positive and score-negative subpopulations without anakinra treatment. For this analysis, they studied patients receiving only standard-of-care treatment (with or without placebo) during the same period. This analysis included 421 patients who were screened for eligibility for the SAVE-MORE trial and for whom the 14-day outcome was known. These patients either had suPAR < 6 ng/mL and failed screening (n = 240) or had suPAR ≥ 6 ng/mL and were enrolled in the SAVE-MORE trial and allocated to SoC + placebo treatment (n = 181). Figure 8.9 shows the time to progression to SRF by Day 14 for patients who are positive or negative for SCORE 2. Despite the limitations of this exploratory analysis, it appears that SCORE 2 is able to identify patients at risk for progression to SRF.

TABLE 8.9

Summary of Performance of SCORE 2 to Predict suPAR Levels of 6 ng/ml or Higher

Training Dataset (SAVE-MORE Trial)				External Validation Dataset (SAVE Trial)			
	suPAR ≥ 6	**suPAR < 6**	**Total**		**suPAR ≥ 6**	**suPAR < 6**	**Total**
SCORE 2 (+)	231 (PPV = 0.95, sensitivity = 0.41)	12 (FPR = 0.04)	243	SCORE 2 (+)	95 (PPV = 0.94, sensitivity = 0.37)	6 (FPR = 0.07)	101
SCORE 2 (-)	338	256 (NPV = 0.43, specificity = 0.96)	243	SCORE 2 (–)	159	76 (NPV = 0.32, specificity = 0.93)	235

Source: FDA Anakinra EUA review (FDA 2022b).
Note: PPV: Positive predictive value; NPV: Negative predictive value; FPR: False positive rate.

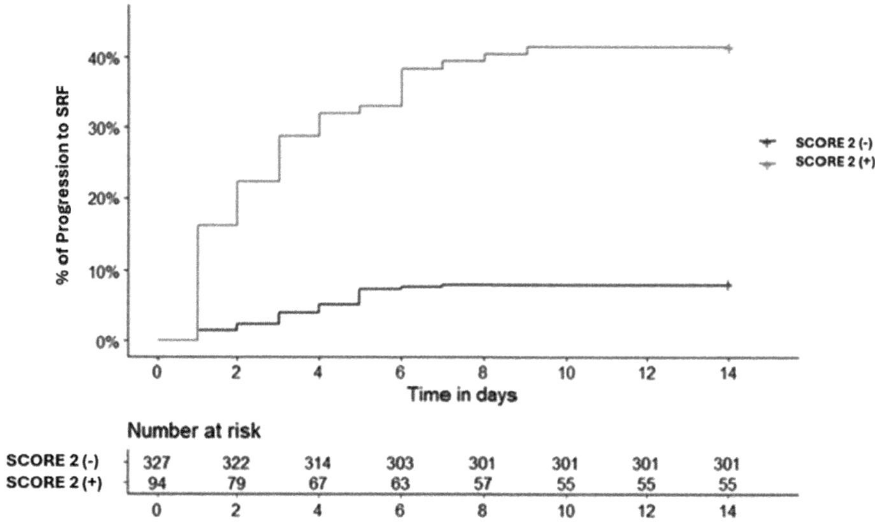

FIGURE 8.9
Probability of progression to severe respiratory failure. The curves on the top and the bottom correspond to Score 2 (−) and Score 2 (+) groups, respectively. (Source: FDA Anakinra EUA review, FDA 2022b.)

8.6.3 Forecasting Abbreviated New Drug Application

8.6.3.1 Background

Generic drug products undergo approval by the U.S. Food and Drug Administration (FDA) through the submission of Abbreviated New Drug Applications (ANDAs). The review and approval process for ANDAs involves coordination among various offices within the FDA. Accurately forecasting the timing of ANDA submissions is crucial for effective resource allocation and workload management within the agency. A group of researchers at the FDA employed ML methodologies to predict the time to the first ANDA submission referencing new chemical entities (NCEs) subsequent to their earliest lawful ANDA submission dates (Hu et al. 2019). The predictive modeling approach based on supervised learning utilized a combination of drug product information, regulatory factors, and pharmacoeconomic factors as input variables. A random survival forest ML method was developed and its performance compared to the conventional Cox regression model. The results indicated that the ML method exhibited superior predictive performance compared to the conventional Cox regression model. This conclusion was robustly supported by thorough internal and external validation procedures. The findings by Hu et al. (2019) suggest that the ML approach holds promise as an effective forecasting tool for strategic workload management and research planning concerning generic drug applications.

8.6.3.2 Data

To build the predictive models for the time to the first ANDA submission for NCEs, drug product information, regulatory factors, and pharmacoeconomic information were collected and curated from various sources. These variables are presented in Table 8.10 along with associated data sources.

8.6.3.2.1 Drug Information

Notably, several variables derived from the GDUFA II commitment letter were created to describe drug product information, reflecting the difficulty in developing a generic drug product. These variables were created by FDA subject matter experts. For example, the variable "Complex API" indicates if the drug product API includes peptides, heterogeneous mixtures of small molecules, natural and synthetic polymers, and/or macromolecular complexes. The variable "Complex Dosage Form" reflects whether the drug product assumes a nonoral complex formulation/dosage form, often with two or more discrete states of matter within the formulation. The variable "Complex Delivery Route" indicates a nonsolution drug product with a nonsystemic site of action (e.g., topical, ophthalmic, local gastrointestinal action). The variable "Complex Drug-Device Combination" refers to products with two constituents, a drug constituent preloaded in a product-specific device constituent or specifically cross-labeled for use with a specific device, where the device design affects drug delivery to the site of action and/or absorption. The variable "Abuse Deterrent Formulation" refers to solid oral opioid drug products with abuse deterrence properties in their FDA-approved labeling.

All variables in the "Drug Product Information" category, along with the variable "Risk Evaluation and Mitigation Strategy" (REMS), were directly used as predictive variables. REMS is a safety strategy to manage a known or potential serious risk associated with a medicine and to enable patients to have continued access to such medicines by managing their safe use. The FDA may require a REMS as part of the approval of a new product or for an approved product when new safety information arises.

8.6.3.2.2 Regulatory Factors

The regulatory information includes several critical regulatory dates: "NDA Approval Date," "NCE Exclusivity Expiration Date," "Patent Expiration Date," "First ANDA Submission Date," and "First Product-Specific Guidance Publishing Date." The ELASD is defined as 4 years from the approval date of NCE. In contrast, the ELASD for the "First Product-Specific Guidance Publishing Date" is used to generate a variable "Has PSG 1-Year before the ELASD," indicating if the corresponding PSG had been published more than 1 year before the ELASD.

The time span between the "Patent Expiration Date" and the ELASD is calculated as a predictive variable, while the time span between the "First

TABLE 8.10

Variables Used for Predictive Analysis for Time to First ANDA Submission

Information Category	Variable	Denotation	Data Type	Data Source
Drug Product Information	Complex API	Whether drug product contains complex active ingredients	1-Yes, 0-No	Internal Classification
	Complex dosage form	Whether drug product uses complex dosage form (e.g., cream and ointment)	1-Yes, 0-No	Internal Classification
	Complex drug-device combination	Whether drug product uses complex route of delivery (e.g., topical cream)	1-Yes, 0-No	Internal Classification
	Abuse deterrent formulation	Whether drug product is the complex drug-device combination	1-Yes, 0-No	Internal Classification
	Oral MR	Whether drug product is the complex drug-device combination	1-Yes, 0-No	Internal Classification
	ATC	Anatomic Therapeutic Chemical Classification	Categorical	Internal Classification
	Acute/chronic disease	Disease conditions: (1) acute; (2) chronic; (3) both	Categorical	IQVIA database
Regulatory Information	NDA approval date	NDA approval date	Date	Red Book IQVIA database
	NCE exclusivity expiration date	NCE exclusivity expiration date	Date	Orange Book
	Patent expiration date	Patent expiration date	Date	Orange Book
	First ANDA submission date	First ANDA submission date	Date since July 2019	FDA internal database
	First PSG publishing date	First PSG publishing date	Date	FDA PSG website
	REMS	First PSG publishing date	1-Yes, 0-No	REMS data files
Pharmacoeconomic Information	Quarterly sales from 2011 to 2017	Quarterly sales from 2011 to 2017	Continuous	IQVIA database

Source: Hu et al. (2019).

Note: ANDA = Abbreviated New Drug Application; API = active pharmaceutical ingredient; MR = modified release; NCE = new chemical entity; NDA = New Drug Application; PSG = product-specific guidance; REMS = risk evaluation and mitigation strategy.

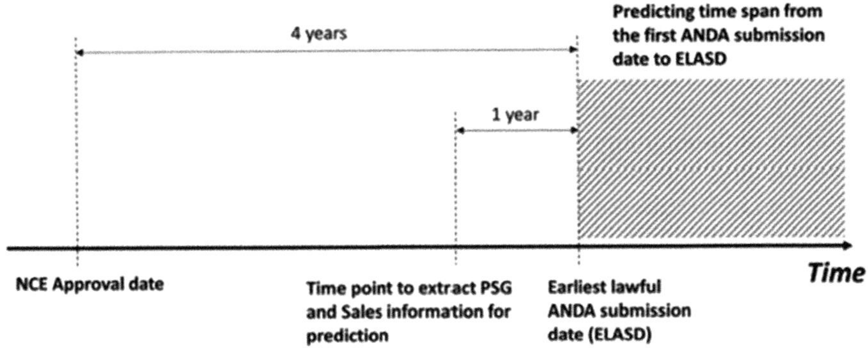

FIGURE 8.10
The data model delineates how the collected variables are utilized to establish a predictive model. The earliest lawful Abbreviated New Drug Application (ANDA) submission date (ELASD) is calculated as four years after the new chemical entity (NCE) approval date. (Adapted from Hu et al. 2019.)

ANDA Submission Date" and the ELASD serves as the dependent variable time in survival analysis. Data censoring occurs when no ANDA has been received by the FDA by the time of analysis. For pharmacoeconomic information, drug sales 1 year before ELASD were retrieved as a predictive variable.

The relative scales for all time-relevant variables are depicted in Figure 8.10.

8.6.3.2.3 Pharmacoeconomic Information

The quarterly sales data from 2011 to 2017 were sourced from an IQVIA database. The data were adjusted for inflation using the 2011 Producer Price Index data.

8.6.3.2.4 Final Dataset

The originally collected dataset contained 887 NCEs. Among those, 156 were retained for model building after applying two selection criteria: (i) only include NCEs whose ELASDs are after July 2009, when ANDA submissions became electronically available and (ii) only include NCEs with available sales data (sales data are available only for 2011–2017).

8.6.3.3 Model Training and Validation

8.6.3.3.1 Random Survival Forest Model

Random survival forest (RSF), a well-known ML algorithm for survival analysis, is a direct methodological extension of the random forest (RF) algorithm. The basic unit of RF is a binary tree, constructed using the classification and regression tree methodology, where binary splits recursively partition the tree into homogeneous or near-homogeneous terminal nodes, known as "leaves."

Given the greedy nature of one-step-at-a-time node splitting in binary trees, RF addresses the overfitting issue (i.e., the inability to generalize to unseen data) by implementing two forms of randomization: (i) creating a collection of binary decision trees (e.g., hundreds of trees) grown independently based on bootstrap samples of the original data and (ii) selecting a random subset of variables as candidates for splitting at each node. This approach allows RF to approximate complex functions while maintaining low generalization error.

In survival analysis, RSF applies the same principles as RF, with two key differences: (i) each split is made by maximizing survival differences between the daughter nodes and (ii) the outcomes of RSF are the ensemble estimates for the cumulative hazard function (CHF).

8.6.3.3.2 Model Training

Building the RSF model involved bootstrap sampling to create a series of survival trees and using the out-of-bag (OOB) technique to fine tune model parameters (Figure 8.11). Specifically, N sets of bootstrap samples are randomly drawn from the original data. For each bootstrap sample, approximately 37% of the data (known as the out-of-bag [OOB] data) are set aside for prediction testing. A survival tree is grown for each of the bootstrap samples. This process encompasses randomly selecting a subset of predictors and each node of the tree, and splitting the node into two daughter nodes, using the candidate variable that maximally differentiates the two daughter nodes. The tree is grown to full size with each terminal node providing a predicted CHF. The final RSF consists of this collection of trained trees. When the trained

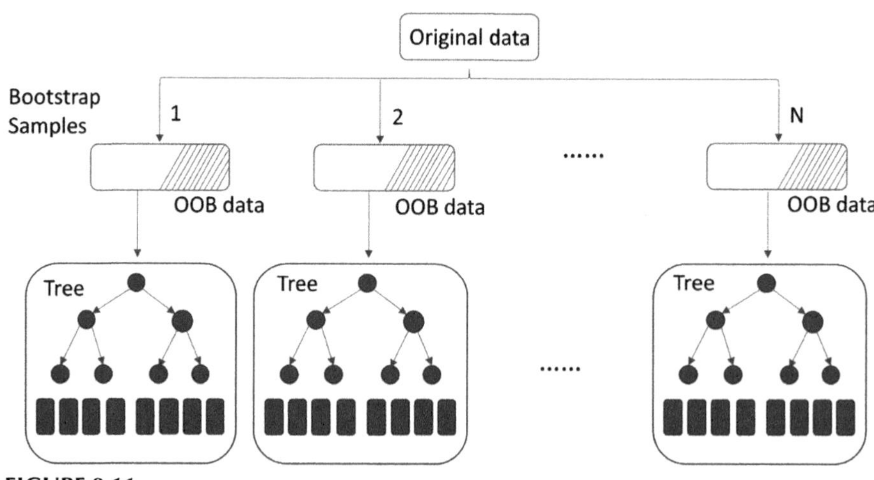

FIGURE 8.11
RSF model training. (Adapted from Hu et al. 2019.)

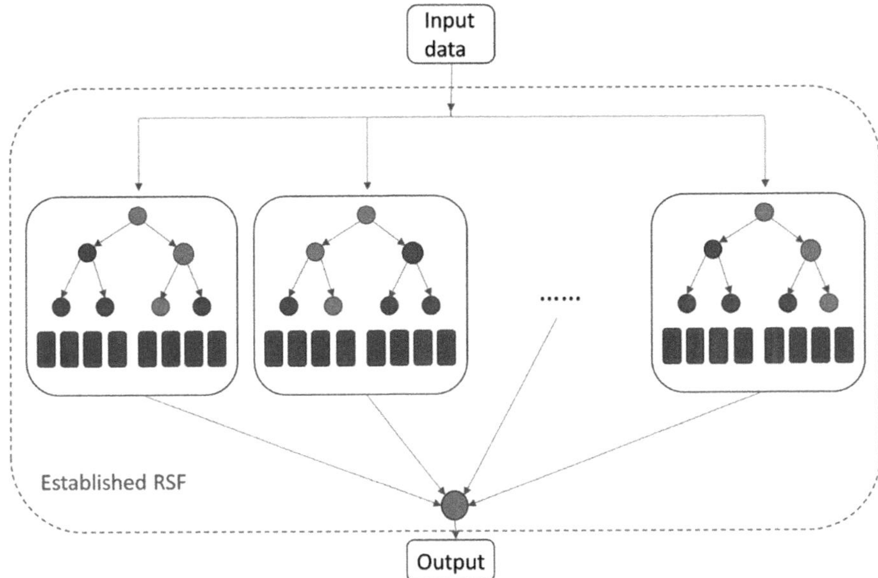

FIGURE 8.12
Prediction by the established RSF model. (Adapted from Hu et al. 2019.)

trees analyze the OOB data, the OOB prediction error, measured in terms of the C-index, can be estimated. This estimation is primarily used for parameter fine-tuning.

As shown in Figure 8.12, with the established RSF model, for each NCE, the associated drug, regulatory and pharmacoeconomic predictive factors will pass through each individual tree to generate predicted CHFs. The RSF then combines these predictions to provide the ensemble CHF as the output.

8.6.3.3.3 Performance Evaluation

Evaluating predictive performance is essential for validating a predictive model's success. This effort requires a suitable metric be selected to assess the survival predictions provided by the model. Because of the continuous nature of the predicted outcomes of the RSF, a concordance measure between the actual and predicted outcomes, known as C-index, was employed to assess the survival predictions of the established model. This index assumes a value between 0 and 1 with 1 corresponding to a perfect prediction.

8.6.3.3.4 Results

Hu et al. (2019) evaluated the performance of the RSF model with the same dataset that was used to train the model and obtained a C-index value of 0.894. Additionally, Hu et al. (2019) employed the Variable Importance (VIMP) algorithm to generate a list of predictive variables ranked by their prediction power (Figure 8.13). The VIMP value of a variable (e.g., Complex Active Pharmaceutical Ingredient, API) signifies the difference in out-of-bag (OOB) prediction errors between the RSF model with and without the inclusion of this predictive variable. The sales before the earliest lawful submission date (ELSD), the availability of product-specific guidance (PSG) before ELSD, and the complexity of API were most influential factors for predicting time to first ANDA submission.

Figure 8.14a depicts Kaplan–Meier (KM) plots stratified by the median value of the variable "Sales 1-Year before ELASD" (US $16.6 million). It illustrates that higher sales 1-year before the ELASD are associated with earlier first ANDA submission, with a p-value < 0.001 based on the log-rank test. KM plots were also generated, stratified by a randomly selected less important

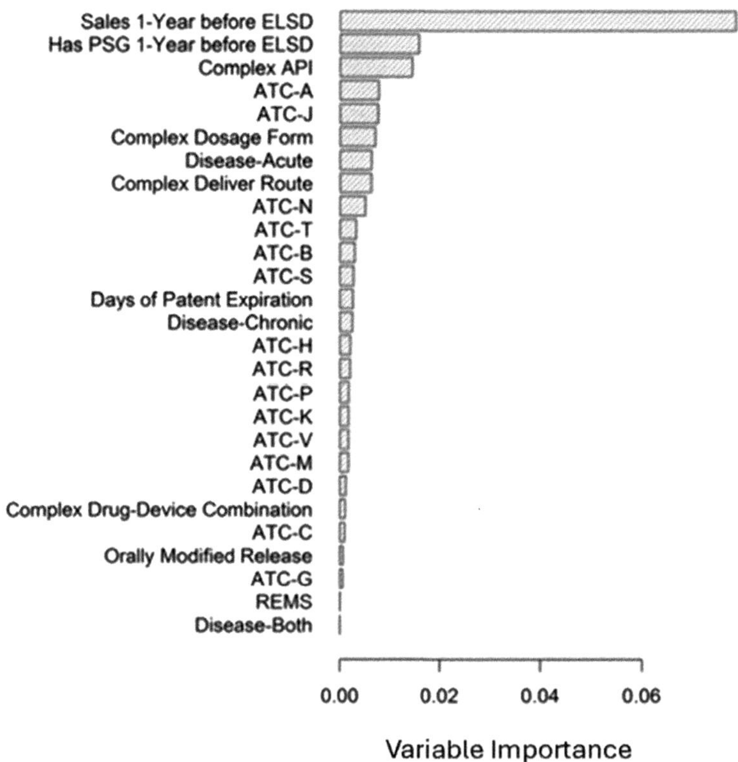

FIGURE 8.13

Variable importance (VIMP) of predictive variables. Bar length indicates the VIMP value of the variable, which represents difference of the out-of-bag prediction errors 17 between the random survival forest model with and without this predictive variable being permuted. API, active pharmaceutical ingredient; ATC, anatomic therapeutic classification; ELSD, earliest lawful submission dates; PSG, product-specific guidance; REMS, risk evaluation and mitigation strategy. (Adapted from Hu et al. 2019.)

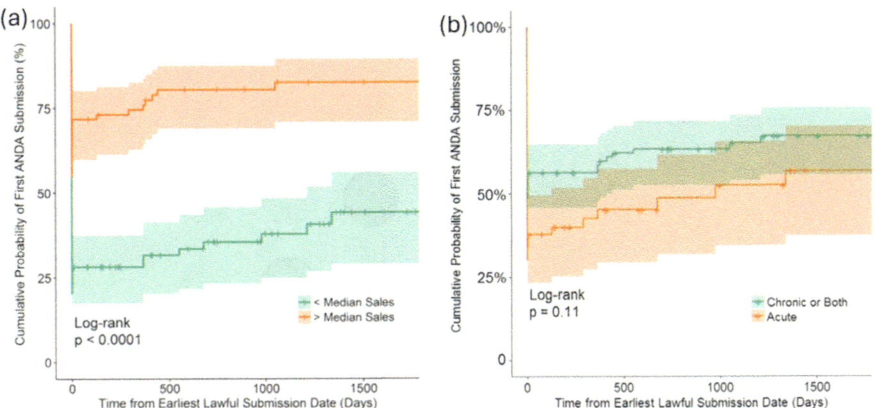

FIGURE 8.14
Kaplan–Meier plots stratified by (a) the median value of sales data (the curves at the top and the bottom correspond to "< Median Sales" and "> Median Sales," respectively), and (b) variable "Acute Disease," a less important variable for prediction (the curves at the top and the bottom correspond to "Not Acute Disease"" and "> Acute Disease," respectively). (Adapted from Hu et al. 2019.)

variable, "Acute Disease" (Figure 8.14b). The log-rank test indicated no significant difference between them.

8.6.3.4 Alternate RSF Models

The RSF model discussed previously uses the same data for both model training and testing. This may impede its generalizability. Recognizing this potential issue with the model, Hu et al. (2019) evaluated two additional methods to build the predictive models.
Leave-One-Out Method

In the Leave-One-Out (LOO) approach, the original dataset was split into two sets, one for training and the other for validation. This was done by taking one sample from the original data and setting it aside as the testing data, while the remaining samples are used to train the RSF model. This process was repeated for each sample in turn, where the chosen sample serves as the testing data. Consequently, survival predictions for all samples were generated. Figure 8.15 presents a hypothetical example for illustration. In the example, there are 5 data records. Each data record is selected as the testing data in one iteration, while the other 4 data records form the training dataset. Using independent datasets for model training and validation enhances the predictive power for new data, thus its generalizability. The new RSF model based on the LOO method achieved a C-index value of 0.785.

The KM survival was used to model the time to the first ANDA submission, with the first ANDA submission as the "event." Figure 8.16 displays the KM curve

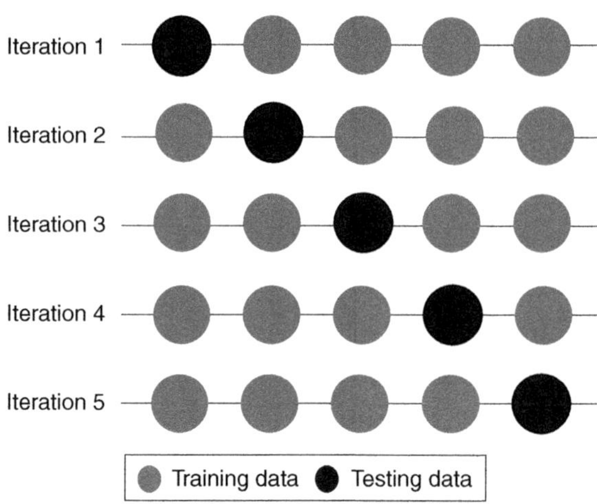

FIGURE 8.15
An example to illustrate the leave-one-out method.

based on all eligible data for model building. The KM survival plot illustrates the cumulative probability of first ANDA submission. Notably, approximately 40% of the NCE drug products have ANDA submissions at their earliest lawful ANDA submission dates (ELASDs), and over 50% of the NCE drug products have the first ANDA submissions within a few days after the ELASD.

Applying the RSF model based on the LOO method, the predicted KM curve was also obtained. Both the actual and predicted survival curves are presented in Figure 8.16 along with their respective 95% confidence intervals (CIs). The 95% CI of the predicted KM curve being entirely contained within that of the actual KM curve, indicating a good overall prediction by the new RSF model. It is also noticeable that the two curves trend well together in the first 1,000 days or approximately 3 years.

Hu et al. (2019) further explored a different method for building the predictive RSF model, by splitting data into a training set consisting of all data up to 2017, and testing set comprising data from 2018. This approach simulates the task of "predicting the future" for an RSF model established using past data. The predictive RSF model was trained on the data before the end of 2017, and subsequently evaluated for predicting the outcomes for data with ELASDs within 2018 ($N = 27$). Notably, variables such as "Sales 1-Year before ELASD" and "PSG Availability 1-Year before ELASD" would be unavailable if predicting outcomes for NCEs approved within a year. The predictive performance of this approach was assessed using the C-index, yielding a value of 0.861, indicating reasonably accurate predictions.

Among the 27 products with ELASDs within 2018, 11 had generic submissions to the FDA during that year. Figure 8.17a illustrates the average predicted cumulative probabilities of ANDA submission for these 11 products with generic submissions and the 16 products without submissions,

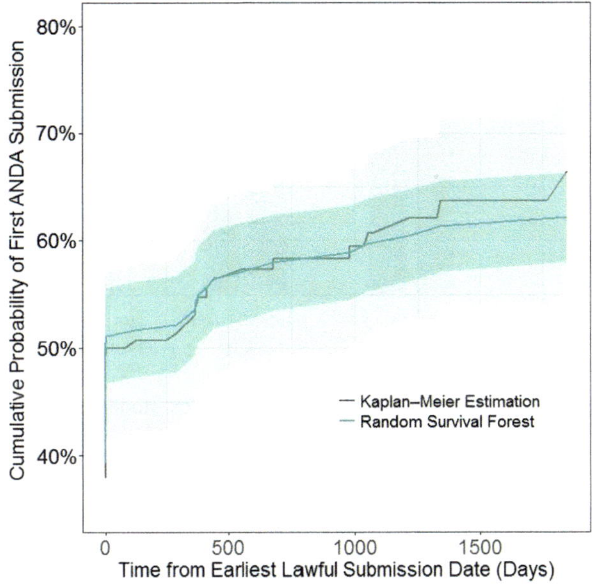

FIGURE 8.16
Kaplan–Meier survival plot of all eligible data (black) and the predicted survival plot (gray) based on the random survival forest using the leave-one-out method. The shaded area represents the 95% confidence intervals. (Adapted from Hu et al. 2019.)

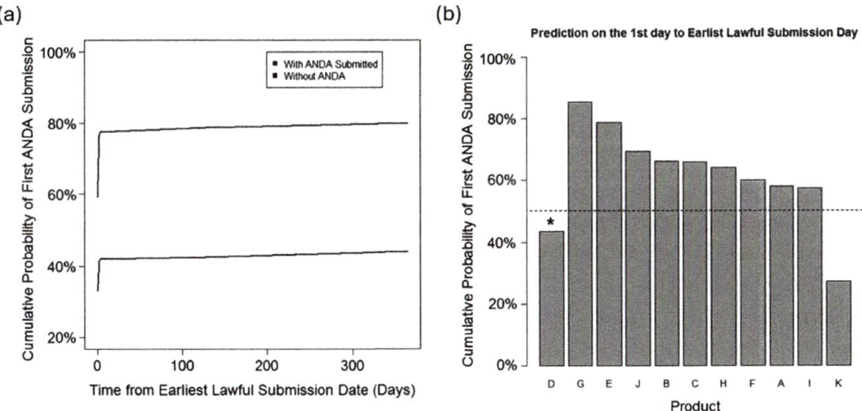

FIGURE 8.17
Results from the modeling with external validation using products whose earliest lawful ANDA submission dates (ELASDs) were within 2018 as the test dataset are presented. (a) shows the predicted cumulative probabilities of first Abbreviated New Drug Application (ANDA) submission for products with submissions (red) and without submissions (blue). The curves at the top and the bottom correspond to " With ANDA Submitted" and Without ANDA," respectively. (b) illustrates the predicted probabilities of the products with submissions within 1 day of their ELASDs. Note: Product D received its first ANDA submission on day 291 post-ELASD. (Adapted from Hu et al. 2019.).

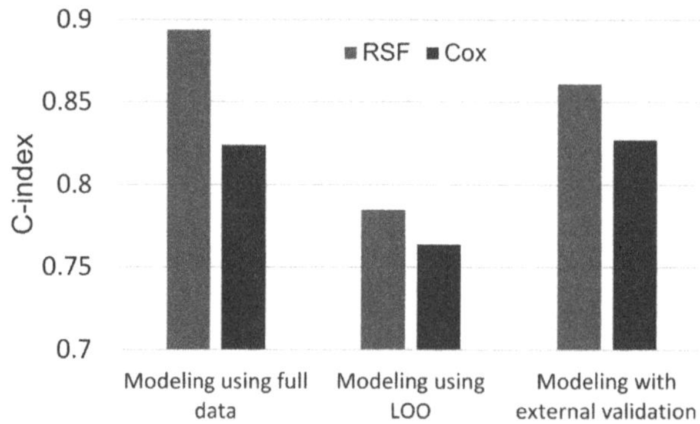

FIGURE 8.18
Performance comparisons between the RSF model and Cox model using full data, LOO, and truncated dataset by year.

plotted against their ELASDs. The results demonstrate that the model accurately reflects ANDA submissions in 2018, with low predicted probabilities for products without submissions and vice versa. Notably, 10 out of the 11 products with ANDA submissions received their first generic submissions within 1 day post their ELASDs, while only product D received its first generic submission 291 days post its ELASD.

Figure 8.17b shows the predicted probabilities of ANDA submission for the 11 products (A–K) within 1 day post their ELASDs. These results further support the predictive performance of the proposed model, with product D exhibiting a lower probability compared to other products except for product K. The relatively low predicted probability for product K may be attributed to unrecognized factors not captured by the current model input variables.

8.6.3.4.1 Model Comparison

Hu et al. (2019) conducted a comparison of model performances between the proposed RSF model and the conventional Cox proportional hazards (PH) regression model. Both models were applied to the same dataset using identical evaluation procedures. The resulting C-index values indicate that the proposed RSF model generally outperforms the conventional Cox model in terms of predictive performance (Figure 8.18). Specifically, while the Cox model identified only two significant variables ($p < 0.05$), the RSF model exhibited superior predictive capabilities across a wider range of variables.

8.6.4 Leveraging Large Language Model for Developing Guidance on Food Effect

8.6.4.1 Background

The U.S. Food and Drug Administration (FDA) releases product-specific guidance (PSG) to delineate its current directives on establishing bioequivalence between a proposed test drug product and its corresponding reference standard drug product. These PSGs serve as vital tools in guiding and expediting generic drug product development, facilitating the submission and approval of Abbreviated New Drug Applications (ANDAs), and ultimately ensuring public access to safe, effective, high-quality, and affordable generic drugs. However, the assessment process associated with PSG can be labor-intensive, particularly in gathering supportive information. Consequently, there is a burgeoning necessity to automate the retrieval of information from diverse data sources. Automating the collection of pharmacokinetic information, such as absorption, distribution, metabolism, and excretion (ADME) of Reference Listed Drug (RLD) products, stands as a pivotal advancement in enhancing PSG assessments (Shi et al. 2023).

A significant hurdle in streamlining the burden of PSG assessment lies in automatically summarizing key elements from extensive documents. For instance, food effect studies play a critical role in understanding drug absorption, as they can impact pharmacokinetics through various mechanisms, including delaying gastric emptying, altering gastrointestinal pH, and interacting with the dosage form or drug itself. Within the context of review documents for RLD products, food effect studies elucidate whether there exists a clinically significant influence of food on the product's bioavailability or pharmacokinetics. Nonetheless, these documents often span dozens of pages and can prove time-consuming to summarize manually. To tackle this challenge, there is a pressing need to automatically generate an initial summary that FDA staff can subsequently refine to ensure accuracy and relevancy.

Recent advancements in Large Language Models (LLMs) have showcased remarkable performance in zero/few-shot tasks in Natural Language Processing (NLP), prompting interest in their potential for automatic text summarization (Bai et al. 2022; Brown et al. 2020; Goyal et al. 2022; Liu et al. 2022). AI-powered chatbots, exemplified by ChatGPT (OpenAI 2022) and GPT-4 (OpenAI 2023), represent the cutting-edge in LLM technology. While these models were not initially developed for the biomedical domain, their capabilities have exhibited significant promise in biomedical research (Gao et al. 2023; Holmes et al. 2023; Lee et al. 2023).

Despite this progress, prior research has yet to explore the potential of harnessing LLM outputs to iteratively refine text summaries. In this study by Shi et al. (2023), the authors proposed a straightforward strategy termed iterative prompting, which involves engaging in iterative conversations with chatbots. Specifically, leveraging two state-of-the-art chatbots, ChatGPT and GPT-4, they developed a method to train the chatbots to generate summaries of food effect studies within NDA review files. This was accomplished through an iterative conversation between humans and chatbots with the aim of enhancing the accuracy and efficiency of food effect study summarization. Their approach is illustrated through an example in Figure 8.19, highlighting how ChatGPT collaborates with humans to enhance the summary.

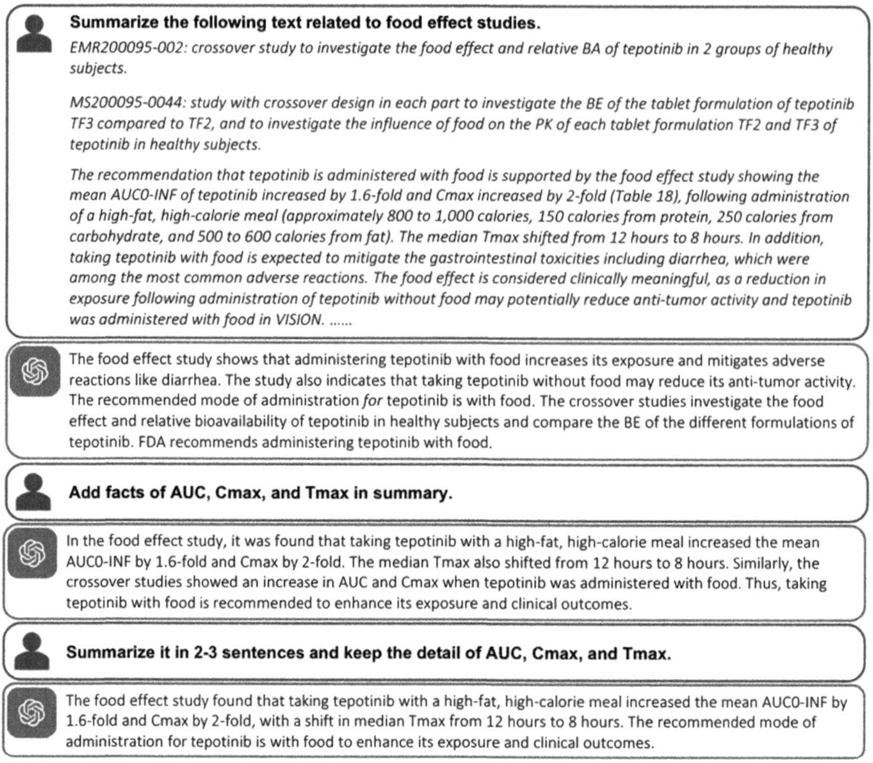

Summarize the following text related to food effect studies.

EMR200095-002: crossover study to investigate the food effect and relative BA of tepotinib in 2 groups of healthy subjects.

MS200095-0044: study with crossover design in each part to investigate the BE of the tablet formulation of tepotinib TF3 compared to TF2, and to investigate the influence of food on the PK of each tablet formulation TF2 and TF3 of tepotinib in healthy subjects.

The recommendation that tepotinib is administered with food is supported by the food effect study showing the mean AUC0-INF of tepotinib increased by 1.6-fold and Cmax increased by 2-fold (Table 18), following administration of a high-fat, high-calorie meal (approximately 800 to 1,000 calories, 150 calories from protein, 250 calories from carbohydrate, and 500 to 600 calories from fat). The median Tmax shifted from 12 hours to 8 hours. In addition, taking tepotinib with food is expected to mitigate the gastrointestinal toxicities including diarrhea, which were among the most common adverse reactions. The food effect is considered clinically meaningful, as a reduction in exposure following administration of tepotinib without food may potentially reduce anti-tumor activity and tepotinib was administered with food in VISION. ......

The food effect study shows that administering tepotinib with food increases its exposure and mitigates adverse reactions like diarrhea. The study also indicates that taking tepotinib without food may reduce its anti-tumor activity. The recommended mode of administration *for* tepotinib is with food. The crossover studies investigate the food effect and relative bioavailability of tepotinib in healthy subjects and compare the BE of the different formulations of tepotinib. FDA recommends administering tepotinib with food.

Add facts of AUC, Cmax, and Tmax in summary.

In the food effect study, it was found that taking tepotinib with a high-fat, high-calorie meal increased the mean AUC0-INF by 1.6-fold and Cmax by 2-fold. The median Tmax also shifted from 12 hours to 8 hours. Similarly, the crossover studies showed an increase in AUC and Cmax when tepotinib was administered with food. Thus, taking tepotinib with food is recommended to enhance its exposure and clinical outcomes.

Summarize it in 2-3 sentences and keep the detail of AUC, Cmax, and Tmax.

The food effect study found that taking tepotinib with a high-fat, high-calorie meal increased the mean AUC0-INF by 1.6-fold and Cmax by 2-fold, with a shift in median Tmax from 12 hours to 8 hours. The recommended mode of administration for tepotinib is with food to enhance its exposure and clinical outcomes.

FIGURE 8.19
An example illustrating the interactive conversation with ChatGPT for text summarization of food effect study (NDA214096). (Adapted from Shi et al. 2023.)

8.6.4.2 Interactive Prompting Approach

ChatGPT and GPT-4 are AI-powered chatbots capable of conversational interaction with humans. The underlying AI model uses its "working memory" to recall previous outputs, enabling it to respond to follow-up questions and adapt based on human feedback. Shi et al. (2023) proposed a three-turn interactive conversation between a PSG assessor and a chatbot for the summarization of food effect studies (Figure 8.20). Each iteration incorporates a query or constraint in the prompt reflecting human feedback from the previous output.

8.6.4.3 Data

The dataset for this study has two segments: detailed food effect study documents from NDA review files, accessible through the Drugs@FDA website, which serve as the source material for summarization, and the

Turn 1 (task instruction):
Summarize the following text related to food effect studies. {article}
Turn 2 (keyword-constrained prompt):
Add facts of AUC, Cmax, and Tmax in the summary.
Turn 3 (length-constrained prompt):
Summarize it in 2-3 sentences and keep the detail of AUC, Cmax, and Tmax.

FIGURE 8.20
Three-turn iterative prompting approach to utilizing chatbot to provide food effect summarization. (Adapted from Shi et al. 2023.)

corresponding concise food effect sections from drug labeling, serving as the ground-truth reference summaries.

The FDA NDA review files encompass a variety of documents, including multidisciplinary reviews, integrated reviews, and clinical pharmacology reviews. A team of FDA assessors selected new drug applications from 2009 to 2023. Excluded were drugs without publicly available reviews or lacking a food effect section in their labeling. The final dataset consisted of 100 drugs, representing a wide range of clinical categories, drug substances, and dosage forms. The NDA review files of these drugs were meticulously reviewed by the accessors, and sections relevant to food effect studies were annotated. These sections were used as inputs (articles) to the chatbots for summarization. The corresponding golden reference summaries were manually collected from the food effect sections of FDA-approved drug labeling available on DailyMed.

8.6.4.4 Method

Two large language models, ChatGPT-3.5-urbo and GPT4 were evaluated for their capabilities in generating text summaries of food effect studies. First, can multi-turn iterative conversations with a chatbot, enhanced by human feedback, improve summary quality? Second, which model – ChatGPT or GPT-4 – excels in this specific task? Third, considering that GPT-4 is the latest iteration in the GPT series with advanced features and capabilities, to what extent can GPT-4 produce summaries that are factually consistent with the reference summaries found in drug labeling? To address these three questions, the following evaluations methods were used.

8.6.4.4.1 Comparison between Summaries of Different Iterations

To determine if iterative prompts to ChatGPT, guided by human feedback, can improve summary quality, we devised an evaluation task involving preference annotations. For each drug, assessors were presented with summaries from all three turns in a randomized order to eliminate presentation bias. They were asked to select their most and least preferred summaries and optionally provide a free-text justification for their choices.

8.6.4.4.2 Comparison between ChatGPT and GPT-4

We conducted a blinded pairwise comparison to evaluate the summarization capabilities of ChatGPT and GPT-4. Assessors were given pairs of summaries, with one summary generated by ChatGPT and the other by GPT-4, both from the final iteration. The order of these summary pairs was randomized to reduce bias. Assessors had to choose the better summary from each pair or indicate if the summaries were equally good if they couldn't differentiate between them.

8.6.4.4.3 GPT-4 Consistency Evaluation

Despite substantial improvements over previous GPT models, GPT-4 can still produce summaries with factual inaccuracies or hallucinations. This task aimed to assess the factual consistency of GPT-4 generated summaries against reference summaries.

8.6.4.5 Evaluation Metrics

Shi et al. (2023) used three evaluation methods to assess model performance:

Automated Metrics: We use ROUGE to evaluate the summaries generated by ChatGPT and GPT-4 against reference summaries. This helps us measure the impact of iterative prompting and compare the two models' performance across each turn.

Human Evaluation: Three independent FDA assessors, each with over three years of experience in PSG assessment, evaluate the summaries. They are not given specific quality definitions to avoid bias, instead relying on their professional judgment.

GPT-4 Evaluation: Leveraging GPT-4's advanced capabilities, we use it to assess the relevance of model-generated summaries relative to the reference summaries. GPT-4's evaluation tasks include comparing ChatGPT and GPT-4's performance and checking the factual consistency of GPT-4-generated summaries. Recent studies have shown GPT-4 can produce reliable rankings and detailed assessments, suggesting it might be comparable to human evaluators.

8.6.4.6 Results

8.6.4.6.1 Effectiveness of Three-Turn Prompting

Figure 8.21 illustrates the distribution of assessor votes for the most and least preferred summaries at each iteration. Summaries from Turn 3, the final iteration, received the highest number of "Most Preferred Summary" votes, making them the top choice. These summaries were favored by all three assessors in nearly half of the cases. Conversely, Turn 1 summaries were most frequently voted as the "Least Preferred Summary," indicating they were the least favored. However, 42 drugs received zero "Least Preferred Summary"

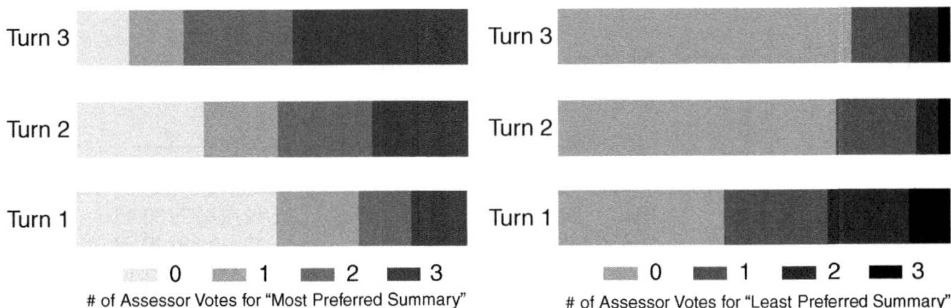

FIGURE 8.21
Distribution of assessor votes for the most and least preferred summaries at each turn during the iterative prompting process. (Adapted from Shi 2023.)

votes in Turn 1, suggesting ChatGPT can produce satisfactory summaries even with initial task instructions.

Additionally, we used Krippendorff's alpha (Castro 2017) to measure inter-annotator agreement for the most and least preferred summaries. The results showed moderate to substantial agreement among assessors, with alpha values of 0.49 for the most preferred and 0.55 for the least preferred summaries.

Table 8.11 presents the ROUGE-1/2/L evaluation results for summaries generated at each turn. It shows that summaries increasingly resemble the reference as the iterative process progresses. Turn 3 outperforms other turns in both ROUGE-1 and ROUGE-L scores for ChatGPT and GPT-4, aligning with human evaluation results. This demonstrates that ROUGE metrics effectively capture the improvements across turns.

8.6.4.6.2 Performance of ChatGPT vs GPT-4

Figure 8.22 highlights that both human assessors and GPT-4 favored summaries generated by GPT-4 over those by ChatGPT. Human votes showed GPT-4 winning 43% of the time, with a tie in 45% of cases. The assessors had moderate agreement (Krippendorff's alpha = 0.35). However, agreement between human votes and GPT-4 was low (Krippendorff's alpha = 0.07), indicating GPT-4's difficulty in identifying ties where both models performed equally well.

8.6.4.6.3 Consistency of GPT-4

Three assessors compared GPT-4 summaries with the food effect sections from drug labeling. The majority vote indicated that 85% of GPT-4 summaries were factually consistent with the reference, with reasonable agreement among assessors (Krippendorff's alpha = 0.41). GPT-4's evaluation showed a 72% consistency rate. Despite this being lower than the human rating, there was a strong overlap of 69% between human and GPT-4 ratings.

TABLE 8.11
ROUGE Scores (1/2/L) of Summaries at Each Turn

	ROUGE-1/2/L	
	ChatGPT	**GPT-4**
Turn1	29.4/11.24/19.48	31.44/12.72/30.45
Turn2	32.67/**14.06**/21.75	31.37/**12.88**/20.98
Turn3	**34.04**/13.36/22.60	**33.64**/11.44/**22.12**

Source: Shi et al. (2023).

Note: Bold indicates the best score among the turns for each metric.

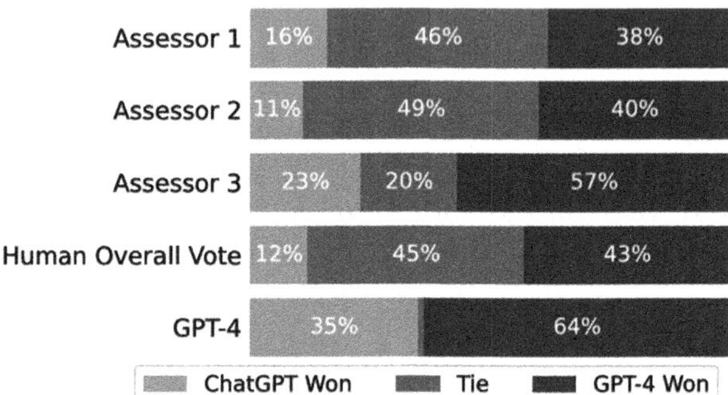

FIGURE 8.22
Comparison of text summaries generated by ChatGPT and GPT-4. (Adapted from Shi et al. 2023.)

8.7 Concluding Remarks

The modernization of clinical trials is pivotal in advancing drug development and ensuring the timely availability of safe and effective medicines. Regulatory initiatives have been at the forefront of this transformation, emphasizing the need for innovative trial designs, the integration of digital technologies, and streamlined regulatory processes. These efforts aim to address the challenges of lengthy development times, large patient populations, and significant risks and costs associated with traditional clinical trials.

The adoption of novel clinical trial designs, such as adaptive trials and decentralized trials, has shown promise in increasing the efficiency and flexibility of clinical research. These designs not only reduce the burden on patients but also enable more dynamic and responsive studies, adapting in real-time to emerging data and trends. Furthermore, the integration of digital technologies, including AI and ML, has revolutionized the way trials are conducted. These technologies facilitate sophisticated data analysis, enhance patient monitoring, and improve the overall execution of trials, thereby expediting the clinical development process.

Regulatory authorities have recognized the potential of these advancements and are actively developing guidelines and frameworks to support their implementation. The FDA's proactive approach in utilizing AI and ML for regulatory review, decision-making, and administrative automation serves as a prime example of this commitment. By leveraging these technologies, the FDA aims to enhance the accuracy, efficiency, and transparency of its processes, ultimately benefiting patients and the healthcare system as a whole.

The case studies presented in this chapter highlight the tangible impact of these regulatory initiatives. From product-specific guidance to the automation of administrative tasks, these efforts demonstrate a clear trajectory toward a more modernized and efficient clinical trial landscape. As regulatory authorities continue to embrace and refine these innovations, the future of clinical trials holds great promise for more effective and expeditious drug development.

In conclusion, the regulatory initiatives discussed in this chapter underscore the critical role of innovative trial designs and digital technologies in modernizing clinical trials. By fostering a collaborative and forward-thinking regulatory environment, we can overcome traditional barriers and accelerate the development of life-saving therapies. The ongoing evolution of regulatory frameworks will ensure that clinical trials remain robust, ethical, and patient-centric, ultimately improving public health outcomes and advancing medical science.

References

Abatemarco, D., Perera, S., et al. (2018). Training augmented intelligent capabilities for pharmacovigilance: Applying deep-learning approaches to individual case safety report processing. *Pharmaceutical Medicine*, 32(3). https://doi.org/10.1007/s40290-018-0251-9

Ali, J., and Luqman, S. (2023). Using machine learning for risk-based monitoring and data integrity in clinical trial. www.researchgate.net/publication/373214571_Using_Machine_Learning_for_Risk-based_Monitoring_and_Data_Integrity_in_Clinical_Trials. Accessed January 18, 2025.

Aliper, A., Kudrin, R., et al. (2023). Prediction of clinical trial outcomes based on target choice and clinical trial design with multi-modal artificial intelligence. *Clinical Pharmacology & Therapeutics*, 114(5), 972–980.

Allen-Ramey, F.C., Gupta, S., and DiBonaventura, M.D. (2012). Patient characteristics, treatment patterns, and health outcomes among COPD phenotypes. *International Journal of Chronic Obstructive Pulmonary Disease*, 7, 779–787.

Altexsoft. (2020). Clinical trial software: Understanding EDC, CTMS, ePRO, RTSM, and addressing industry challenges. www.altexsoft.com/blog/clinical-trial-software-edc-cdms-ctms-rtsm/. Accessed August 31, 2024.

Anagnostopoulos, C., Champagne, D., et al. (2023). How artificial intelligence can power clinical development. www.mckinsey.com/industries/life-sciences/our-insights/how-artificial-intelligence-can-power-clinical-development. Accessed September 15, 2024.

Andrews, J.A., Berry, J.D., Baloh, R.H., Carberry, N., Cudkowicz, M.E., Dedi, B., et al. (2020). Amyotrophic lateral sclerosis care and research in the United States during the COVID-19 pandemic: Challenges and opportunities. *Muscle Nerve*, 62, 182–186.

Anisimov, V.V., and Fedorov, V.V. (2007). Modelling, prediction and adaptive adjustment of recruitment in multicentre trials. *Statistics in Medicine*, 26(27), 4958–4975.

Arbour, K.C, Luu, A.T., et al. (2021). Deep learning to estimate RECIST-defined outcomes from radiology reports in patients with non-small cell lung cancer. *Cancer Discovery*, 11(1), 59–67.

Arkenau, H.T., Olmos, D., Ang, J.E., De Bono, H., Judson, I., and Kaye, S. (2008). Clinical outcome and prognostic factors for patients treated within the context of a phase I study: The Royal Marsden Hospital experience. *British Journal of Cancer*, 98, 1029–1033.

Array. (2023). Mitigating clinical trial enrollment challenges with information and communication. www.arraylive.com/blog/how-to-increase-clinical-trial-enrollment-with-information-and-communication. Accessed September 1, 2024.

Arrowsmith, H. (2012). A decade of change. *Nature Review Drug Discovery*, 11, 17–18.

Arrowsmith, J. (2011). Trial watch: Phase II failures: 2008–2010. *Nature Reviews Drug Discovery*, 10(5), 328–329. https://doi.org/10.1038/nrd3439

Arrowsmith, J., and Miller, P. (2013). Trial watch: Phase II and phase III attrition rates 2011–2012. *Nature Reviews Drug Discovery*, 12(8), 569.

Askin, S., Burkhalter, D., et al. (2023). Artificial intelligence applied to clinical trials: Opportunities and challenges. *Health and Technology*, 13(2), 203–213.

Athey, S., Tibshirani, J., and Wager, S. (2019). Generalized random forests. *The Annals of Statistics*, 47(2), 1148–1178.

Austin, C.P., Mount, B.A., and Colvis, C.M. (2021). Envisioning an actionable research agenda to facilitate repurposing of off-patent drugs. *Nature Review Drug Discovery*, 20(10), 723–724.

AVOCA (2020). Webinar RBQM – The AQC's risk-based monitoring framework and monitoring plans consistent with new FDA draft guidance. www.theavocagroup.com/news_events/webinar-rbqm-the-aqcs-risk-based-monitoring-framework-and-monitoring-plans-consistent-with-new-fda-draft-guidance/. Accessed January 12, 2025.

Azam, T.U., Shadid, H., Blakely, P., O'Hayer, P., Berlin, H., and Pan, M. (2020). Soluble urokinase receptor in COVID-19 related acute kidney injury. *Journal of the American Society of Nephrology*, 31, 2725–2735. https://doi.org/10.1681/ASN.2020060829

Bafadhel, M., McCormick, M., Saha, S., McKenna, S., Shelley, M., Hargadon, B., et al. (2012a). Profiling of sputum inflammatory mediators in asthma and chronic obstructive pulmonary disease. *Respiration*, 83(1), 36–44.

Bafadhel, M., McKenna, S., Terry, S., Mistry, V., Pancholi, M., Venge, P., et al. (2012b). Blood eosinophils to direct corticosteroid treatment of exacerbations of chronic obstructive pulmonary disease: A randomized placebo-controlled trial. *American Journal of Respiratory and Critical Care Medicine*, 186(1), 48–55.

Bafadhel, M., McKenna, S., Terry, S., Mistry, V., Reid, C., Haldar, P., et al. (2011). Acute exacerbations of chronic obstructive pulmonary disease: Identification of biologic clusters and their biomarkers. *American Journal of Respiratory and Critical Care Medicine*, 184(6), 662–671.

Bagiella, E., and Heitjan, D.F. (2001). Predicting analysis times in randomized clinical trials. *Statistics in Medicine*, 20(24), 2055–2063.

Bai, Y., Jones, A., Ndousse, K., Askell, A., Chen, A., DasSarma, N., Drain, D., Fort, S., Ganguli, D., Henighan, T., Joseph, N., Kadavath, S., Kernion, J., Conerly, T., El-Showk, S., Elhage, N., Hatfield-Dodds, Z., Hernandez, D., Hume, T., … and Kaplan, J. (2022). Training a helpful and harmless assistant with reinforcement learning from human feedback (arXiv:2204.05862). *arXiv*. https://doi.org/10.48550/arXiv.2204.05862

Bain, E.E., Shafner, L., Walling, D.P., Othman, A.A., Chuang-Stein, C., Hinkle, J., and Hanina, A. (2017). Use of a novel artificial intelligence platform on mobile devices to assess dosing compliance in a Phase 2 clinical trial in subjects with schizophrenia. *JMIR mHealth and uHealth*, 5, e18. https://doi.org/10.2196/mhealth.7030

Barker, A.D., et al. (2009). I-SPY 2: An adaptive breast cancer trial design in the setting of neoadjuvant chemotherapy. *Clinical Pharmacology & Therapeutics*, 86(1), 97–100.

Barnes, P.J., and Celli, B.R. (2009). Systemic manifestations and comorbidities of COPD. *European Respiratory Journal*, 33(5), 1165–1185.

Barrett, J.S., Goyal, R.K., Gobburu, J., Bara, S., and Varshney, J. (2023). An AI approach to generating MIDD assets across the drug development continuum. *The APPS Journal*, 25, 70.

Barton, S. (2000). Which clinical studies provide the best evidence: The best RCT still trumps the best observational. *BMJ*, 321(7256), 255–256.

Bateman, R.J., Benzinger, T.L., Berry, S., Clifford, D.B., Duggan, C., Fagan, A.M., Fanning, K., Farlow, M.R., Hassenstab, J., McDade, E.M., Mills, S., Paumier, K., Yuintana, M., Salloway, S.P., Santacruz, A., Schneider, L.S., Wang, G., Xiang, C., and the Dian-TU Pharma Consortium for the Dominantly Inherited Alzheimer network. (2017). The DIAN-TU Next Generation Alzheimer's prevention trial: Adaptive design and disease progression model. *Alzheimer's & Dementia*, 13(1), 8–19.

Beacher, F.D., Mujica-Parodi, L.R., et al. (2021). Machine learning predicts outcomes of phase III clinical trials for prostate cancer. *Algorithms*, 14, 147.

Beam, A.L., and Kohane, I.S. (2018). Big data and machine learning in health care. *JAMA*, 319(13), 1317–1318.

Beaulieu, D., Berry, J.D., Glass, D.J., et al. (2021). Development and validation of a machine-learning ALS survival model lacking vital capacity (VC Free) for use in clinical trials during the COVID-19 pandemic. *Amyotrophic Lateral Sclerosis and Frontotemporal Degeneration*, 22, 22–32.

Beauregard, A., Lavoie, L., and Labrie, F. (2015). A centralized monitoring approach using Excel for the quality management of clinical trials. https://d1wqtxts1xzle7.cloudfront.net/42617803/A_Centralized_Monitoring_Approach_Using_Excel_for_the_Quality_Management_of_Clinical_Trials-libre.pdf?1455293897=&response-content-disposition=inline%3B+filename%3DA_Centralized_Monitoring_Approach_Using.pdf&Expires=1729448958&Signature=Vh4m7VV~KK5RgHm4tVL4Lyd9XDjS5bW4~8~g-nnE458QwHWnYZ8vY4qPySZLG3X~LuWEOGsOZO4fnirrPP4zgizh6cJGHaYIWLOHmR1W2Bm9Ff7ObsglSmQg3ymhLu0POv9sS9kQB79tlxpaV69V8CDgd8autaky5HGb01KKeCSy0v22VQqCtyTKg7Jm0Bbc0YnMQch5MoYwhtG~oFQEWSXCNIpoXd0r2oYRyxxOFEoIWsJs5NmYivYgC76CGBzCcQerbVA9EegcA9Fa2j5h8K1I5K8I1tMybQRcfd-PCn--1sErBxe0NnswiGG~8xyS9RSmgKlXI8eZZfaUMtFMyA__&Key-Pair-Id=APKAJLOHF5GGSLRBV4ZA. Accessed October 20, 2024.

Beck, J.A., Singer, P.M., et al. (2018). Improving data for behavioral health workforce planning: Development of a minimum data set. *American Journal of Preventive Medicine*, 54(6 Suppl 3), S192–S198.

Beck, J.T., et al. (2020). *Artificial Intelligence Tool for Optimizing Eligibility Screening for Clinical Trials in a Large Community Center*. JCO Clinical Cancer Informatics.

Beckman, R.A., Clark, J., and Chen, C. (2011). Integrating predictive biomarkers and classifiers into oncology clinical development programs. *Nature Reviews Drug Discovery*, 10(10), 735–748.

Beinse, G., Teller, V., Charver, V., Deutsh, E., et al. (2019). Prediction of drug approval after phase I clinical trials in oncology resolved. *JCO Clinical Cancer Informatics*, 3, 1–10.

Benda, N., Branson, M., Maurer, W., and Friede, T. (2010). Aspects of modernizing drug development using clinical scenario planning and evaluation. *Regulatory Affairs*, 44, 299–315.

Benitez, J.C., Recondo, G., Rassy, E., and Mezquita, L. (2020). The LIPI score and inflammatory biomarkers for selection of patients with solid tumors treated with checkpoint inhibitors. *The Quarterly Journal of Nuclear Medicine and Molecular Imaging*, 64(2), 162–174. https://doi.org/10.23736/S1824-4785.20.03250-1

Berger, J.O. (1985). *Statistical Decision Theory and Bayesian Analysis*. 2nd ed. Springer.

Berger, M., and Doban, V. (2014). Big data, advanced analytics and the future of comparative effectiveness research. *Journal of Comparative Effectiveness Research*, 3(2), 167–176.

Bergmeir, C., and Benítez, J.M. (2012). On the use of cross-validation for time series predictor evaluation. *Information Sciences*, 191, 192–213. https://doi.org/10.1016/j.ins.2011.12.028

Bernardez-Pereira, S., et al. (2014). Prevalence, characteristics, and predictors of early termination of cardiovascular clinical trials due to low recruitment: Insights from the ClinicalTrials.gov registry. *American Heart Journal*, 168(2), 213–219.e1.

Berry, J., Lloyd-Jones, D.M., Garside, D.B., and Greenland, P. (2007). Framingham risk score and prediction of coronary heart disease death in young men. *American Heart Journal*, 154(1), 80–86.

Berry, D.A., and Stangl, D.K. (Eds). (1996). *Bayesian Biostatistics*. Marcel Dekker.

Berry, S.M., Carlin, B.P., Lee, J.J., and Muller, P. (2015). *Bayesian Adaptive Methods for Clinical Trials*. CRC Press.

Bhagat, R., Bojarshi, L., Chevalier, S., et al. (2021). Quality tolerance limits: Framework for successful implementation in clinical development. *Therapeutic Innovation & Regulatory Science*, 55(2), 251–261.

Bhatt, D.L., and Mehta, C. (2016). Adaptive designs for clinical trials. *New England Journal of Medicine*, 375(1), 65–74. https://doi.org/10.1056/NEJMra1510061

Biau, G., and Scornet, E. (2016). A random forest guided tour. *Test*, 25(2), 197–227.

Bieganek, C., Alferis, C., and Ma, S. (2022). Prediction of clinical trial enrollment rates. *PLOS ONE*, 17(2), e0263193.

Bigot, F., Castanon, E., Baldini, C., Hollebecque, A., Carmona, A., Postel-Vinay, S., et al. (2017). Prospective validation of a prognostic score for patients in immunotherapy phase I trials: The Gustave Roussy Immune Score (GRIm-Score). *European Journal of Cancer*, 84, 212–218.

BIO, Informa Pharma Intelligence, and QLS Advisors. (2021). *Clinical development success rates and contributing factors 2011–2022*. www.bio.org/clinical-development-success-rates-and-contributing-factors-2011-2020. Accessed May 1, 2024.

Bitton, A., and Gaziano, T. (2010). The Framingham heart study's impact on global risk assessment. *Progress in Cardiovascular Diseases*, 53(1), 68–78.

Black, N. (1996). Why we need observational studies to evaluate effectiveness of health care. *BMJ*, 312, 1215–1218.

Blumenthal, D., and Tavenner, M. (2010). The "meaningful use" regulation for electronic health records. *New England Journal of Medicine*, 363, 501–504.

Blumenthal, G., Jain, L., and Loeser, A. (2021). Optimizing dosing in oncology drug development. In *Friends of Cancer Research Annual Meeting 2021*. https://friendsofcancerresearch.org/wp-content/uploads/Optimizing_Dosing_in_Oncology_Drug_Development.pdf. Accessed January 14, 2025.

Bolland, K., Sooriyarachchi, M.R., and Whitehead, J. (1998). Sample size review in a head injury trial with ordered categorical responses. *Statistics in Medicine*, 17(24), 2835–2847.

Bothwell, L.E., Avorn, J., Khan, N.F., and Kesselheim, A.S. (2018). Adaptive design clinical trials: a review of the literature and ClinicalTrials.gov. *BMJ Open*, 8(2), 08320.

Böttcher, S., et al. (2022). Data quality evaluation in wearable monitoring. *Scientific Reports*, 12, 21412.

Boxer, A.L., Lang, A.E., Grossman, M., Knopman, D.S., Miller, B.L., Schneider, L.S., et al. (2014). Davunetide in patients with progressive supranuclear palsy: A randomised, double-blind, placebo-controlled phase 2/3 trial. *Lancet Neurology*, 13(7), 676–685.

Breiman, L. (2001). Random forests. *Machine Learning*, 45(1), 5–32.

Brightling, C.E., Monteiro, W., Ward, R., et al. (2000). Sputum eosinophilia and short-term response to prednisolone in chronic obstructive pulmonary disease: A randomized controlled trial. *Lancet*, 356, 1480–1485.

Brock, K., et al. (2015). Modelling clinical trial recruitment using Poisson processes. In *3rd International Clinical Trials Methodology Conference*, vol. 16, article number P85. https://trialsjournal.biomedcentral.com/articles/10.1186/1745-6215-16-S2-P85. Accessed January 14, 2025.

Brown, M.J., Chuang-Stein, C., and Kirby, S. (2012). Designing studies to find early signals of efficacy. *Journal of Biopharmaceutical Statistics*, 22(6), 1097–1108.

Brown, T.B., Mann, B., Ryder, N., Subbiah, M., Kaplan, J., Dhariwal, P., Neelakantan, A., Shyam, P., Sastry, G., Askell, A., Agarwal, S., Herbert-Voss, A., Krueger, G., Henighan, T., Child, R., Ramesh, A., Ziegler, D.M., Wu, J., Winter, C., ... and Amodei, D. (2020). Language models are few-shot learners. ArXiv:2005.14165 [Cs]. http://arxiv.org/abs/2005.14165

Bryant, J., and Day, R. (1995). Incorporating toxicity considerations into the design of two-stage phase II clinical trials. *Biometrics*, 51(4), 1372–1383.

Bunnage, M.E. (2011). Getting pharmaceutical R&D back on target. *Nature Chemical Biology*, 7, 335–339.

Butryn, T., et al. (2016). Keys to success in clinical trials: A practical review. *International Journal of Academic Medicine*, 2(2), 203–216.

Cai, C., Liu, S., and Yuan, Y. (2014). A Bayesian design for phase II clinical trials with delayed responses based on multiple imputation. *Statistics in Medicine*, 33(23), 4017–4028.

Calverley, P.M.A., Rabe, K.F., Goehring, U.-M., Kristiansen, S., Fabbri, L.M., and Martinez, F.J. (2009). Roflumilast in symptomatic chronic obstructive pulmonary disease: Two randomised clinical trials. *Lancet*, 374(9691), 685–694.

Cantor, S.B., Sun, C.C., Tortolero-Luna, G., Richards-Kortum, R., and Follen, M. (1999). A comparison of C/B ratios from studies using receiver operating characteristic curve analysis. *Journal of Clinical Epidemiology*, 52(9), 885–892.

Capozzo, R., Zoccolella, S., Musio, M., Barone, R., Accogli, M., and Logroscino, G. (2020). Telemedicine is a useful tool to deliver care to patients with amyotrophic lateral sclerosis during COVID-19 pandemic: Results from Southern Italy. *Amyotrophic Lateral Sclerosis and Frontotemporal Degeneration*, 21, 542–548.

Cardiac Arrhythmia Suppression Trial (CAST) Investigators. (1989). Preliminary report: Effect of encainide and flecainide on mortality in a randomized trial of arrhythmia suppression after myocardial infarction. *The New England Journal of Medicine*, 321(6), 406–412.

Carlisle, B., et al. (2015). Unsuccessful trial accrual and human subjects protections: An empirical analysis of recently closed trials. *Clinical Trials*, 12(1), 77–83.

Carroll, K.J. (2013). Decision making from phase II to phase III and the probability of success: Reassured by "assurance"? *Journal of Biopharmaceutical Statistics*, 23(5), 1188–1200.

Cawley, G.C., and Talbot, N.L.C. (2010). On over-fitting in model selection and subsequent selection bias in performance evaluation. *Journal of Machine Learning Research*, 11(70), 2079–2107.

Centers for Disease Control and Prevention. (2021). Lesson1: Introduction to Epidemiology. https://archive.cdc.gov/www_cdc_gov/csels/dsepd/ss1978/lesson1/section9.html. Accessed July 23, 2023.

Cerqueira, V., Torgo, L., and Mozetič, I. (2020). Evaluating time series forecasting models: An empirical study on performance estimation methods. *Machine Learning*, 109(11), 1997–2028. https://doi.org/10.1007/s10994-020-05910-7

Champiat, S., Ferrara, R., Massard, C., Besse, B., Marabelle, A., Soria, J.C., et al. (2018). Hyperprogressive disease: Recognizing a novel pattern to improve patient management. *Nature Reviews Clinical Oncology*, 15(12), 748–762.

Chang, M. (2008). *Adaptive Decision Theory and Implementation using ASA and R*. Chapman & Hall/CRC, Taylor & Francis Group.

Chang, M. (2010). *Monte Carlo Simulation for the Pharmaceutical Industry: Concepts, Algorithms, and Case Studies*. CRC Press.

Chekroud, A.M., Zotti, R.J., et al. (2016). Cross-trial prediction of treatment outcome in depression: A machine learning approach. *Lancet Psychiatry*, 3(3), 243–250.

Chen, Y.J., Gesser, R., and Luxembourg, A. (2015). A seamless Phase IIB/III adaptive outcome trial: Design rationale and implementation challenges. *Clinical Trials*, 12(1), 84–90.

Cheung, Y.K., and Chappell, R. (2000). Sequential designs for phase I clinical trials with late-onset toxicities. *Biometrics*, 56(4), 1177–1182.

Chevret, S. (1993). The continual reassessment method in cancer phase I clinical trials: A simulation study. *Statistics in Medicine*, 12, 1093–1108.

Chevret, S. (2006). *Statistical Methods for Dose-Finding Experiments*. Wiley.

Choi, E., et al. (2017). Using recurrent neural network models for early detection of heart failure onset. *Journal of the American Medical Informatics Association*, 24(2), 361–370.

Chow, S.C., and Chang, M. (2008). *Adaptive Design Methods in Clinical Trials*. PRC/Press.

Christensen, A., Melgaard, H., and Iwersen, J. (2003). Environmental monitoring based on a hierarchical Poisson-Gamma model. *Journal of Quality Technology*, 35(3), 11.

Chuang-Stein, C., and Kirby, S. (2017). *Quantitative Decisions in Drug Development*. Springer.

Chuang-Stein, C., Kirby, S., French, J., Kowalski, K., Marshall, S., Smith, M.K., et al. (2011). A quantitative approach for making Go/No-Go decisions in drug development. *Drug Information Journal*, 45(2), 187–202.

Clinical Trials Transformation Initiative (CTTI). (2024). *Quality by Design Toolkit*. https://ctti-clinicaltrials.org/our-work/quality/qbd-quality-by-design-toolkit/. Accessed September 14, 2024.

CMR Institute. (n.d.). *Understanding Target Product Profiles (TPPs) and Their Value in Drug Development*. Retrieved from www.cmrinstitute.org

Cohen, A.M., Chamberlin, S., Deloughery, T., et al. (2020). Detecting rare diseases in electronic health records using machining learning and knowledge engineering: Case study of acute hepatic porphyria. *PLOS ONE*. https://journals.plos.org/plosone/article?id=10.1371/journal.pone.0235574

Cohen, E., et al. (2023). Comparative effectiveness of eConsent: Systematic review. *Journal of Medical Internet Research*, 25, e43883.

Colin, P. (n.d.). *Predictive probability of success (PPoS) application to an interim analysis*. www.efspi.org/Documents/Events/Events%202018/Decision_Making_Dec_18/Presentations/Poster%20-%20Colin%20-%20Predictive%20Proba%20of%20Success%20Application%20to%20an%20interim%20analysis.pdf. Accessed May 1, 2023.

Collignon, O., Gartner, C., Haidich, A.B., James Hemmings, R., Hofner, B., Schiel, A., and Moes, D. (2020). Current statistical considerations and regulatory perspectives on the planning of confirmatory basket, umbrella, and platform trials. *Clinical Pharmacology & Therapeutics*, 107(5), 1059–1067.

Collins, F.S., and Varmus, H. (2015). A new initiative on precision medicine. *New England Journal of Medicine*, 372(9), 793–795. Available at: www.nejm.org/doi/full/10.1056/NEJMp1500523

Collins, R., and MacMahon, S. (2001). Reliable assessment of the effects of treatment on mortality and major morbidity, I: Clinical trials. *Lancet*, 357, 373–380.

Colosimo, B.M., and del Castillo, E. (2007). *Bayesian Process Monitoring, Control and Optimization*. Chapman & Hall/CRC.

Conover, W.J. (1999). *Practical Nonparametric Statistics*. 3rd ed. New York: Wiley.

Conti, P., Ronconi, G., Caraffa, A., Gallenga, C.E., Ross, R., Frydas, I., and Kritas, S.K. (2020). Induction of pro-inflammatory cytokines (IL-1 and IL-6) and lung inflammation by coronavirus-19 (COVI-19 or SARS-CoV-2): Anti-inflammatory strategies. *Journal of Biological Regulators and Homeostatic Agents*, 34, 327–331.

Corrigan-Curay, J., Sacks, L., and Woodcock, J. (2018). Real-world evidence and real-world data for evaluating drug safety and effectiveness. *JAMA*, 320(9), 867–868.

Corrigan, B.W., Lockwood, P.A., Marshall, S.A., and Benincosa, L.J. (2007). Model-based drug development. *Clinical Pharmacology & Therapeutics*, 82(1), 21–32.

Crowson, C.S., Atkinson, E.J., and Therneau, T.M. (2016). Assessing calibration of prognostic risk scores. *Statistical Methods in Medical Research*, 25, 1692–1706.

Cutillo, C.M., et al. (2020). Machine intelligence in healthcare-perspectives on trustworthiness, explainability, usability, and transparency. *npj Digital Medicine*, 3, 47.

Dallow, N., Best, N., and Montague, T.H. (2017). Better decision making in drug development through adoption of formal prior elicitation. *Pharmaceutical Statistics*, 17(4), 301–316.

De Marchi, F., Cantello, R., Ambrosini, S., Mazzini, L., and CANPALS Study Group (2020). Telemedicine and technological devices for amyotrophic lateral sclerosis in the era of COVID-19. *Neurological Sciences*, 41, 1365–1367.

DeMets, D.L., and Lan, K.K.G. (1994). Interim analysis – the alpha spending function approach. *Statistics in Medicine*, 13, 1341–1352.

Deng, Y., et al. (2017). Bayesian modeling and prediction of accrual in multi-reginal clinical trials. *Statistical Methods in Medical Research*, 26(2), 752–765.

DiCicco, R., Illis, B., and Evans, D. (2023). From documents to digital data: Clinical drug development's burning platform. *DIA Global Forum*, July Issue. https://globalforum.diaglobal.org/issue/july-2023/from-documents-to-digital-data-clinical-drug-developments-burning-platform/. Accessed August 31, 2024.

Dietrich, J., Forrest, A., and Meeker-O'Connell, A. (2018). QbD in clinical trials: A focused approach to quality assurance and risk management. *Clinical Leader*.

www.clinicalleader.com/doc/qbd-in-clinical-trials-a-focused-approach-to-quality-assurance-and-risk-management-0001. Accessed September 9, 2024.

DiMasi, J.A., Feldman, L., Sheckler, A., and Wilson, A. (2010). Trends in risks associated with new drug development: Success rates for investigational drugs. *Clinical Pharmacology & Therapeutics*, 87(3), 272–277. https://doi.org/10.1038/clpt.2009.295

DiMasi, J.A., Grabowski, H.G., and Hansen, R.W. (2016). Innovation in the pharmaceutical industry: New estimates of R&D costs. *Journal of Health Economics*, 47, 20–33.

DiMasi, J.A., Hermann, J., Twyman, K., Kondru, R.K., et al. (2015). A tool for predicting regulatory approval after phase II testing of new oncology compounds. *Clinical Pharmacology & Therapeutics*, 98(5), 506–513.

Dmitrienko, A., Muysers, C., Fritsch, A., and Lipkovich, I. (2016). General guidance on exploratory and confirmatory subgroup analysis in late-stage clinical trials. *Journal of Biopharmaceutical Statistics*, 26, 71–98.

Dmitrienko, A., and Pulkstenis, E. (2017). *Clinical Trial Optimization Using R*. CRC Press.

Dodd, L.E., Proschan, M.A., Neuhaus, J., Koopmeiners, J.S., Neaton, J., Beigel, J.D., Barrett, K., Lane, H.C., and Dvey, Jr., R.T. (2016). Design of a randomized controlled trial for Ebola virus disease medical countermeasures: PREVAIL II, the Ebola MCM study. *Journal of Infectious Diseases*, 213(12), 1906–1913.

Dorsey, E.R., and Topol, E.J. (2020). Telemedicine 2020 and the next decade. *The Lancet*, 395(10227), 859.

EC. (2020). *White Paper on Artificial Intelligence – A European Approach to Excellence and Trust*. https://commission.europa.eu/publications/white-paper-artificial-intelligence-european-approach-excellence-and-trust_en. Accessed September 22, 2024.

EC. (2021). *Proposal for a Regulation Laying Down Harmonised Rules on Artificial Intelligence*. https://digital-strategy.ec.europa.eu/en/library/proposal-regulation-laying-down-harmonised-rules-artificial-intelligence. Accessed September 22, 2024.

Eddy, D.M., and Schlessinger, L. (2003). Validation of the Archimedes diabetes model. *Diabetes Care*, 26(11), 3102–3110.

EFPIA. (2020). *Innovation in Clinical Trial Design: A Review of the Clinical Trial Design Landscape*. www.efpia.eu/media/547507/efpia-position-paper-innovation-in-clinical-trial-design-white-paper.pdf. Accessed September 18, 2024.

EMA. (2007). *Reflection Paper on Methodological Issues in Confirmatory Clinical Trials Planned with an Adaptive Design*. EMA/CHMP/EWP/2459/02, 2007. www.ema.europa.eu/en/documents/scientific-guideline/reflection-paper-methodological-issues-confirmatory-clinical-trials-planned-adaptive-design_en.pdf. Accessed September 22, 2024.

EMA. (2010). *Guideline on the Investigation of Bioequivalence*. European Medicines Agency (not available any more, find a paper instead: The revised EMA guideline for the investigation of bioequivalence for immediate release oral formulations with systemic action).

EMA. (2013). *Reflection Paper on Risk Based Quality Management in Clinical Trials*. European Medicines Agency (not available any more but have put into database).

EMA. (2017). *Concept Paper on the Development and Lifecycle of Personalised Medicines and Companion Diagnostics that Measure Predictive Biomarkers Which Help to Assess the Most Likely Response to a Particular Treatment*. July 2017. www.ema.europa.eu/en/news/concept-paper-development-lifecycle-personalisedmedicines-companion-diagnostic

EMA. (2020). *Regulatory Science Strategy to 2025*. www.ema.europa.eu/en/about-us/how-we-work/regulatory-science-strategy. Accessed June 19, 2024.

EMA. (2022). *ICH M11 Guideline, Clinical Study Protocol Template and Technical Specifications – Scientific Guideline*. www.ema.europa.eu/en/ich-m11-guideline-clinical-study-protocol-template-and-technical-specifications-scientific-guideline. Accessed August 31, 2024.

EMA. (2023a). *Guideline on Computerized Systems and Electronic Data in Clinical Trials*. www.ema.europa.eu/en/documents/regulatory-procedural-guideline/guideline-computerised-systems-and-electronic-data-clinical-trials_en.pdf. Accessed June 30, 2024.

EMA. (2023b). *Reflection Paper on the Use of Artificial Intelligence (AI) in Medicinal Product Lifecycle*. EMA. Available at: www.ema.europa.eu/en/documents/scientific-guideline/reflection-paper-use-artificial-intelligence-ai-medicinal-product-lifecycle_en.pdf

Emanuel, E.J., Wendler, D., and Grady, C. (2000). What makes clinical research ethical? *JAMA*, 283(20), 2701–2711. https://doi.org/10.1001/jama.283.20.2701

Emmanuel, T., Maupong, T., et al. (2021). A survey on missing data in machine learning. *Journal of Big Data*, 8(1), 140. https://doi.org/10.1186/s40537-021-00516-9

Evan, S.R., et al. (2021). Real-world data for planning eligibility criteria and enhancing recruitment: Recommendations from the Clinical Trials Transformation Initiative. *Therapeutic Innovation & Regulatory Science*, 55(3), 545–553.

Fard, M.J., Chawla, S., and Reddy, C. (2016). Early-stage event prediction for longitudinal data. *Lecture Notes in Computer Science*, PAKDD Part I, LNAI 9651, 139–151. https://doi.org/10.1007/978-3-319-31753-3_2

FDA. (1997). 21 CFR Part 11, electronic records; electronic signatures; final rule. *Federal Register*, 62(54), 13429. www.govinfo.gov/content/pkg/FR-1997-03-20/html/97-6833.htm. Accessed August 31, 2024.

FDA. (1999). Guidance for Industry Providing Regulatory Submissions to the Center for Biologics Evaluation and Research (CBER) in Electronic Format—Biologics Marketing Applications. www.fda.gov/media/77755/download. Accessed January 18, 2025.

FDA. (2003). *Guidance for Industry Part 11, Electronic Records; Electronic Signatures — Scope and Application*. www.fda.gov/media/75414/download. Accessed June 30, 2024.

FDA. (2004a). *Innovation/Stagnation: Challenge and Opportunity on the Critical Path to New Medical Products*. https://c-path.org/wp-content/uploads/2013/08/FDACPIReport.pdf. Accessed June 7, 2024.

FDA. (2004b). Innovation/Stagnation: Challenge and Opportunity on the Critical Path to New Medical Products. U.S. Food and Drug Administration. https://wayback.archive-it.org/7993/20180125035414/https:/www.fda.gov/ScienceResearch/SpecialTopics/CriticalPathInitiative/ucm076689.htm. Accessed September 18, 2024.

FDA. (2007a). *FDA Guidance for Industry: Computerized Systems Used in Clinical Investigations*. www.fda.gov/media/70970/download. Accessed August 31, 2024.

FDA. (2007b). *Guidance for Industry and Review Staff: Target Product Profile — A Strategic Development Process Tool. U.S. Department of Health and Human Services*. Retrieved from www.fda.gov/media/72566/download

FDA. (2009a). *FDA Guidance for Industry: Drug-Induced Liver Injury: Premarketing Clinical Evaluation*. www.fda.gov/media/116737/download. Accessed September 15, 2024.

FDA. (2009b). *Guidance for Industry: Drug-Induced Liver Injury: Premarketing Clinical Evaluation*. Rockville, MD: FDA.

FDA. (2010). *Guidance for Industry and FDA Staff: Guidance for the Use of Bayesian Statistics in Medical Device Clinical Trials*. www.fda.gov/media/71512/download. Accessed January 14, 2025.

FDA. (2013a). *Guidance for Industry: Electronic Source Data in Clinical Investigations*. U.S. Food and Drug Administration.

FDA. (2013b). *Guidance for Industry: Oversight of Clinical Investigations—A Risk-Based Approach to Monitoring*. www.fda.gov/media/116754/download. Accessed January 14, 2025.

FDA. (2016a). *21st Century Cures Act*. www.fda.gov/regulatory-information/selected-amendments-fdc-act/21st-century-cures-act. Accessed June 7, 2024.

FDA. (2016b). *Non-Inferiority Clinical Trials. Guidance for Industry*. www.fda.gov/media/78504/download. Accessed January 23, 2025.

FDA. (2016c). Principles for Co-development of an In Vitro Companion Diagnostic Device with a Therapeutic Product. 2016. Guidance for Industry and FDA Staff. www.fda.gov/regulatory-information/search-fda-guidance-documents/principlescodevelopment-vitro-companion-diagnostic-device-therapeutic-produc

FDA. (2017a). Use of Real-World Evidence to Support Regulatory Decision-Making for Medical Devices. www.fda.gov/regulatory-information/search-fda-guidance-documents/use-real-world-evidence-support-regulatory-decision-making-medical-devices. Accessed June 7, 2024.

FDA. (2017b). Digital Health Innovation Action Plan. www.fda.gov/media/106331/download. Accessed January 11, 2025.

FDA. (2017c). Use of Real-World Evidence to Support Regulatory Decision-Making for Medical Devices. *Guidance for Industry and Food*. www.fda.gov/media/99447/download. Accessed June 19, 2024.

FDA. (2018a). Use of Electronic Health Record Data in Clinical Investigations. www.fda.gov/media/97. Accessed June 11, 2024.

FDA. (2018b). Statement from FDA Commissioner Scott Gottlieb, M.D., on FDA's New Strategic Framework to Advance Use of Real-world Evidence to Support Development of Drugs and Biologics. www.fda.gov/news-events/press-announcements/statement-fda-commissioner-scott-gottlieb-md-fdas-new-strategic-framework-advance-use-real-world. Accessed June 7, 2024.

FDA. (2018c). Complex Innovative Trial Design (CID) Pilot Meeting Program. https://public4.pagefreezer.com/content/FDA/23-11-2021T07:28/https://www.fda.gov/drugs/development-resources/complex-innovative-trial-design-pilot-meeting-program. Accessed September 19, 2024.

FDA. (2018d). Draft guidance: Biomarker Qualification: Evidentiary Framework Guidance for Industry and FDA Staff. December 2018. www.fda.gov/regulatory-information/search-fdaguidance-documents/biomarker-qualification-evidentiary-framework

FDA. (2018e). Guidance for Industry: Clinical Trial Endpoints for the Approval of Cancer Drugs and Biologics. www.fda.gov/media/71195/download. Accessed January 14, 2025.

FDA. (2019a). Guidance for Industry: Enrichment Strategies for Clinical Trials to Support Determination of Effectiveness of Human Drugs and Biological Products. www.fda.gov/media/121320/download. Accessed June 19, 2024.

FDA. (2019b). Submitting Documents Using Real-World Data and Real-World Evidence to FDA for Drugs and Biologics Guidance for Industry. www.fda.gov/media/124795/download. January 7, 2020.

FDA. (2019c). Guidance for Industry; Adaptive Designs for Clinical Trials of Drugs and Biologics. www.fda.gov/media/78495/download. Accessed June 16, 2024.

FDA. (2019d). Rare Diseases: Natural History Studies for Drug Development. www.fda.gov/media/122425/download. January 7, 2020.

FDA. (2019e). Draft guidance: Developing and Labeling In Vitro Companion Diagnostic Devices for a Specific Group or Class of Oncology Therapeutic Product. Guidance for Industry. December 2019. www.fda.gov/regulatory-information/search-fda-guidance-documents/developingand-labeling-vitro-companion-diagnostic-devices-specific-group-or-class-oncology

FDA. (2019f). Enrichment Strategies for Clinical Trials to Support Approval of Human Drugs and Biological Products: Draft Guidance. *FDA*. Available at: www.fda.gov/media/121320/download. Accessed September 18, 2024.

FDA. (2019g). Impact Story: Using Innovative Statistical Approaches to Provide the Most Reliable Treatment Outcomes Information to Patients and Clinicians. Food and Drug Administration. June 10, 2019. www.fda.gov/drugs/regulatory-science-action/impact-story-using-innovative-statistical-approaches-provide-most-reliable-treatment-outcomes

FDA. (2020a). About the Digital Health Center of Excellence. www.fda.gov/medical-devices/digital-health-center-excellence. Accessed June 19, 2024.

FDA. (2020b). *Digital Health Innovation Action Plan.* www.fda.gov/media/106331/download. Accessed June 19, 2024.

FDA. (2020c). Part 11, Electronic Records; Electronic Signatures – Scope and Application. Retrieved from www.fda.gov/regulatory-information/search-fda-guidance-documents/part-11-electronic-records-electronic-signatures-scope-and-application. Accessed September 3, 2024.

FDA. (2020d). Guidance for Industry: Enhancing the Diversity of Clinical Trial Populations — Eligibility Criteria, Enrollment Practices, and Trial Designs. www.fda.gov/media/127712/download. Accessed July 8, 2024.

FDA. (2020e). Guidance for Industry: Interacting with the FDA on Complex Innovative Trial Designs for Drugs and Biological Products. www.fda.gov/media/130897/download. Accessed September 19, 2024.

FDA. (2021a). Artificial Intelligence/Machine Learning (AI/ML)-Based Software as a Medical Device (SaMD) Action Plan. FDA. Available at: www.fda.gov/medical-devices/software-medical-device-samd/artificial-intelligencemachine-learning-aiml-based-software-medical-device-samd-action-plan

FDA. (2021b). Determination Letter. www.fda.gov/media/155363/download. Accessed July 12, 2024.

FDA. (2021c). Project Optimus: Reforming the Dose Optimization and Dose Selection Paradigm in Oncology. www.fda.gov/about-fda/oncology-center-excellence/project-optimus. Accessed September 19, 2024.

FDA. (2021d). Providing Regulatory Submissions in Electronic Format — Standardized Study Data. www.fda.gov/media/82716/download. Accessed August 31, 2024.

FDA. (2021e). Software Precertification (Pre-Cert) Program: A Working Model.

FDA. (2022a). Complex Innovative Trial Design Meeting Program. Food and Drug Administration. 27 October 2022. www.fda.gov/drugs/development-resources/complex-innovative-trial-design-meeting-program. Accessed July 4, 2024.

FDA. (2022b). Fact Sheet for Healthcare Providers: Emergency Use Authorization for Kineret. KINERET HCP FS 11082022 (fda.gov). Accessed July 6, 2024.

FDA. (2023a). Guidance for Industry: Master Protocols for Drug and Biological Product Development. www.fda.gov/media/174976/download. Accessed June 16, 2024.

FDA. (2023b). Guidance for Industry, Investigators, and Other Stakeholders: Decentralized Clinical Trials for Drugs, Biological Products, and Devices. www.fda.gov/media/167696/download. Accessed September 1, 2024.

FDA. (2023c). Guidance for Industry: A Risk-Based Approach to Monitoring of Clinical Investigations Questions and Answers. www.federalregister.gov/documents/2023/04/12/2023-07687/a-risk-based-approach-to-monitoring-of-clinical-investigations-questions-and-answers-guidance-for. Accessed January 14, 2025.

FDA. (2023d). Guidance for Industry: Optimizing the Dosage of Human Prescription Drugs and Biological Products for the Treatment of Oncologic Diseases. www.fda.gov/media/164555/download. Accessed September 19, 2024.

FDA. (2023e). Using Artificial Intelligence & Machine Learning in the Development of Drug & Biological Products. www.federalregister.gov/documents/2023/05/11/2023-09985/using-artificial-intelligence-and-machine-learning-in-the-development-of-drug-and-biological. Accessed July 3, 2024.

FDA. (2023f). *Using Artificial Intelligence & Machine Learning in the Development of Drug & Biological Products*. FDA Discussion Paper.

FDA. (2024a). FDA establishes CDER Center for Clinical Trial Innovation (C3TI). www.fda.gov/drugs/drug-safety-and-availability/fda-establishes-cder-center-clinical-trial-innovation-c3ti. Accessed June 19, 2024.

FDA. (2024b). FDA Workshop: Advancing the Use of Complex Innovative Designs in Clinical Trials: From Pilot to Practice. www.fda.gov/news-events/advancing-use-complex-innovative-designs-clinical-trials-pilot-practice-03052024. Accessed September 19, 2024.

FDA. (2025). Guidance from Industry and Other Interested Parties: Considerations for the Use of Artificial Intelligence to Support Regulatory Decision-Making for Drug and Biological Products. www.fda.gov/media/184830/download. Accessed January 19, 2025.

FDA and AACR. (2024). FDA-AACR Public Workshop on Optimizing Dosage for Oncology Products: Quantitative Approaches to Select Dosages for Clinical Trials. February 15–16, 2024, Washington, D.C. www.aacr.org/professionals/policy-and-advocacy/regulatory-science-and-policy/events/fda-aacr-workshop-optimizing-dosages-for-oncology-drug-products/. Accessed September 19, 2024.

FDA-ASCO. (2022). Getting the Dose Right: Optimizing Dose Selection Strategies in Oncology – An FDA-ASCO Virtual Workshop May 3–5, 2022.
Feijoo, F., Palopoli, M., Bernstein, J., Siddiqui, S., and Albright, T.E. (2020). Key indicators of phase transition for clinical trials through machine learning. *Drug Discovery Today*, 25(2), 414–421. https://doi.org/10.1016/j.drudis.2019.12.014
Feroze, F.F. (2021). Application to Automate Clinical Study Report Generation Using AI. PharmaSUG 2021, www.lexjansen.com/pharmasug/2021/AI/PharmaSUG-2021-AI-012.pdf. Accessed August 31, 2024.
Fisher, B., Costantino, J.P., Wickerham, D.L., Redmond, C.K., Kavanah, M., Cronin, W.M., … and Wolmark, N. (1998). Tamoxifen for prevention of breast cancer: Report of the National Surgical Adjuvant Breast and Bowel Project P-1 Study. *Journal of the National Cancer Institute*, 90(18), 1371–1388. https://doi.org/10.1093/jnci/90.18.1371
Fisher, C.K., Smith, A.M., and Walsh, J.R. (2019). Machine learning for comprehensive forecasting of Alzheimer's disease progression. *Scientific Report*, 9. www.nature.com/articles/s41598-019-49656-2. Accessed July 7, 2024.
Flaherty, K.T., et al. (2020). The Molecular Analysis for Therapy Choice (NCI-MATCH) trial: Lessons for genomic trial design. *Journal of the National Cancer Institute*, 112(10), 1021–1029.
Fleming, T.R. (1982). One-sample multiple testing procedure for phase II clinical trials. *Biometrics*, 38(1), 143–151.
Fogel, D.B. (2018). Factors associated with clinical trials that fail and opportunities for improving the likelihood of success: A review. *Contemporary Clinical Trials Communications*, 11, 156–164. https://doi.org/10.1016/j.conctc.2018.08.001
Freidlin, B., McShane, L.M., and Korn, E.L. (2010). Randomized clinical trials with biomarkers: Design issues. *Journal of the National Cancer Institute*, 102(3), 152–160.
Freidlin, B., and Simon, R. (2005). Adaptive signature design: An adaptive clinical trial design for generating and prospectively testing a gene expression signature for sensitive patients. *Clinical Cancer Research*, 11(21), 7872–7878.
Frewer, P., Mitchell, P., Watkins, C., and Matcham, J. (2016). Decision-making in early clinical drug development. *Pharmaceutical Statistics*, 15(3), 255–263.
Friedman, L.M., Furberg, C.D., and DeMets, D.L. (2010). *Fundamentals of Clinical Trials*. Springer.
Gaddale, J.R. (2015). Clinical data acquisition standards harmonization importance and benefits in clinical data. *Perspectives in Clinical Research*, 6(4), 179–183.
Gail, M., and Simon, R. (1985). Testing for qualitative interactions between treatment effects and patient subsets. *Biometrics*, 41(2), 361–372.
Gail, M.H., Brinton, L.A., Byar, D.P., Corle, D.K., Green, S.B., Schairer, C., and Mulvihill, J.J. (1989). Projecting individualized probabilities of developing breast cancer for white females who are being examined annually. *Journal of the National Cancer Institute*, 81(24), 1879–1886. https://doi.org/10.1093/jnci/81.24.1879
Gambo, M.A., Tiwari, R.C., and LaVange, L.M. (2014). Bayesian approach to the design and analysis of non-inferiority trials for anti-infective products. *Pharmaceutical Statistics*, 13(1), 25–40.
Gao, C.A., Howard, F.M., Markov, N.S., Dyer, E.C., Ramesh, S., Luo, Y., and Pearson, A.T. (2023). Comparing scientific abstracts generated by ChatGPT to real abstracts with detectors and blinded human reviewers. *npj Digital Medicine*, 6(1), Article 1. https://doi.org/10.1038/s41746-023-00819-6

Gazali, S.K., and Singh, I. (2017). Artificial intelligence based clinical data management systems: A review. *Informatics in Medicine Unlocked*, 9, 219–229.

Gehan, E.A. (1961). The determination of the number of patients required in a preliminary and a follow-up trial of a new chemotherapeutic agent. *Journal of Chronic Diseases*, 13, 346–353.

Gerlinger, C., Evers, T., Rassen, J., and Wyss, R. (2020). Using real-world data to predict clinical and economic benefits of a future drug based on its target product profile. *Drugs – Real World Outcomes*, 7, 221–227.

Ghosh, S.K., Mukhopadhyay, P., and Lu, J.-C.J. (2006). Bayesian analysis of zero-inflated regression models. *Journal of Statistical planning and Inference*, 136(4), 1360–1375.

Girard, P. (2005). Clinical trial simulation: A tool for understanding study failures and preventing them. *Basic & Clinical Pharmacology & Toxicology*, 96(3), 228–234.

Gkioni, E., Rius, R., Dodd, S., and Gamble, C. (2019). A systematic review describes models for recruitment prediction at the design stage of a clinical trial. *Journal of Clinical Epidemiology*, 115, 141–149.

Gloy, V., Speich, B., Griessbach, A., Heravi, A.T., Schulz, A., Fabbro, T., Magnus, C.P., McLennan, S., Bertram, W., and Briel, M. (2022). Scoping review and characteristics of publicly available checklists for assessing clinical trial feasibility. *BMC Medical Research Methodology*, 22, 1–8. https://doi.org/10.1186/s12874-022-01617-6.

Godfrey, A., et al. (2020). BioMeT and algorithm challenges: A proposed digital standardized evaluation framework. *IEEE Journal of Translational Engineering in Health and Medicine*, 8, 0700108.

Godfried, I. (2018). A review of recent reinforcement learning applications to healthcare – taking machine learning beyond diagnosis to find optimal treatments. https://towardsdatascience.com/a-review-of-recent-reinforcment-learning-applications-to-healthcare-1f8357600407. Accessed June 17, 2024.

Gold, D.L., Dawson, M., Yang, H., Paker, J., and Gossage, D.L. (2014). Clinical trial simulation to assist in COPD trial planning and design with a biomarker-based diagnostic: When to pull the trigger? *Journal of Chronic Obstructive Pulmonary Disease*, 11(2), 226–235.

Goldman, B., LeBlanc, M., and Crowley, J. (2008). Interim futility analysis with intermediate endpoints. *Clinical Trials*, 5(1), 14–22.

Goldsack, J.C., et al. (2020). Verification, analytical validation, and clinical validation (V3): The foundation of determining fit-for-purpose for Biometric Monitoring Technologies (BioMeTs). *npj Digital Medicine*, 3, 55.

Goldstein, H., and Rigdon, J. (2019). Using machine learning to identify heterogeneous effects in randomized clinical trials – moving beyond the forest plot and into the forest. *JAMA Network*, 2(3), e190004. https://pubmed.ncbi.nlm.nih.gov/30848801/. Accessed September 20, 2024.

Gómez, F.P., and Rodriguez-Roisin, R. (2002). Global Initiative for Chronic Obstructive Lung Disease (GOLD) guidelines for chronic obstructive pulmonary disease. *Current Opinion in Pulmonary Medicine*, 8(2), 81–86.

Gong, P., Wang, Z., Han, S., Bai, M., Li, X., Zhang, C., Wang, Y., Yang, W., Wei, Y., and Zhang, Y. (2022). Decoding kinase-adverse event associations for small molecule kinase inhibitors. *Nature Communications*, 13, Article 6606. https://doi.org/10.1038/s41467-022-34173-4

Gong, X., Hu, M., Basu, M., and Zhao, L. (2021). Heterogenous treatment effect analysis based on machine-learning methodology. *CPT Pharmacometrics & Systems Pharmacology*, 10, 1433–1445.

Goodman, S., Zahurak, M.L., and Piantadosi, S. (1995). Some practical improvements in the continual reassessment method for phase I studies. *Statistics in Medicine*, 14, 1149–1161.

Gossage, D.L., Khatry, D.B., Geba, G.P., Molfino, N., and Parker, J.M. (2015). *Methods of Diagnosing and Treating Pulmonary Diseases or Disorders*. United States Patent, Patent No.: US 8,961,965 B2, February 24, 2015.

Götte, H., Kieser, M., and Häring, D.A. (2014). Optimal designs for phase II proof-of-concept trials with adaptive treatment selection. *Statistics in Medicine*, 33(28), 4981–4993.

Govil, K., and Vundela, S. (2019). Utilizing artificial intelligence for efficient CRF design. *PhUSE US Connect, Pager ML12*. www.lexjansen.com/phuse-us/2019/ml/ML12.pdf. Accessed August 31, 2024.

Govindarajan, R., Berry, J.D., Paganoni, S., Pulley, M.T., and Simmons, Z. (2020). Optimizing telemedicine to facilitate amyotrophic lateral sclerosis clinical trials. *Muscle Nerve*, 62, 321–326.16.

Goyal, T., Li, J.J., and Durrett, G. (2022). News Summarization and Evaluation in the Era of GPT-3 (arXiv:2209.12356). arXiv. https://doi.org/10.48550/arXiv.2209.12356

Grady, C., Cummings, S.R., Rowbotham, M.C., McConnell, M.V., Ashley, E.A., and Kang, G. (2017). Informed consent. *New England Journal of Medicine*, 376(9), 856–867. https://doi.org/10.1056/NEJMra1603773

Graf, A.C., Posch, M., König, F., and Bauer, P. (2015). Adaptively enriched designs in clinical trials. *Statistics in Medicine*, 34(17), 2331–2351.

Grahnén, A., and Karlsson, M.O. (2001). Concentration-controlled or effect-controlled trials. *Clinical Pharmacokinetics*, 40, 317–325.

Green, S.J., and Dahlberg, S. (1992). Planned versus attained design in phase II clinical trials. *Statistics in Medicine*, 11, 853–862.

Grömping, U. (2009). Variable importance assessment in regression: Linear regression versus random forest. *The American Statistician*, 63, 308–319. www.tandfonline.com/doi/abs/10.1198/tast.2009.08199.

Guenther, W. (1972). Simple approximation to the negative binomial (and regular binomial). *Technometrics*, 14, 385–398.

Guthrie, Carpenter, J., et al. (2019). Emergence of digital biomarkers to predict and modify treatment efficacy: Machine learning study. *BMJ*, e030710. https://doi.org/10.1136/bmjopen-2019-030710

Haddad, T., et al. (2021). Accuracy of an artificial intelligence system for cancer clinical trial eligibility screening: Retrospective pilot study. *JMIR Medical Informatics*, 9(3). https://medinform.jmir.org/2021/3/e27767/. Accessed January 14, 2025.

Haight, F.A. (1967). *Handbook of the Poisson Distribution*. John Wiley & Sons.

Hamburg, M.A. (2010). Shattuck lecture. Innovation, regulation, and the FDA. *New England Journal of Medicine*, 363(23), 2228–2232.

Hampson, L.V, Björn, H., et al. (2021). Improving the assessment of the probability of success in late stage drug development. https://arxiv.org/abs/2102.02752. Accessed January 13, 2025.

Hampson, L.V, Björn, H., et al. (2022). A new comprehensive approach to assess the probability of success of development programs before pivotal trials. *Clinical Pharmacology & Therapeutics*, 111(5), 1050–1060.

Harrell, F.E., Lee, K.L., and Mark, D.B. (1996). Multivariable prognostic models: Issues in developing models, evaluating assumptions and adequacy, and measuring and reducing errors. *Statistics in Medicine*, 15, 361–387.

Harrer, S., Shah, P., Antony, B., and Hu, J. (2019). Artificial intelligence for clinical trial design. *Trends in Pharmacological Sciences*, 40(8), 577–591. https://doi.org/10.1016/j.tips.2019.06.004

Hassanzadeha, H., Karimib, S., and Nguyena, A. (2023). Matching patients to clinical trials using semantically enriched document representation. *Journal of Biomedical Informatics*. https://doi.org/10.1016/j.jbi.2020.103406. Accessed June 22, 2024.

Hay, M., Thomas, D.W., Craighead, J.L., Economides, C., and Rosenthal, J. (2014). Clinical development success rates for investigational drugs. *Nature Biotechnology*, 32(1), 40–51. https://doi.org/10.1038/nbt.2786

He, W., Liu, J., Binkowitz, B., and Quan, H. (2006). A model-based approach in the estimation of the maximum tolerated dose in phase I cancer clinical trials. *Statistics in Medicine*, 25(12), 2027–2042.

He, Z., Tang, X., Yang, X., Guo, Y., George, T.J., Charness, N., Quan Hem, K.B., Hogan, W., and Bian, J. (2020). Clinical trial generalizability assessment in the big data era: A review. *Clinical and Translational Science*, 13(4), 675–684.

Hee, S.W., Hamborg, T., Day, S., Madan, J., Miller, F., Posch, M., and Stallard, N. (2016). Decision-theoretic designs for small trials and pilot studies: A review. *Statistical Methods in Medical Research*, 25(3), 1027–1047.

Hegge, S., Thunecke, M., et al. (2020). Predicting success of Phase III trials in oncology, medRxiv. www.medrxiv.org/content/10.1101/2020.12.15.20248240v1.full. Accessed September 22, 2024.

Heitjan, D.F. (1997). Bayesian interim analysis of phase II cancer clinical trials. *Statistics in Medicine*, 16, 1791–802.

Heitjan, D.F., Ge, Z., and Ying, G.-S. (2015). Real-time prediction of clinical trial enrollment and event counts: A review. *Contemporary Clinical Trials*, 45, 26–33.

Hernán, M.A., and Robins, J.M. (2020). *Causal Inference: What If*. Chapman & Hall/CRC.

Heyd, J.M., and Carlin, B. (1999). Adaptive design improvements in continual reassessment method for phase I studies. *Statistics in Medicine*, 18(11), 1307–1321.

Hirakawa, A., Asano, J., Sato, H., and Teramukai, S. (2018). Master protocol trials in oncology: Review and new trial designs. *Contemporary Clinical Trials Communications*, 12, 1–8.

Hoffman, D. (2004). Negative binomial control limits for count data with extra-Poisson variation. *Pharmaceutical Statistics*, 2(2), 127–132.

Höglinger, G.U., Litvan, I., Mendonca, N., Wang, D., Zheng, H., RendenbachMueller, B., et al. (2021). Safety and efficacy of tilavonemab in progressive supranuclear palsy: A phase 2, randomised, placebo-controlled trial. *Lancet Neurology*, 20(3), 182–192.

Holford, N., Ma, S.C., and Ploeger, B.A. (2010). Clinical trial simulation: A review. *Clinical Pharmacology & Therapeutics*, 88(2), 166–182.

Holmes, J., Liu, Z., Zhang, L., Ding, Y., Sio, T.T., McGee, L.A., Ashman, J.B., Li, X., Liu, T., Shen, J., and Liu, W. (2023). Evaluating large language models on a highly-specialized topic, radiation oncology physics (arXiv:2304.01938). arXiv. https://doi.org/10.48550/arXiv.2304.01938

Holmgren, E.B. (2014). *Theory of Drug Development*. CRC Press.

Hsu, Y-C., and Shen, S. (2019). Testing treatment effect heterogeneity in regression discontinuity designs. *Journal of Economics*, 208(2), 468–486.

Hu, M., Babiskin, A., Wittayanukorn, S., Schick, A., Rosenberg, M., Gong, X., Kim, M.-J., Zhang, L., Lionberger, R., and Zhao, L. (2019). Predictive analysis of first abbreviated new drug application submission for new chemical entities based on machine learning methodology. *Clinical Pharmacology and Therapeutics*. https://doi.org/10.1002/cpt.1479

Huang, X., and Chiu, Y.-L. (2023). Machine learning for precision medicine. In *Data Science, AI, and Machine Learning in Drug Development*, edited by Yang, H. CRC Press, pp. 145–176.

Huque, M., and Mushti, S. (2015). Alpha-cycling for the analyses of primary and secondary endpoints of clinical trials. www.bassconference.org/tutorials/BASS%202015%20Huque%20Mushti.pdf. Accessed September 14, 2024.

Hurst, J.R., Vestbo, J., Anzueto, A., Locantore, N., Müllerova, H., Tal-Singer, R., et al. (2010). Susceptibility to exacerbation in chronic obstructive pulmonary disease. Evaluation of COPD longitudinally to identify predictive surrogate endpoints (ECLIPSE) investigators. *The New England Journal of Medicine*, 363(12), 1128–1138.

Hutson, M. (2024). Cutting to the chase. *Nature*, 627, s2–s5.

Hyman, D.M., et al. (2015). Vemurafenib in multiple nonmelanoma cancers with BRAF V600 mutations. *New England Journal of Medicine*, 373(8), 726–736.

ICH. (1995). ICH Guideline E3: Structure and Content of Clinical Study Reports. https://database.ich.org/sites/default/files/E3_Guideline.pdf. Accessed August 31, 2024.

ICH. (1998). ICH E9 Statistical Principles for Clinical Trials. ICH E9 statistical principles for clinical trials – Scientific guideline | European Medicines Agency (europa.eu). Accessed July 4, 2024.

ICH. (2016). ICH E6(R2) Good Clinical Practice: Integrated Addendum to ICH E6(R1). International Council for Harmonisation of Technical Requirements for Pharmaceuticals for Human Use. www.hhs.gov/guidance/document/e6r2-good-clinical-practice-integrated-addendum-ich-e6r1. Accessed September 9, 2024.

ICH. (2021a). ICH E8(R1) General Considerations for Clinical Studies. https://database.ich.org/sites/default/files/E8-R1_Guideline_Step4_2021_1006.pdf. Accessed January 14, 2025.

ICH. (2021b). Guidance for Industry: E2F Development Safety Update Report. www.fda.gov/media/71255/download. Accessed September 1, 2024.

ICH. (2023). E6(R3) Good Clinical Practice (GCP). https://database.ich.org/sites/default/files/ICH_E6%28R3%29_DraftGuideline_2023_0519.pdf. Accessed September 1, 2024.

ICMRA. (2021). Horizon scanning assessment report – artificial intelligence. www.icmra.info/drupal/sites/default/files/2021-08/horizon_scanning_report_artificial_intelligence.pdf. Accessed January 14, 2025.

Infosys BPM. (n.d.) Adapting to procurement changes with AI. www.infosysbpm.com/blogs/sourcing-procurement/adapting-to-procurement-changes-with-ai.html?cmpid=pm_ser_Gen_AI_Focus_google_sourcingprocurement_textlink_06082024_7fa5161384883ac9c76294b8d500da8b_ai%20contract%20lifecycle%20management_g_1018688_21543998831&gad_source=1&gclid=EAIaIQobChMIoLSs6pWjiAMVY09HAR1u8BeoEAAYASAAEgJsK_D_BwE. Accessed September 1, 2024.

Ionan, A.C., Clark, J., Travis, J., Amatya, A., Scott, J., Smith, J.P., Chattopadhyay, S., Salerno, M.J., and Rothmann, M. (2023). Bayesian methods in human drug and biological products development in CDER and CBER. *Therapeutic Innovation & Regulatory Science*, 57(3), 436–444. https://doi.org/10.1007/s43441-022-00483-0. Epub December 2, 2022.

IQVIA. (2020). Modernizing the Nature History of Disease Research – IQVIA Perspectives from Human Data Science Lab. www.iqvia.com/library/publications/evolving-the-understanding-of-the-natural-history-of-disease. Accessed July 10, 2024.

Ironclad. (n.d.). What is AI for Contract Management. https://ironcladapp.com/journal/contract-management/ai-contract-management/. Accessed September 1, 2024.

Ivanova, A., Paul, B., Marchenko, O., Song, G., Patel, N., and Moschos, S.J. (2016). Nine-year change in statistical design, profile, and success rates of Phase II oncology trials. *Journal of Biopharmaceutical Statistics*, 26(1), 141–149.

Izmailova, E.S., Wagner, J.A., and Perakslis, E.D. (2018). Wearable devices in clinical trials: Hype and hypothesis. *Clinical Pharmacology & Therapeutics*, 104(1), 42–52.

Jacobson, I.M., McHutchison, J.G., Dusheiko, G., Di Bisceglie, A.M., Reddy, K.R., Bzowej, N.H., … and Zeuzem, S. (2011). Telaprevir for previously untreated chronic hepatitis C virus infection. *New England Journal of Medicine*, 364(25), 2405–2416. https://doi.org/10.1056/NEJMoa1012912

James, G., et al. (2013). *An Introduction to Statistical Learning*. Springer.

Jeffreys, H. (1961). *Theory of Probability*. 3rd ed, Oxford University Press.

Jha, A. K., et al. (2009). Use of electronic health records in U.S. hospitals. *New England Journal of Medicine*, 360(6), 1628–1638.

Ji, Y., Li, Y., and Bekele, B.N. (2007). Dose-finding in oncology clinical trials based on toxicity probability intervals. *Clinical Trials*, 4, 235–244.

Johnson, O. (2015). An evidence-based approach to conducting clinical trial feasibility assessments. *Clinical Investigation*, 5(5), 491–499.

Jonker, A.H., Day, S., Gabaldo, M., Stone, H., de Kort, M., O'Connor, D.J., and Pasmooij, A.M.G. (2023). IRDiRC Drug Repurposing Guidebook: Making better use of existing drugs to tackle rare diseases. *Nature Review Drug Discovery*, 22(12), 937–938.

Jung, S.H., Lee, T., Kim, K.M., and George, S.L. (2004). Admissible two-stage designs for phase II cancer clinical trials. *Statistics in Medicine*, 23, 561–569.

Jung, Y.L., Kim, K.M., et al. (2022). Artificial intelligence-based decision support model for new drug development planning. *BMC Bioinformatics*, 21(Suppl 4), 145.

Jung, Y.L., Yoo, H.S., and Hwang, J. (2020). Artificial intelligence-based decision support model for new drug development planning. *Expert Systems with Applications*, 198. https://doi.org/10.1016/j.eswa.2022.116825

Juran, J.M. (1992). *Quality by Design: The New Steps for Planning Quality into Goods and Services*. The Free Press.

Kalscheur, M.M., Kipp, R.T., Tattersall, M.C., Mei, C., Buhr, K.A., DeMets, D.L., et al. (2018). Machine learning algorithm predicts cardiac resynchronization therapy outcomes: Lessons from the companion trial. *Circulation: Arrhythmia and Electrophysiology*, 11(1), e005499.

Kang, T., et al. (2017). EliE: An open-source information extraction system for clinical trial eligibility criteria. *Journal of the American Informatics Association*, 24(6), 1062–1071.

Kaplan, R., et al. (2017). FOCUS4: A seamless trial design for biomarker-driven trials of multiple agents in colorectal cancer. *Annals of Oncology*, 28(5), 1007–1013.

Karia, S., and Taylor, B. (2020). Using AI to Drive Automation for Clinical Data, from Protocol to Submission. www2.deloitte.com/content/dam/Deloitte/us/Documents/technology/us-deloitte%E2%80%99s-automated-clinical-life-cycle-management-product.pdf. Accessed August 31, 2024.

Kasenda, B., et al. (2024). Prevalence, characteristics, and publication of discontinued randomized trials. *JAMA*, 311(10), 1045–1051.

Kavalci, E., and Hartshorn, A. (2023). Improving clinical trial design using interpretable machine learning based prediction of early trial termination. *Scientific Reports*, 13, 121.

Kazandijan, D.G., Gong, Y., Kazandijian, H., Pazdur, R., and Blumenthal, G.M. (2018). Exploration of baseline derived neutrophil to lymphocyte ration (dNLR) and lactate dehydrogenase (LDH) in patients (pts) with metastatic non-small cell lung cancer (mNSCLC) treated with immune checkpoint inhibitors (ICI) or cytotoxic chemotherapy (CCT). *Journal of Clinical Oncology*, 36(15_suppl), 3035.

Khatry, D.B., Gossage, D.L., Geba, G.P., Parker, J.M., Jarjour, N.N., Busse, W.W., and Molfino, N.A. (2015). Discriminating sputum-eosinophilic asthma: Accuracy of cutoffs in blood eosinophil measurements versus a composite index, ELEN. *Journal of Allergy and Clinical Immunology*, 136(3), 812–814.e2.

Khin, N.A., Francis, G., Mulinde, J., et al. (2020). Data Integrity in global clinical trials: Discussions from Joint US FDA and MHRA UK Good Clinical Practice Workshop. *Clinical Pharmacology & Therapeutics*, 108(5), 1–18.

Kidwell, K.M., Roychoudhury, S., et al. (2022). Application of Bayesian methods to accelerate rare disease drug development: Scopes and hurdles. *Orphanet Journal of Rare Disease*, 17, 86.

Kihara, Y., Heeren, T.F.C., Lee, C.S., et al. (2019). Estimating retinal sensitivity using optical coherence tomography with deep-learning algorithms in macular telangiectasia type 2. *JAMA Network Open*, 2, e188029.

Kikawa, M., and Nakajima, Y. (2020). Hot to let machine learning clinical data review as it can support reshaping the future of clinical data cleaning process. *PharmaSUG* 2020-paper AI-224.

Kim, J.H., and Scialli, A.R. (2011). Thalidomide: The tragedy of birth defects and the effective treatment of disease. *Toxicological Sciences*, 122(1), 1–6.

Kim, J.H., Ta, C.N., et al. (2020). Towards clinical data driven eligibility criteria optimization for interventional COVID-19 clinical trials. *Journal of the American Medical Informatics Association*, 28(1), 14–22.

Kirby, J.C., et al. (2016). PheKB: A catalog and workflow for creating electronic phenotype algorithms for transportability. *JAMA*, 23(6), 1046–1052. https://doi.org/10.1093/jamia/ocv202

Kitaguchi, Y., Komatsu, Y., Fujimoto, K., Hanaoka, M., and Kubo, K. (2012). Sputum eosinophilia can predict responsiveness to inhaled corticosteroid treatment in patients with overlap syndrome of COPD and asthma. *International Journal of Chronic Obstructive Pulmonary Disease*, 7, 283–289.

Koesmahargyo, V., Abbas, A., Zhang, L., Guan, Feng, S., Yadav, V., and Galatzer-Levy, I.R. (2020). Accuracy of machine learning-based prediction of medication adherence in clinical research. *Psychiatry Research*, 294, 1–7.

Kola, I., and Landis, J. (2004). Can the pharmaceutical industry reduce attrition rates? *Nature Reviews Drug Discovery*, 3(8), 711–716.

Kolla, L., Gruber, F.K., Khalid, O., Hill, C., and Parikh, R.B. (2021). The case for AI-driven cancer clinical trials - the efficacy arm in silico. *Biochim et Biophys acta Reviews Cancer*, 1876(1). https://doi.org/10.1016/J.BBCAN.2021.188572.

Korn, E.L., Midthune, D., Chen, T.T., Rubinstein, L.V., Christain, M.C., and Simon, R. (1994). A comparison of two phase I trial designs. *Statistics in Medicine*, 13, 1799–1806.

Kraiczi, H., Jang, T., Ludden, T., and Peck, C.C. (2003). Randomized concentration-controlled trials: Motivations, use, and limitations. *Clinical Pharmacology & Therapeutics*, 74(3), 203–214.

Krisam, J., and Kieser, M. (2014). Decision rules for subgroup selection based on a predictive biomarker. *Journal of Biopharmaceutical Statistics*, 24(1), 188–202.

Krishnankutty, B., et al. (2012). Data management in clinical research: An overview. *Indian Journal of Pharmacology*, 42(2), 168–172.

Kulkarni, V.S., Alagarsamya, V., Solomona, V.R., Josea, P.A., and Murugesan, S. (2023). Drug repurposing: An effective tool in modern drug discovery. *Russian Journal of Bioorganic Chemistry*, 49(2), 157–166.

Kumar-Sinha, C., and Chinaiyan, M.A. (2018). Precision oncology in the age of integrative genomics. *Nature Biotechnology*, 36(1), 46–60.

Kyriazopoulou, E., Panagopoulos, P., et al. (2021). An open label trial of anakinra to prevent respiratory failure in COVID-19. *eLife*, 10, e66125.

Labovitz, D.L., Shafner, L., Reyes Gil, M., Virmani, D., and Hanina, A. (2017). Using Artificial intelligence to reduce the risk of nonadherence in patients on anticoagulation therapy. *Stroke*, 48, 1416–1419. https://doi.org/10.1161/STROKEAHA.116.016281

Lalonde, R.L., Kowalski, K.G., Hutmacher, M.M., Ewy, W., Nichols, D.J., Milligan, P.A., et al. (2007). Model-based drug development. *Clinical Pharmacology & Therapeutics*, 82(1), 21–32.

Lalonde, R.L., and Peck, C.C. (2022). Probability ofsuccess: A crucial concept to inform decisionmaking in pharmaceutical research anddevelopment. *Clinical Pharmacology & Therapeutics*, 111,1001–1003.

Lan, Y., Tang, G., and Heitjan, D.F. (2019). Statistical modeling and prediction of clinical trial recruitment. *Statistics in Medicine*, 38(6), 945–955.

Landray, M.J., Gradinetti, C., Kramer, J.M., Morrison, B.W., Ball, L., et al. (2012). Clinical trials: Rethinking how we ensure quality. *Drug Information Journal*, 46, 657–660.

Le Tourneau, C., Lee, J.J., and Siu, L.L. (2009). Dose escalation methods in phase I cancer clinical trials. *Journal of the National Cancer Institute*, 101(10), 708–720.

Leacy, F.P., and Stuart, E.A. (2014). On the joint use of propensity and prognostic scores in estimation of the Average Treatment Effect on the Treated: A simulation study. *Statistics in Medicine*, 33(20), 3488–3508.

LeCun, Y., Bengio, Y., and Hinton, G. (2015). Deep learning. *Nature*, 521(7553), 436–444.

Lee, C.S., and Lee, A.Y. (2020). How artificial intelligence can transform randomized controlled trials. *Translational Vision Science & Technology*, 9(2). https://doi.org/10.1167/TVST.9.2.9

Lee, H., et al. (2018). Prediction of bispectral index during target-controlled infusion of propofol and remifentanil. *Anesthesiology*, 128(3), 492–501.

Lee, J.J., and Liu, D.D. (2008). A predictive probability design for phase II cancer trials. *Clinical Trials*, 5, 93–106.

Lee, P., Bubeck, S., and Petro, J. (2023). Benefits, limits, and risks of GPT-4 as an AI chatbot for medicine | NEJM. *The New England Journal of Medicine*, 388, 1233–1239.

Leigh, R., Pizzichini, M.M.M., Morris, M.M., Maltais, F., Hargreave, F.E., and Pizzichini, E. (2006). Stable COPD: Predicting benefit from high-dose inhaled corticosteroid treatment. *European Respiratory Journal*, 27(5), 946–971.

Leon, A.C., Mallinckrodt, C.H., Chuang-Stein, C., Archibald, D.G., Archer, G.E., and Chartier, K. (2007). Attrition in randomized controlled clinical trials: Methodological issues in psychopharmacology. *Biological Psychiatry*, 59, 1001–1005.

Li, J., Chen, G., Lin, J., Chi, A., and Daviess, S. (2021). External control using RWE and historical data in clinical development. In *Real-World Evidence in Drug Development and Evaluation*, edited by Yang, H. CRC Press, pp. 71–100.

Li, Q.H., Deng, Q., and Ting, N. (2021). Proof of concept: Drug selection? Or dose selection? Thoughts on multiplicity issues. *Therapeutic Innovation & Regulatory Science*, 55, 1001–1005.

Li, X., and Tang, Q. (2021). Introduction to artificial intelligence and deep learning with a case study in analyzing electronic health records for drug development. In *Real-World Evidence in Drug Development and Evaluation*, edited by Yang, H., and Yu, B., CRC Press, pp. 151–172.

Li, X., Liu, C., Mao, Z., Xiao, M., and Wang, W. (2020). Predictive value of the neutrophil-to-lymphocyte ratio on mortality in patients with COVID-19. *Journal of Clinical Virology*, PMC7667659.

Li, X., Sharan, S., Wang, H., and Hu, M. (n.d.) Deep learning-based pharmacokinetic modeling for drugs with complex pharmacokinetic profiles. Unpublished manuscript.

Liang, M., Ye, T., and Fu, H. (2018). Estimating individualized optimal combination therapies through outcome weighted deep learning algorithms. *Statistics in Medicine*, 37(27), 3869–3886.

Lin, Y., and Shih, W.J. (2001). Statistical properties of the traditional algorithm-based designs for phase I cancer clinical trials. *Biostatistics*, 2(2), 203–215.

Lipkovich, I., Dmitrienko, A., and D' Agostino Sr, R.B. (2017). Tutorial in biostatistics: Data-driven subgroup identification and analysis in clinical trials. *Statistics in Medicine*, 36(1), 136–196. https://doi.org/10.1002/sim.7064

Lipsky, M.S., and Sharp, L.K. (2001). From idea to market: The drug approval process. *Journal of American Board of Family Medicine*, 14(5), 362–367.

Little, R.J.A., and Rubin, D.B. (2002). *Statistical Analysis with Missing Data*. Wiley Series in Probability and Statistics.

Liu, J., et al. (2020). Bayesian accrual modeling and prediction in multicenter clinical trials with varying center activation times. *Pharmaceutical Statistics*, 19(5), 692–703.

Liu, R., Rizzo, S., Whipple, S., Pal, N., Pineda, A.L., Lu, M., Arnieri, B., Lu, Y., Capra, W., Copping, R., et al. (2021). Evaluating eligibility criteria of oncology trials using real-world data and AI. *Nature*, 592(7855), 629–633.

Liu, S., and Yuan, Y. (2015). Bayesian optimal interval designs for phase I clinical trials. *Journal of the Royal Statistical Society: Series C*, 64, 507–523.

Liu, X., Shi, C., et al. (2021). A scalable AI approach for clinical trial cohort optimization. https://arxiv.org/abs/2109.02808. Accessed July 2024.

Liu, X., Shi, C., Deore, U., Wang, Y., Tran, M., Kalil, I., and Devarakonda, M. (2021). A scalable AI approach for clinical trial cohort optimization. In *Machine Learning and Principles and Practice of Knowledge Discovery in Databases*, pp. 479–489. https://arxiv.org/abs/2109.02808. Accessed July 2024.

Liu, Y., Liu, P., Radev, D., and Neubig, G. (2022). BRIO: Bringing Order to Abstractive Summarization (arXiv:2203.16804). arXiv. https://doi.org/10.48550/arXiv.2203.16804

Lo, A.W., et al. (2018). Machine Learning with Statistical Imputation for Predicting Drug Approvals. Available at SSRN: https://ssrn.com/abstract=2973611 or https://doi.org/10.2139/ssrn.2973611. Accessed September 15, 2024.

Lo, A.W., Siah, K.W., and Wong, C.H.W. (2017). Machine-learning Models for Predicting Drug Approvals and Clinical-phase Transitions. https://economics.harvard.edu/files/economics/files/lo-andrew_paper-4_sbbi-4-20-18_predictive_15.pdf. Accessed September 22, 2024.

Lopez, A.D., Shibuya, K., Rao, C., Mathers, C.D., Hansell, A.L., Held, L.S., Schmid, V., and Buist, S. (2006). Chronic obstructive pulmonary disease: Current burden and future projections. *European Respiratory Journal*, 27, 397–412.

Lu, M., Sadiq, S., Feaster, D.J., and Ishwaran, H. (2018). Estimating individual treatment effect in observational data using random forest methods. *Journal of Computational and Graphical Statistics*, 27(1), 209–219.

Madabushi, R., Seo, P., Zhao, L., et al. (2022). Review: Role of model-informed drug development approaches in the lifecycle of drug development and regulatory decision-making. *Pharmaceutical Research*, 39, 1669–1680.

Magda, S. (2010). Susceptibility to exacerbation in chronic obstructive pulmonary disease – Data from the ECLIPSE study. *Maedica*, 5(3), 223–224.

Mahajan, P., Uddin, S., et al. (2023). Ensemble learning for disease prediction: A review. *Healthcare (Basel)*, 11(12), 1808. https://doi.org/10.3390/healthcare11121808

Mahlich, J., Bartol, A., and Dheban, S. (2021). Can adaptive clinical trials help to solve the productivity crisis of the pharmaceutical industry? – A scenario analysis. *Health Economics Review*, 11(1), 4. https://doi.org/10.1186/s13561-021-00302-6

Makady, A., Ham, R.T., de Boer, A., Hillege, H., Klungel, O., and Goettsch, W. (2017). Policies for use of real-world data in health technology assessment (HTA): A comparative study of six HTA agencies. *Value in Health*, 20(4), 520–532.

Makady, A., van Veelen, A., Jonsson, P., Moseley, O., D'Andon, A., and de Boer, A. (2017). Using real-world data in health technology assessment (HTA) practice: A comparative study of five HTA agencies. *PharmacoEconomics*, 35(1), 89–103. https://doi.org/10.1007/s40273-017-0559-5

Maleki, F., Muthukrishnan, N., et al. (2020). Machine learning algorithm validation: From essentials to advance applications and implications for regulatory certification and deployment. *Neuroimaging Clinics of North America*, 30, 433–445.

Mannino, D.M., and Buist, A.S. (2007). Global burden of COPD: Risk factors, prevalence, and future trends. *Lancet*, 370(9589), 765–773.

Massard, C., et al. (2017). High-Throughput genomics and clinical outcome in hard-to-treat advanced cancers: Results of the MOSCATO 01 trial. *Cancer Discovery*, 7(6), 586–595.

Mayo, C.S., Matuszak, M.M., Schipper, M.J., Jolly, S., Hayman, J.A., and Haken, R.K.T. (2017). Big data in designing clinical trials: Opportunities and challenges. *Frontiers in Oncology*, 7, 187. https://doi.org/10.3389/fonc.2017.00187

McElvaney, O.J., McEvoy, N.L., et al. (2020). Characterization of the inflammatory response to severe COVID-19 illness. *American Journal of Respiratory and Critical Care Medicine*, 202, 812–821.

McMurray, J.J., Packer, M., Desai, A.S., Gong, J., Lefkowitz, M.P., Rizkala, A.R., ... and PARADIGM-HF Investigators and Committees. (2014). Angiotensin–neprilysin inhibition versus enalapril in heart failure. *New England Journal of Medicine*, 371(11), 993–1004. https://doi.org/10.1056/NEJMoa1409077

McNeil, B.J., Keller, E., and Adelstein, S.J. (1975). Primer on certain elements of medical decision making. *The New England Journal of Medicine*, 293(5), 211–215. https://doi.org/10.1056/NEJM197507312930501

MedDRA. (n.d.) What is MedDRA and how is it used? www.meddra.org/training/schedule/1677/What-is-MedDRA-and-How-is-it-Used%3F. Accessed August 31, 2024.

Meeker-O'Connell, A., Glessner, C., Behm, M., et al. (2016). Enhancing clinical evidence by integrating quality by design (QbD) principles in clinical trials: A review. *Therapeutic Innovation & Regulatory Science*, 50(1), 22–31.

Meeker-O'Connell, A., Sam, L.M., et al. (2016). TransCelerte's clinical quality management systems: From a vision to a conceptual framework. *Therapeutic Innovation & Regulatory Science*, 50(4), 397–413.

Ménard, T., Barmaz, Y., Koneswarakantha, B., Bowling, R., and Popko, L. (2019). Enabling data-driven clinical quality assurance: Predicting adverse event reporting in clinical trials using machine learning. *Drug Safety*, 42, 1045–1053. https://doi.org/10.1007/s40264-019

Metz, C.E. (1978). Basic principles of ROC analysis. *Seminars in Nuclear Medicine*, 8(4), 283–298.

Mezquita, L., Auclin, E., Ferrara, R., Charrier, M., Remon, J., et al. (2018). Association of the Lung Immune Prognostic Index with immune checkpoint inhibitor outcomes in patients with advanced non-small cell lung cancer. *JAMA Oncology*, 4(3), 351–357.

Mitsi, G., et al. (2022). Implementing digital technologies in clinical trials: Lessons learned. *Innovations in Clinical Neuroscience*, 19(4), 65 69.

Mittermaier, M., Venkatesh, K.P., and Kvedar, J.C. (2023). Digital health technology in clinical trials. *npj Digital Medicine*, 6, 88.

Molfino, N.A., Gossage, D., Kolbeck, R., Parker, J.M., and Geba, G.P. (2012). Molecular and clinical rationale for therapeutic targeting of interleukin-5 and its receptor. *Clinical & Experimental Allergy*, 42(5), 712–737.

Moller, S. (1995). An extension of the continual reassessment methods using a preliminary up-and-down design in a dose finding study in cancer patients, in order to investigate a greater range of doses. *Statistics in Medicine*, 14, 911–922.

Montgomery, D.C. (2013). *Introduction to Statistical Quality Control*. 7th edition. Wiley.

Morgan, S.L., and Winship, C. (2007). *Counterfactuals and Causal Inference: Methods and Principles for Social Research*. Cambridge University Press.

Morita, S., Thall, P., and Muller, P. (2008). Determining the effective sample size of a parametric prior, *Biometrics*, 64, 595–602.

Munsaka, M., Liu, M., Xing, Y., and Yang, H. (2022). Leveraging machine learning, natural language processing, and deep learning in drug safety and pharmacovigilance. In *Data Science, AI, and Machine Learning in Drug Development*, edited by Yang, H. CRC Press, pp. 193–230.

Murphy, S.N., Mendis, M.E., Hackett, K., Kuttan, R., Pan, W., Phillips, L.C., … and Kohane, I.S. (2007). Architecture of the open-source clinical research chart from informatics for integrating biology and the bedside. *AMIA Annual Symposium Proceedings*, 548–552.

Nantz, E., Liu-Sheifert, H., and Skyljarevski, V. (2009). Predictors of premature discontinuation of treatment in multiple disease states. *Patient Preference and Adherence*, 3, 31–43.

National Research Council. (2011). *Toward Precision Medicine: Building a Knowledge Network for Biomedical Research and a New Taxonomy of Disease*. The National Academies Press.

NCI (2021). *ALCHEMIST: Adjuvant Lung Cancer Enrichment Marker Identification and Sequencing Trial*. ClinicalTrials.gov Identifier: NCT02194738.

Nilsson, A., Bonander, C., Strömberg, U., et al. (2019). Assessing heterogeneous effects and their determinants via estimation of potential outcomes. *European Journal of Epidemiology*, 34(9), 823–835.

Nova In Silico. (2023). Conquer the complexity of biology with clinical trial simulations. www.novainsilico.ai/. Accessed July 7, 2024.

O'Brian, P.C., and Fleming, T.R. (1979). A multiple testing procedure for clinical trials. *Biometricka*, 35, 549–556.

O'Hagan, A. (2019). Expert knowledge elicitation: Subjective but scientific. *The American Statistician*, 73(1), 69–81.

O'Quigley, J. (1992). Estimating the probability of toxicity at the recommended dose following a phase I clinical trial in cancer. *Biometrics*, 48, 885–893.

O'Quigley, J., and Chevret, S. (1991). Methods for dose-finding studies in cancer clinical trials: A review and results of a Monte Carlo study. *Statistics in Medicine*, 10, 1647–1664.

O'Quigley, J., Pepe, M., and Fisher, L. (1990). Continual reassessment method: A practical design for phase I clinical trials in cancer. *Biometrics*, 2(2), 203–215.

O'Quigley, J., and Zhen, L.Z. (1996). Continual reassessment methods: A likelihood approach. *Biometrics*, 52, 673–684.

Obermeyer, Z., and Emanuel, E.J. (2016). Predicting the future – big data, machine learning, and clinical medicine. *NEJM*, 375(13), 1216–1219.

Ondra, T., Jobjörnsson, S., Beckman, R.A., Posch, M., and Graf, A. (2016). Optimizing trial designs for targeted therapies. *Journal of Biopharmaceutical Statistics*, 26(1), 120–135.

OpenAI. (2022, November 30). *Introducing ChatGPT*. https://openai.com/blog/chatgpt

OpenAI. (2023). GPT-4 Technical Report (arXiv:2303.08774). arXiv. https://doi.org/10.48550/arXiv.2303.08774

Padmaja, A.V., Vemulapally, A., and Srikanth, J. (2024). Personalized eProtocol design: A data-driven approach with AI. *International Journal of Trend in Scientific Research and Development (IJTSRD)*, 8(1), 555–558.

Papadimitrakopoulou, V.A., et al. (2018). Lung-MAP: A biopharma/academic partnership for genomically driven precision medicine in squamous cell lung cancer. *Journal of Clinical Oncology*, 36(15), 1405–1406.

Park, J.J., Siden, E., Zoratti, M., et al. (2019). Systematic review of basket trials, umbrella trials, and platform trials: A landscape analysis of master protocols. *Trials*, 20(57). https://trialsjournal.biomedcentral.com/articles/10.1186/s13063-019-3664-1. Accessed January 14, 2025.

Parulekar, W.R., and Eisenhauer, E.A. (2004). Phase I trial design for solid tumor studies of targeted, non-cytotoxic agents: Theory and practice. *Journal of the National Cancer Institute*, 96(13), 990–997.

Paul, S.M., Mytelka, D.S., Dunwiddie, C.T., Persinger, C.C., Munos, B.H., Lindborg, S., R., and Schacht, A.L. (2010). How to improve R&D productivity: The pharmaceutical industry's grand challenge. *Nature Review Drug Discovery*, 9, 203–214.

Peck, C. (1990). The randomized concentration-controlled clinical trial (CCT): An information-rich alternative to the randomized placebo controlled clinical Trial (PCT). *Clinical Pharmacology & Therapeutics*, 47, 2–148.

Pedersen, T.R., Faergeman, O., et al. (2004). Design and baseline characteristics of incremental decrease in end points through aggressive lipid lowering (IDEAL) study group. (1994). Randomised trial of cholesterol lowering in 4444 patients with coronary heart disease: The Scandinavian Simvastatin Survival Study (4S). *Lancet*, 344(8934), 1383–1389. https://doi.org/10.1016/S0140-6736(94)90566-5

Peng, Y., Wang, N., et al. (2020). Effeteness of mobile applications on medication adherence in adults with chronic diseases: A systematic review and meta-analysis. *Journal of Managed Care & Specialty Pharmacy*, 26(4), 550–561.

Perperoglou et al. (2023). Modeling time-varying recruitment rates in multicenter clinical trials. *Biometrical Journal*, 65(6), e2100377. https://pubmed.ncbi.nlm.nih.gov/36287068/. Accessed January 14, 2025.

Piantadosi, S. (1997). *Clinical Trials: A Methodologic Perspective*. Wiley.

Piantadosi, S., and Liu, G. (1996). Improved design for dose escalation studies using pharmacokinetic measurements. *Statistics in Medicine*, 15, 1605–1618.

Piccart-Gebhart, M.J., Procter, M., Leyland-Jones, B., Goldhirsch, A., Untch, M., Smith, I., … and Herceptin Adjuvant (HERA) Trial Study Team. (2005). Trastuzumab after adjuvant chemotherapy in HER2-positive breast cancer. *New England Journal of Medicine*, 353(16), 1659–1672. https://doi.org/10.1056/NEJMoa052306

Pixley, R.A., Espinola, R.G., Ghebrehiwet, B., Joseph, K., Kao, A., Bdeir, K., Cines, D.B., and Colman, R.W. (2011). Interaction of high-molecular weight kininogen with endothelial cell binding proteins suPAR, gC1qR and cytokeratin 1 determined by surface plasmon resonance (BiaCore). *Thrombosis and Haemostasis*, 105, 1053–1059. https://doi.org/10.1160/TH10-09-0591, PMID: 21544310

Ploquin, A., et al. (2012). Prediction of early death among patients enrolled in phase I trials: Development and validation of a new model based on platelet count and albumin. *British Journal of Cancer*, 107(7), 1025–1030.

PMDA (2007). Basic Principles on Global Clinical Trials. PFSB/ELD Notification No. 0921001, 2007. www.pmda.go.jp/files/000157900.pdf

Pocock, S.J. (1977). Group sequential methods in the design and analysis of clinical trials. *Biometrica*, 64, 191–199.

Pocock, S.J. (1982). Interim analyses for randomized clinical trials: The group sequential approach. *Biometrics*, 38, 153–162.

Poynton, M.R., et al. (2009). Machine learning methods applied to pharmacokinetic modelling of remifentanil in healthy volunteers: A multi-method comparison. *Journal of International Medical Research*, 37(6), 1680–1691.

PREVAIL II Writing Group (2016). A Randomized, Controlled Trial of ZMapp for Ebola Virus Infection. *NEJM*, 375, 1448–1456.

Price, D., and Scott, J. (2021). The U.S. Food and Drug Administration's complex innovative trial design pilot meeting program: Progress to date. *Clinical Trials*, 18(6), 706–710. https://doi.org/10.1177/17407745211050580

Pulkstenis, E., Patra, K., and Zhang, J. (2017). A Bayesian paradigm for decision-making in proof-of-concept trials. *Journal of Biopharmaceutical Statistics*, 27(3), 442–456.

Pulkstenis, E., Sfarjal, K., Dimitrienko, A., and Milligan, P.A. (2017). Decision-making in drug development: Integrating quantitative approaches into the TPP. In *Clinical Trial Optimization Using R*, edited by Dimitrienko, K. and Pulkstenis, E CRC Press pp. 63–92.

Qi, Y., and Tang, Q. (2019). Predicting phase 3 clinical trial results by modeling phase 2 clinical trial subject level data using deep learning. *Proceeding of Machine Learning Research*, 106, 1–14.

Rajadhyaksha, V. (2010). Conducting feasibilities in clinical trials: An investment to ensure a good study. *PICR*, 1(3), 106–109.

Rajkomar, A., Oren, E., Chen, K., et al. (2018). Scalable and accurate deep learning with electronic health records, *npj Digital Medicine*, 1(1), 18, https://doi.org/10.1038/s41746-018-0029-1

Ramsey, B.W., Davies, J., McElvaney, N.G., Tullis, E., Bell, S.C., Dřevínek, P., … and VX08-770-102 Study Group. (2011). A CFTR potentiator in patients with cystic fibrosis and the G551D mutation. *New England Journal of Medicine*, 365(18), 1663–1672. https://doi.org/10.1056/NEJMoa1105185

Ratain, M.J., Mick, R., Shilsky, R.L., and Siegler, M. (1993). Statistical and ethical issues in the design and conduct of phase I and II clinical trials of new anticancer agents. *Journal of the National Cancer Institute*, 85, 1637–1643.

Reiners, F., Sturm, J., Bouw, L.J.W., and Wouters, E.J.M. (2019). Sociodemographic factors influencing the use of ehealth in people with chronic diseases. *International Journal of Environmental Research and Public Health*, 16, 645.

Richard, R. R., Anzueto, A., and Weisman, I. (2011). Optimizing management of chronic obstructive pulmonary disease in the upcoming decade. *International Journal of Chronic Obstructive Pulmonary Disease*, 6 , 47–61.

Robert, C. (2013). Bayesian Computational Tools. https://arxiv.org/pdf/1304.2048.pdf. Accessed June 30, 2024.

Roberts, S.W. (1959). Control chart tests based on geometric moving averages. *Technometrics*, 1, 239–250.

Rogatko, A., Schoneneck, D., Jonas, W., Tighinouaet, M., Khuri, F., and Porter, A. (2007). Translation of innovative designs into phase II trials. *Journal of Clinical Oncology*, 25(31), 4982–4986.

Rosenbaum, P.R. (2020). *Design of Observational Studies*. Springer Science & Business Media.

Rostami-Hodjegan, A. (2012). Physiologically based pharmacokinetics joined with in vitro-in vivo extrapolation of ADME: A marriage under the arch of systems pharmacology. *Clinical Pharmacology & Therapeutics*, 92(1), 50–61.

Rothwell, P.M. (2005). Subgroup analysis in randomized controlled trials: Importance, indications, and interpretation. *The Lancet*, 365(9454), 176–186. https://doi.org/10.1016/S0140-6736(05)17709-5

Rothwell, R., Nikolov, N.P., Maynard, J.W., and Levin, G. (2020). Noninferiority trials to evaluate drug effects in rheumatoid arthritis. *Arthritis & Rheumatology*, 72(8), 1258–1265. https://doi.org/10.1002/art.41257

Rubin, D.B. (2005). Causal inference using potential outcomes: Design, modeling, decisions. *Journal of the American Statistical Association*, 100(469), 322–331.

Rubin, L.A. (1990). The soluble Interleukin-2 receptor: Biology, function, and clinical application. *Annals of Internal Medicine*, 113, 619–627. https://doi.org/10.7326/0003-4819-113-8-619

Ruifbach, K., Burger, H.U., and Abt, M. (2016). Bayesian predictive power: Choice of prior and some recommendations for its use as probability of success in drug development. *Pharmaceutical Statistics*, 15(5), 438–446.

Saama (2024). Accelerate Your Data Review Processes – Today. www.saama.com/wp-content/uploads/SDQ_Fact_Sheet_Jan_2024.pdf. Accessed October 20, 2024.

Saeed, S.A., and Masters, R.M. (2021). Disparities in health care and the digital divide. *Current Psychiatry Reports*, 23, 61.

Saha, S., and Brightling, C.E. (2006). Eosinophilic airway inflammation in COPD. *International Journal of Chronic Obstructive Pulmonary Disease*, 1(1), 39–47.

Sahni, N., et al. (2018). Development and validation of machine learning models for prediction of 1-year mortality utilizing electronic medical record data available at the end of hospitalization in multicondition patients: A proof-of-concept study. *Journal of General Internal Medicine*, 33, 921–928.

Saint-Hilary, G. (2019). Probability of success. PSI/EFSPI SIG QDM – Webinar 2, December 10, 2019. https://slideplayer.com/slide/17810602/. Accessed September 21, 2024.

Sakamoto, J., and Buyse, M. (2016). Fraud in clinical trials: Complex problem, simple solutions? *International Journal of Clinical Oncology*, 21(1), 13–14.

Sanathanan, L.P., and Peck, C.C. (1991). The randomized concentration-controlled trial: An evaluation of its sample size efficiency. *Contemporary Clinical Trials*, 12, 780–794.

Sanathanan, L., and Peck, C. (1993). Concentration-controlled rials: Basic concepts, design, and implementation. In *Integration of Pharmacokinetics, Pharmacodynamics, and Toxicokinetics in Rational Drug Development*, edited by Yacobi, A., Skelly, J.P., Shah, V.P. et al. Springer Nature Link, pp. 239–247. https://link.springer.com/book/10.1007/978-1-4757-1520-0. Accessed January 23, 2025.

Sanathanan, L.P., Peck, C.C., et al. (1991). Randomization, PK-controlled dosing, and titration: An integrated approach for designing clinical trials. *Drug Information Journal*, 25, 425–431.

Sangari, N., and Qu, Y. (2020). A comparative study on machine learning algorithms for predicting breast cancer prognosis in improving clinical trials. In *Proceedings – 2020 International Conference on Computational Science and Computational Intelligence, CSCI 2020*, pp. 813–818. https://doi.org/10.1109/CSCI51800.2020.00152

Sangwan, P., and Singha, P. (2023). Digital data flow (DDF) and technological solution providers. *PharmaSUG. Paper SI-006.* www.pharmasug.org/proceedings/2023/SI/PharmaSUG-2023-SI-006.pdf. Accessed August 31, 2024.

Sargent, D.J., Chan, V., and Goldberg, R.M. (2001). A three-outcome design for phase II clinical trials. *Controlled Clinical Trials*, 22(2), 117–125.

Saville, B.R., and Berry, S.M. (2016). Efficiencies of platform clinical trials: A vision of the future. *Clinical Trials*, 13(3), 358–366.

SCDM (2013). Good Clinical Data Management Practices (GCDMP). *Society for Clinical Data Management*. www.jscdm.org/collections/4/. Accessed September 9, 2024.

Scheuermann, R.H., Ceusters, W., and Smith, B. (2009). Toward an ontological treatment of disease and diagnosis. *Summit on Translational Bioinformatics*, 116–120. https://pubmed.ncbi.nlm.nih.gov/21347182/. Accessed January 25, 2025.

Schneider, L.S. (2003). Clinical antipsychotic trials of intervention effectiveness (CATIE): Alzheimer's disease trial. *Schizophrenia Bulletin*, 29(1), 57–72.

Schork, N.J. (2015). Personalized medicine: Time for one-person trials. *Nature*, 520, 609–611.

Schuhmacher, A., Gassmann, O., Hinder, M., and Kuss, M. (2021). The present and future of project management in pharmaceutical R&D. *Drug Discovery Today*, 26(1), 1–4. https://doi.org/10.1016/j.drudis.2020.07.020

Scott, M.B., Connor, J.T., and Lewis, J.R. (2015). The platform trial: An efficient strategy for evaluating multiple treatments. *JAMA*, 313(16), 1619–1620.

Sechidis, K., Papangelou, K., et al. (2018). Distinguishing prognostic and predictive biomarkers: An information theoretic approach. *Bioinformatics*, 34(19), 3365–3376.

Sen, A., Ryan, P., Goldstein, A., Chakrabarti, S., Wang, S., Koski, E., and Weng, C. (2017). Correlating eligibility criteria generalizability and adverse events using Big Data for patients and clinical trials. *Annals of the New York Academy of Sciences*, 1387(1), 34–43. https://doi.org/10.1111/nyas.13195

Shefner, J.M., Cudkowicz, M.E., Hardiman, O., Cockroft, B.M., Lee, J.H., Malik, F.I., VITALITY-ALS Study Group, et al. (2019). A phase III trial of tirasemtiv as a potential treatment for amyotrophic lateral sclerosis. *Amyotrophic Lateral Sclerosis and Frontotemporal Degeneration*, 20, 584–594.

Shelton, R. (2001). Steps following attainment of remission: Discontinuation of antidepressant therapy. *Primary Care Companion to The Journal of Clinical Psychiatry*, 3, 168–174.

Sherman, R.E., Anderson, S.A., Dal Pan, G.J., Gray, G.W., Gross, T., Hunter, N.L., … and Califf, R.M. (2016). Real-world evidence—what is it and what can it tell us. *New England Journal of Medicine*, 375(23), 2293–2297.

Shi, Y., Ren, P., Wang, J., Han, B., ValizadehAslni, T., Agbavor, F., Zhang, Y., Hu, M., Zhao, L., and Liang (2023). Leveraging GPT-4 for food effect summarization to enhance product-specific guidance development via iterative prompting. *Journal of Biomedical Informatics*, 148(12), 104533. https://doi.org/10.1016/j.jbi.2023.104533

Shiah, K.W., Kelley, N.W., et al. (2021). *Predicting Drug Approvals: The Novartis Data Science and Artificial Intelligence Challenge*. Cell Press.

Shickel, B., Tighe, P.J., Bihorac, A., and Rashidi, P. (2018). Deep HER: A survey of recent advances in deep learning techniques for electronic health record (EHR) analysis. *IEEE Journal of Biomedical and Health Informatics*, 22(5), 1589–1604. https://doi.org/10.1109/JBHI.2017.2767063

Shord, S.S., Zhu, H., Liu, J., et al. (2023). US Food and Drug Administration embraces using innovation to identify optimized dosages for patients with cancer. *CPT Pharmacometrics & Systems Pharmacology*, 12, 1573–1576.

Silverman, E.K., Vestbo, J., Agusti, A., Anderson, W., Bakke, P.S., Barnes, K.C., et al. (2011). Opportunities and challenges in the genetics of COPD 2010: An International COPD Genetics Conference report. *COPD*, 8(2), 121–135.

Simon, R. (1989). Optimal two-stage designs for phase II clinical trials. *Controlled Clinical Trials*, 10, 1–10.

Simon, R., and Maitournam, A. (2004). Evaluating the efficiency of targeted designs for randomized clinical trials. *Clinical Cancer Research*, 10(20), 6759–6763.

Simpson, J.L., Scott, R., Boyle, M.J., and Gibson, P.G. (2006). Inflammatory subtypes in asthma: Assessment and identification using induced sputum. *Respirology*, 11(1), 54–61.

Sinha, P., Matthay, M.A., and Calfee, C.S. (2020). Is a "Cytokine Storm" relevant to COVID-19? *JAMA Internal Medicine*, 180, 1152–1154. https://doi.org/10.1001/jamainternmed.2020.3313, PMID: 32602883.

Siva, R., Green, R.H., Brightling, C.E., Shelley, M., Hargadon, B., McKenna, S., et al. (2007). Eosinophilic airway inflammation and exacerbations of COPD: A randomised controlled trial. *European Respiratory Journal*, 29(5), 906–913.

Slamon, D.J., Leyland-Jones, B., Shak, S., Fuchs, H., Paton, V., Bajamonde, A., ... and Norton, L. (2001). Use of chemotherapy plus a monoclonal antibody against HER2 for metastatic breast cancer that overexpresses HER2. *New England Journal of Medicine*, 344(11), 783–792. https://doi.org/10.1056/NEJM200103153441101

SonLa Study, G. (2007). Using a fingerprint recognition system in a vaccine trial to avoid misclassification. *Bulletin of the World Health Organization*, 85, 64–67.

Soter, S., Barta, I., and Antus, B. (2013). Predicting sputum eosinophilia in exacerbations of COPD using exhaled nitric oxide. *Inflammation*. Epub: 10.1007/s10753-013-9653-8

Sperrin, M., Grant, S.W., and Peek, N. (2020). Prediction models for diagnosis and prognosis in Covid-19. *BMJ*, 369. https://doi.org/10.1136/bmj.m1328

Sperry, M., and Ingber, D.E. (2024). Drug repurposing strategies, challenges and successes. www.technologynetworks.com/tn/articles/drug-repurposing-strategies-challenges-and-successes-384263. Accessed June 17, 2024.

Spiegelhalter, D.J., Abrams, K.R., and Myles, J.P. (2004). *Bayesian Approaches to Clinical Trials and Health-Care Evaluation*. Wiley.

Spilker, B. (1991). *Guide to Clinical Trials*. Raven Press.

Stensland, K.D., McBride, R.B., Latif, A., et al. (2014). Adult cancer clinical trials that fail to complete: An epidemic? *Journal of the National Cancer Institute*, 106(9), dju229. https://pubmed.ncbi.nlm.nih.gov/25190726/. Accessed January 14, 2025.

Steyerberg, E.W., and Vergouwe, Y. (2014). Towards better clinical prediction models: Seven steps for development and an ABCD for validation. *European Heart Journal*, 35, 1925–1931.

Sydes, M.R., Parmar, M.K., Mason, M.D., Clarke, N.W., Amos, C., Anderson, J., de Bono, J., Dearnaley, D.P., Dwyer, J., Green, C., Jovic, G., Ritchie, A.W., Russell, J.M., Sanders, K., Thalmann, G., and James, N.D. (2012). Flexible trial design in practice-stopping afor lack-of-benefit and adding contains nonbinding recommendations 33 research arms mid-trial in STAMPEDE: A multi-arm multi-stage randomized controlled rrial. *Trials*, 13(1), 168.

Szefler, S.J., Wenzel, S., Brown, R., Erzurum, S.C., Fahy, J.V., Hamilton, R.G., Hunt, J.F., Kita, H., Liu, A.H., Panettieri Jr, R.A., Schleimer, R.P., and Minnicozzi, M. (2012). Asthma outcomes: Biomarkers. *The Journal of Allergy and Clinical Immunology*, 129(3 Suppl), S9–S23.

Tang, Z. (2015). Optimal futility interim design: A predictive probability of success approach with time-to-event endpoint. *Journal of Biopharmaceutical Statistics*, 25(6), 1313–1319.

Tasneem, A., et al. (2012). The database for aggregate analysis of Clinicaltrials.gov (AACT) and subsequent regrouping by clinical specialty. *PLoS One*, 7, e33677.

Taylor, K., Properzi, F., Cruz, M.J., Ronte, H., and Haughey, J. (2020). Intelligent clinical trials – transforming through AI-enabled engagement. *Deloitte Insights*, 1–32.

Temple, R., and Enas, G.G. (2010). Enrichment of clinical study populations. *Statistics in Medicine*, 28(26), 2898–2912. https://doi.org/10.1002/sim.3688

Thall, P.F., and Lee, S.J. (2003). Practical model-based dose-finding in phase I clinical trials: Methods based on toxicity. *Int J Gynecol Cancer*, 13(3), 251–261.

Thall, P.F., and Wathen, J.K. (2007). Practical Bayesian adaptive randomization in clinical trials. *European Journal of Cancer*, 43(5), 859–866.

Thall, P.F., and Simon, R. (1994a). Practical Bayesian guidelines for phase IIB clinical trials. *Biometrics*, 50, 337–349.

Thall, P.F., and Simon, R. (1994b). A Bayesian approach to establishing ample size and monitoring criteria for phase II clinical trials. *Controlled Clinical Trials*, 15, 463–481.

Thall, P.F., and Simon, R. (1995). Recent developments in the design of phase II clinical trials. *Cancer Treatement and Research*, 75, 49–71.

Thall, P.F., Simon, R.M., and Estey, E.H. (1995). Bayesian sequential monitoring designs for single-arm clinical trials with multiple outcomes. *Statistics in Medicine*, 14, 357–379.

The Cardiac Arrhythmia Suppression Trial (CAST) Investigators. (1989). Preliminary report: Effect of encainide and flecainide on mortality in a randomized trial of arrhythmia suppression after myocardial infarction. *The New England Journal of Medicine*, 321, 406–412.

Thomas, D., et al. (2016). Clinical development success rates 2006–2015. *BIO Industry Analysis*. https://cir.nii.ac.jp/crid/1370572092642204419

Thorlund, K., Haggstrom, J., Park, J., and Mills, E.J. (2018). Key design considerations for adaptive clinical trials: A primer for clinicians. *BMJ*. https://doi.org/10.1136/bmj.k698

ThoughtSphere (n.d.) Clinical data science & statistical programming. https://thoughtsphere.com/. Accessed July 2, 2024.

Tian, L.U., Alizadeh, A.A., Gentles, A.J., et al. (2014). A simple method for estimating interactions between a treatment and a large number of covariates. *Journal of the American Statistical Association*, 109(508), 1517–1532.

Tolosa, E., Litvan, I., Höglinger, G.U., Burn, D., Lees, A., Andrés, M.V., et al. (2014). A phase 2 trial of the GSK-3 inhibitor tideglusib in progressive supranuclear palsy: Tideglusib in PSP. *Movement Disorder*, 29(4), 470–478.

Tomioka, S. (2018). SDTM mapping based on natural language processing and machine learning. 2018 US CDISC Interchange Conference.

Topol, E.J. (2019a). Deep Medicine: How Artificial Intelligence Can Make Healthcare Human Again. *PSNet*. https://psnet.ahrq.gov/issue/deep-medicine-how-artificial-intelligence-can-make-healthcare-human-again. Accessed September 22, 2024.

Topol, E.J. (2019b). High-performance medicine: The convergence of human and artificial intelligence. *Nature Medicine*, 25(1), 44–56.

Toy, E.L., Gallagher, K.F., Stanley, E.L., Swensen, A.R., and Duh, M.S. (2010). Th e economic impact of exacerbations of chronic obstructive pulmonary disease and exacerbation definition: A review. *COPD*, 7(3), 214–228.

TransCelerate (2013). Risk-Based Monitoring Methodology. TransCelerate BioPharma Inc., 2013.

TransCelerate (2020). Demystifying Quality Tolerance Limits webinar. www.transceleratebiopharmainc.com/events/demystifying-quality-tolerance-limits-webinar/. Accessed January 23, 2025..

TransCelerate (2021). Clinical quality management system concept paper. www.transceleratebiopharmainc.com/assets/quality-management-system-solutions/clinical-qms-concept-paper/. Accessed January 23, 2025.

TransCelerate (2023). Modernizing Clinical Trials Using Digitized Protocol Information: An Exploration of New Tools for Digital Transformation. www.transceleratebiopharmainc.com/events/webinar-modernizing-clinical-trials-using-digitized-protocol-information-an-exploration-of-new-tools-for-digital-transformation/. Accessed on August 31, 2024.

TransCelerate (n.d.) Digital Data Flow Solutions. www.transceleratebiopharmainc.com/assets/digital-data-flow-solutions/. Accessed on August 31, 2024.

Travis, J., Tothmann, M., and Thomson, A. (2023). Perspectives on informative Bayesian methods in pediatrics. *Journal of Biopharmaceutical Statistics*, 33(6), 830–843.

Traxinger, K., Kelly, C., Johnson, B.A., Lyles, R.H., and Glass, J.D. (2013). Prognosis and epidemiology of amyotrophic lateral sclerosis: Analysis of a clinic population, 1997-2011. *Neurology Clinical Practice*, 3, 313–320.

U.S. Census Bureau (2019). Metropolitan and Micropolitan Statistical Areas Population Totals and Components of Change: 2010–2019;. www.census.gov/data/datasets/time-series/demo/popest/2010s-totalmetro-and-micro-statistical-areas.html.

Vall, A., Sabnis, Y., Shi, J., Class, R., Hochreiter, S., and Klambauer, G. (2021). The promise of AI for DILI prediction. *Frontiers in Artificial Intelligence*, 4, 638410. www.frontiersin.org/article/10.3389/frai.2021.638410.

van de Schoot, R., Kaplan, D., Denissen, J., Asendorpf, J.B., Neyer, F.J., and Aken, M.A.G. (2013). A gentle introduction to Bayesian analysis: Applications to developmental research. *Children Development*, 85(3), 842–860.

van de Veerdonk, F.L., and Netea, M.G. (2020). Blocking IL-1 to prevent respiratory failure in COVID-19. *Critical Care*, 24, 445. https://doi.org/10.1186/s13054-020-03166-0

van Eijk, P.P.A., Westeneng, H.-J., Nikolakopoulos, S., Verhagen, I.E., van Es, M.A., Eijkemans, M.J.C., and van den Berg, L.H. (2019). Refining eligibility criteria for amyotrophic lateral sclerosis clinical trials Ruben P.A *Neurology*, 92, e451–e460. https://doi.org/10.1212/WNL.0000000000006855

Varadhan, R., and Seeger, J. (2013). Estimating and reporting of heterogeneity of treatment effects. In *Developing a Protocol for Observational Comparative Effectiveness Research: A Users Guide*, edited by Velentgas, P., Dreyer, N.A., Nourjah, P., et al. Agency for Healthcare Research and Quality, pp. 35–45. www.ncbi.nlm.nih.gov/books/NBK126188/. Accessed January 14, 2025.

Varga, A., Bernard-Tessier, Al, Auclin, E., Mezquita, P.L., Baldini, C., et al. (2019). Applicability of the Lung Immune Prognostic Index (LIPI) in patients with metastatic solid tumors when treated with immune checkpoint inhibitors (ICI) in early clinical trials. *Annals of Oncology*, 1, 30(Supplement 1), 2–3.

Vasta, R., Moglia, C., D'Ovidio, F., Di, Pede F., De, Mattei, F., Cabras, S., et al. (2021). Telemedicine for patients with amyotrophic lateral sclerosis during COVID-19 pandemic: An Italian ALS referral center experience. *Amyotrophic Lateral Sclerosis and Frontotemporal Degeneration*, 22, 308–311.

Vestbo, J., Anderson, J.A., and Brook, R.D. (2016). The Salford Lung Study: pragmatic clinical trial design. *Respiratory Medicine*, 117, 61–69.

Vestbo, J., Hurd, S.S., Agusti, A.G., Jones, P.W., Vogelmeier, C., Anzueto, A., et al. (2012). Global strategy for the diagnosis, management and prevention of chronic obstructive pulmonary disease, GOLD executive summary. *American Journal of Respiratory and Critical Care Medicine*; Epub ahead of print.

Vestbo, J., Leather, D., Bakerly, N., et al., (2016). Effectiveness of Fluticasone Furoate–Vilanterol for COPD in clinical practice. *NEJM*, 357, 1253–1260.

Viceconti, M., Henney, A., and Morley-Fletcher, E. (2021). In silico clinical trials: How computer simulation will transform the biomedical industry. *International Journal of Clinical Trials*, 3(2), 78–88.

Viceconti, M., and Hunter, P. (2016). The virtual physiological human: Ten years after. *Annual Review of Biomedical Engineering*, 11(18), 103–123.

Vogelmeier, C., Ramos-Barbon, D., Jack, D., Piggott, S., Owen, R., Higgins, M., et al. (2010). Indacaterol provides 24-hour bronchodilation in COPD: A placebo-controlled blinded comparison with tiotropium. *Respiratory Research*, 11, 135.

VPH Institute (2023). *In Silico trial predict Phase 3 clinical trial result*. www.vph-institute.org/news/in-silico-trial-predict-phase-3-clinical-trial-result.html. Accessed September 15, 2024.

Wager, S., and Athey, S. (2018). Estimation and inference of heterogeneous treatment effects using random forests. *Journal of the American Statistical Association*, 113(523), 1228–1242.

Walsh, J.R., Roumpanis, S., Bertolini, D., and Delmar, P. (2022). Evaluating digital twins for Alzheimer's disease using data from a completed phase 2 clinical trial. *Alzheimer's & Dementia*, 18(S10). https://doi.org/10.1002/alz.065386. https://alz-journals.onlinelibrary.wiley.com/doi/epdf/10.1002/alz.065386. Accessed January 14, 2025.

Walsh, J.R., Smith, A.M., Poulot, Y Li-Bland, D., Loukianov, A., and Fisher, C.K. (2020). Generating digital twins with multiple sclerosis using probabilistic neural networks. arXiv:2002.02779. https://arxiv.org/abs/2002.02779. Accessed September 22, 2024.

Wang, F., Casalino, P.L., and Khullar, D. (2018). Deep learning in medicine-promise, progress, and challenges. *JAMA Internal Medicine*, 179(3), 293–294.

Wang, S.K., and Tsiatis, A.A. (1987). Approximately optimal one-parameter boundaries for a sequential trials. *Biometrics*, 43, 605–611.

Wang, Y., and Li, X. (2022). Personalized medicine and digital twins: A new era of clinical trials. *Personalized Medicine Journal*, 10(3), 301–315.

Wang, Y., Wang, L., Rastegar-Mojarad, M., Moon, S., Shen, F., and Liu, H. (2018). Clinical information extraction applications: A literature review. *Journal of Biomedical Informatics*, 77, 34–49.

Wang, Z., Xiaio, C., and Sun, J. (2023). AutoTrial: Leveraging Large Language Models for Clinical Trial Design. OpenReview.net. https://openreview.net/forum?id=E5r96sfKO0&referrer=%5Bthe%20profile%20of%20Zifeng%20Wang%5D(%2Fprofile%3Fid%3D~Zifeng_Wang3). Accessed September 22, 2024.

Webster, J., and Smith, B.D. (2019). The case for real-world evidence in the future of clinical research on chronic myeloid leukemia. *Clinical Therapeutics*, 41(2), 336–349.

Weidman, J.R., and Belsky, K. (2020). Estimating the probability of regulatory registration success. *Regulatory Focus*. www.raps.org/News-and-Articles/News-Articles/2020/2/Estimating-the-Probability-of-Regulatory-Registrat. Accessed May 1, 2023.

Weinstein, M.C., O'Brien, B., Hornberger, J., Jackson, J., Johannesson, M., McCabe, C., et al. (2003). Principles of good practice for decision analytic modeling in health-care evaluation: Report of the ISPOR Task Force on Good Research Practices--Modeling Studies. *Value Health*, 6(1), 9–17.

Weinstein, J.N., Geller, A., Negussie, Y., and Baciu, A. (2017). *Communities in action: Pathways to health equity*. National Academies Press. https://doi.org/10.17226/24624

Weissler, E.H., Naumann, T., Andersson, T., et al. (2021). The role of machine learning in clinical research: Transforming the future of evidence generation. *Trials*, 22(1), 1–15. https://doi.org/10.1186/s13063-021-05489-x

Westfall, P.H., Tsai, K., Ogenstad, S., Tomoiaga, A., Moseley, S., and Lu, Y. (2008). Clinical trials simulation: A statistical approach. *Journal of Biopharmaceutical Statistics*, 18, 611–630.

Wheler, J., Tsimberidou, A.M., Hong, D., Naing, A., Falchook, G., Pha-Paul, S., et al. (2012). Survival of 1181 patients in a phase I clinic: The MD Anderson Clinical Center for targeted therapy experience. *Clinical Cancer Research*, 18(10), 2992–2999.

WHO (2020). WHO Guidance on Ethics and Governance of Artificial Intelligence in Health. www.who.int/publications/i/item/9789240029200.

Wilkinson, M.D., Dumontier, M., et al. (2016). The FAIR guiding principles for scientific data management and stewardship. *Scientific Data*, 3, 1–27.

Wilson, J.D. (1997). Setting alert and action limits for environmental monitoring. *PDA Journal of Pharmaceutical Science and Technology*, 51(4), 161–162.

Winquist, J. (2012). *Reducing Attrition via Improved Strategies for Pre-clinical Drug Discovery*. Thesis published by Uppsala Universitet.

Wong, C.H., Siah, K.W., and Lo, A.W. (2019). Estimation of clinical trial success rates and related parameters. *Biostatistics*, 20, 273–286.

Woodcock, A., Kakerly, N., et al. (2015). The Salford Lung Study protocol: A pragmatic, randomised phase III real-world effectiveness trial in asthma. *BMC Pulmonary Medicine*. https://bmcpulmmed.biomedcentral.com/articles/10.1186/s12890-015-0150-8. Accessed September 22, 2024.

Woodcock, A., Vestbo, J., Bakerly, N., et al. (2017). Effectiveness of fluticasone furoate plus vilanterol on asthma control in clinical practice: An open-label, parallel group, randomised controlled trial. *Lancet*, 390(10109), 2247–2255.

Woodcock, J., and LaVange, L.M. (2017). Master protocols to study multiple therapies, multiple diseases, or both. *New England Journal of Medicine*, 377(1), 62–70. https://doi.org/10.1056/NEJMra1510062

Xie, Y., Brand, J.E., and Jann, B. (2012). Estimating heterogeneous treatment effects with observational data. *Social Methodology*, 42(1), 314–347.

Yadav, P., Steinbach, M., Kumar, V., and Simon, G. (2017). Mining electronic health records (HER): A survey. *ACM Computing Surveys*, 50(6), 1–40.

Yang, H. (2023a). Transforming pharma with data science, AI, and machine learning. In *Data Science, AI, and Machine Learning in Drug Development*, edited by Yang, H. CRC Press, pp. 1–18.

Yang, H. (2023b). Deep learning with electronic health record. In *Data Science, AI, and Machine Learning in Drug Development*, edited by Yang, H. CRC Press, pp. 265–286.

Yang, H. (2024). Advancing clinical trials through RWD, AI, and machine learning. The 7th International Symposium of International Society for Biopharmaceutical Statistics, March 6–9, 2024, Baltimore, Maryland.

Yang, H., and Khatry, D. (2023). Reinventing medical affairs in the era of big data. In *Data Science, AI, and Machine Learning in Drug Development*, edited by Yang, H. CRC Press, pp. 245–264.

Yang, H., and Novick, S.J. (2019). *Bayesian Analysis with R for Drug Development: Concepts, Algorithms, and Case Studies*. CRC Press.

Yang, H., and Yu, B. (2021). *Real-World Evidence in Drug Development and Evaluation*. CRC Press.

Yang, H., Fu, D., and Roskos, L. (2018). Interchangeability study design and analysis. In *Biosimilar: Regulatory, Clinical, and Biopharmaceutical Development*, edited by Gutka, H.J., Yang, H., and Kakar, S. Springer, pp. 543–570.

Yang, H., Zhang, L., and Cho, I. (2007). Statistical modeling in detecting safety signals for product quality issue. In *Proceedings of 2007 JSM*, 6, 1–12.

Yang, H., Zhao, W., O'Day and Fleming, W. (2013). Environmental monitoring: Setting alert/action limits using zero-inflated models. *PDA Journal of Pharmaceutical Science and Technology*, 67, 2–8.

Youden, W.K (1950). Index for rating diagnostic tests. *Cancer*, 3(1), 32–35.

Yu, L., Yang, H., Dar, M., Roskos, L., Soria, J.-C., Zhao, W., Brokmann, A.G., and Mukhopadhyay, P. (2020). Methods for Determining Treatment for Cancer Patients. Pub. No.: US 2020/0126636 A1. United States Patent Application Publication.

Yuan, Y., Hess, K.R., Hilsenbeck, S.G., and Gilbert, M.R. (2016). Bayesian optimal interval design: A simple and well-performing design for phase I oncology trials, *Clinical Cancer Research*, 22(17), 4291–4301.

Yuan, Y., Lee, J.J., and Hilsenbeck, S.G. (2019). Model-assisted designs for early-phase clinical trials: Simplicity meets superiority. *Journal of Clinical Oncology*, 34(28), 3413–3420.

Zach, N., Ennist, D.L., Taylor, A.A., Alon, H., Sherman, A., Kueffner, R., et al. (2015). Being PRO-ACTive - What can a clinical trial database reveal about ALS? *Neurotherapeutics*, 12, 417–423.

Zaidi, A., et al. (2020). CDIS primer; SDTM and ADaM implementation FAQ. www.cdisc.org/sites/default/files/pdf/CDISC-PHUSE-SlidesMERGED.pdf. Accessed August 31, 2024.

Zarin, D.A., et al. (2016). Trial reporting in ClinicalTrials.gov — the final rule. *The New England Journal of Medicine*, 375, 1998–2004.

Zeng, Y., Pande, A., Zhu, J., and Mohapatra, P. (2017). WearIA: Wearable device implicit authentication based on activity information. *IEEE Xplore* https://doi.org/10.1109/WoWMoM.2017.7974305

Zhang, K., and Demmer-Fushman. (2017). Automated classification of eligibility criteria in clinical trials to facilitate patient-trial matching for specific populations. *Journal of American Medical Informatics Association*, 24(4), 781–787.
Zhang, Y., Zhang, L., Gao, X., et al. (2022). Re-engineering a clinical trial management system using blockchain technology: System design, development, and case studies. *Journal of Medical Internet Research*, 44(6) e36774. https://doi.org/10.2196/36774
Zhang, Y., Yan, L., and Gao, J., et al. (2020). Predicting missing values in medical data via XGBoost regression. *Journal of Healthcare Informatics Research*, 4, 393–394.
Zhang, Z., et al. (2019). Reinforcement learning in clinical medicine: A method to optimize dynamic treatment regime over time. *Annals of Translational Medicine*, 7(14), 1–10.
Zhao, T., et al. (2020). TrueHeart: Continuous authentication on wrist-worn wearables using PPG-based biometrics. *IEEE Xplore* https://doi.org/10.1109/INFOCOM41043.2020.9155526
Zhao, Y., Kosorok, M., and Zeng, D. (2009). Reinforcement learning design for cancer clinical trials. *Statistics in Medicine*, 20(28), 3294–3315.
Zheng, J., Shen, Y., Zhang, Z., Wu, T., Zhang, G., and Lu, H. (2013). Emerging wearable medical devices towards personalized healthcare. In *Proceedings of the 8th International Conference on Body Area Networks*, pp. 427–431. https://eudl.eu/doi/10.4108/icst.bodynets.2013.253725. Accessed June 17, 2024.
Zhong, S., et al. (2023). *Enrollment Forecast for Clinical Trials at the Portfolio Planning Phase Based on Site-level Historical Data*. Pharmaceutical Statistics.
Zhou, H., Yuan, Y., and Nie, L. (2018). Accuracy, safety, and reliability of novel phase I trial designs. *Clinical Cancer Research*, 24(18), 4357–4364.
Zhou, N., and Manser, P. (2020). Does including machine learning predictions in ALS clinical trial analysis improve statistical power? *Annals of Clinical Translational Neurology*, 7(10), 1756–1765.
Zhou, Y., Lin, R., Kuo, Y., Lee, J.J., and Yuan, Y. (2021). BOIN suite: A software platform to design and implement novel early phase clinical trials. *JCO Clinical Cancer Informatics*, 91–101. https://pmc.ncbi.nlm.nih.gov/articles/PMC8462603/. Accessed January 23, 2025.
Zola Matuvanga, T., et al. (2021). Use of iris scanning for biometric recognition of healthy adults participating in an Ebola vaccine trial in the Democratic Republic of the Congo: Mixed methods study. *Journal of Medical Internet Research*, 23, e28573.
ZS (2022). Bring Clinical Trial Optimization into the 21st Century: A Q&A with Dr. Rob Scott, MD. www.zs.com/insights/bringing-clinical trial-optimization-into-the-21st-century. Accessed July 6, 2024.
Zweig, M.H., and Campbell, G. (1993). Receiver-operating characteristic (ROC) plots: A fundamental evaluation tool in clinical medicine. *Clinical Chemistry*, 39(4), 561–577.
ZYLiQ (n.d.) Accelerating medical writing. www.zyliq.ai/index.php. Accessed September 20, 2024.

Index

D

N

O

P